AF870999

ROBOT HANDS AND MULTI-FINGERED HAPTIC INTERFACES

Fundamentals and Applications

ROBOT HANDS AND MULTI-FINGERED HAPTIC INTERFACES

Fundamentals and Applications

Haruhisa Kawasaki

Gifu University, Japan

World Scientific

NEW JERSEY • LONDON • SINGAPORE • BEIJING • SHANGHAI • HONG KONG • TAIPEI • CHENNAI

Published by

World Scientific Publishing Co. Pte. Ltd.
5 Toh Tuck Link, Singapore 596224
USA office: 27 Warren Street, Suite 401-402, Hackensack, NJ 07601
UK office: 57 Shelton Street, Covent Garden, London WC2H 9HE

Library of Congress Cataloging-in-Publication Data
Kawasaki, Haruhisa, 1949–
Robot hands and multi-fingered haptic interfaces : fundamentals and applications / Haruhisa Kawasaki, Gifu University, Japan.
pages cm
Includes bibliographical references and index.
ISBN 978-9814635608 (hardback : alk. paper)
1. Robot hands. 2. Haptic devices. I. Title.
TJ211.43.K39 2015
629.8'933--dc23
2014040799

British Library Cataloguing-in-Publication Data
A catalogue record for this book is available from the British Library.

Copyright © 2015 by World Scientific Publishing Co. Pte. Ltd.

All rights reserved. This book, or parts thereof, may not be reproduced in any form or by any means, electronic or mechanical, including photocopying, recording or any information storage and retrieval system now known or to be invented, without written permission from the publisher.

For photocopying of material in this volume, please pay a copying fee through the Copyright Clearance Center, Inc., 222 Rosewood Drive, Danvers, MA 01923, USA. In this case permission to photocopy is not required from the publisher.

In-house Editor: Amanda Yun

Typeset by Stallion Press
Email: enquiries@stallionpress.com

About the Author

Prof. Haruhisa Kawasaki (born on 27th June 1949) is a Professor in the Department of Mechanical Engineering, Faculty of Engineering (1994–present), and the Kawasaki & Mouri Laboratory in Gifu University, Japan. He is a Senior Member of IEEE, and a Fellow of the Japan Society of Mechanical Engineers, the Robotics Society of Japan, and the Virtual Reality Society of Japan. Prof. Kawasaki is also a member of the Society of Instrument and Control Engineers. His current research interests include Robot Control, Humanoid Robot Hands, Haptic interfaces, Virtual Reality Systems, Hand Rehabilitation Support Systems, and Computer Algebra of Robotics.

Prof. Kawasaki was the National Organization Committee Chair of IFAC SYROCO 2009. Among many other awards and honors, in 2004 he received the Best Paper Award at the World Automation Congress; in 2006, he was honored with a Minister Award from the Ministry of Education, Culture, Sports, Science and Technology of Japan; in 2009, he was the recipient of the Governor Award of the Japan Institute of Invention and Innovation; in 2010, he was given the Industry–Academia–Government Collaboration Promotion Award by the Ministry of Internal Affairs and Communications, as well as a Robotics and Mechatronics Academic Achievement Award from the Robotics and Mechatronics Division of JSME (the Japan Society of Mechanical Engineers); in 2013, he won the IEEE/ASME Transactions on Mechatronics (TMCH) Best Paper Award.

Prior to his academic career, Prof. Kawasaki was a research engineer (1974–1990) at Nippon Telegraph and Telephone Corporation, Japan. From 1990–1994, Prof. Kawasaki served as a Professor in the Kanazawa Institute of Technology Japan. Prof. Kawasaki received his M.Sc. and Ph.D. in Aeronautical Engineering from Nagoya University, Japan, in 1974 and 1986 respectively.

Preface

Multi-fingered robot hands are expected to replace human hands in the dexterous manipulation of objects. When a robot grasps an object, its fingers come into contact with the surface of the object, and its hand motion is constrained by that contact. For dexterous manipulation, a basic understanding of the contact model between a finger and an object, the motion constraint during contact and the dynamics of these constraints is important. Furthermore, the dynamics of robot hands are non-linear, and require control methods that take constraints into consideration.

Haptic interfaces are used to present force and tactile feeling to humans in cases of robot telecontrol, when manipulating objects in a virtual reality (VR) environment, playing games with tactile feeling, and so on. Multi-fingered haptic interfaces have tremendous potential to provide realistic sensations of touch to human fingertips and to cover the workspace of the human arm. Haptic interfaces are one of the major applications for multi-fingered hand robots, but science and technology based on human tactile perception are needed to design and control multi-fingered haptic interfaces.

This book briefly compares human hands and robot hands, discusses the design of robot hands, contact models at grasping, kinematic models of constraint, dynamic models of the multi-fingered hand, the stability theorem of non-linear control systems, robot hand control, design and control of multi-fingered haptic interfaces, application systems using multi-fingered haptic interfaces, and telecontrol of robot hands using a multi-fingered haptic interface.

This book is intended mainly for readers who have already studied basic robot arm engineering. There are many books related to robot arms and non-linear control. To understand robot hand manipulation, however, readers must study kinematic constraint models of fingers, hand dynamics with constraints, stability theorems of non-linear control, and multi-fingered hand control — This book is designed to specifically and comprehensively address these subjects. *Robot Hands and Multi-Fingered Haptic Interfaces* will benefit readers' understanding of the full range of issues regarding robot hand manipulation. Furthermore, the design and control of multi-fingered haptic interfaces are necessary to telecontrol multi-fingered robot hands.

A Brief Rundown of the Chapters

This book is organized into eight chapters:

Chapter 1 — *The Human Hand and the Robotic Hand*: This chapter describes the outline including the structure of the human hand, human grasping, and the structure of a humanoid robot hand called the Gifu Hand. Furthermore, it presents the configuration of a multi-fingered haptic interface called the HIRO, which is a new application of the robot hand.

Chapter 2 — *Kinematics of Multi-Fingered Hands*: Here, we present basic kinematics in terms of the grasp by a multi-fingered robot hand, including contact models, forms of grasps, constraint conditions, manipulable grasps, and so on.

Chapter 3 — *Kinematic Constraint and Controllability*: When a robot grasps an object, the robot hand comes into contact with the surface of the object, and the object in turn, is in contact with its environment and is influenced by it (e.g. friction and gravitational force). Robot hand motion is constrained by these contacts. This chapter presents fundamentals of holonomic and non-holonomic constraints, and the controllability of the constraint system.

Chapter 4 — *Robot Dynamics*: This chapter reviews basic methodologies for dynamics in robot hands. The dynamic model is widely used in the simulation of robot motion, analysis of robot hand structures, and design of control algorithms. When a robot hand with multiple fingers grasps an object, the hand motion is restricted in order to keep contact with the object. A dynamics model with constraint is derived.

Chapter 5 — *Stability Theorem of Non-Linear Control Systems*: Robots are non-linear dynamical systems. This chapter describes the fundamentals of the stability theorem for non-linear autonomous systems and non-linear non-autonomous systems.

Chapter 6 — *Robot Hand Control*: The cooperative control of multiple fingers is a basic technology of dexterous manipulation. In most control laws, it is assumed that the dynamics' parameters are known. However, it is often difficult to find the exact dynamics of the system. In addition, the dynamics parameters will vary depending on the object being manipulated. This chapter describes impedance control for manipulating an object with multiple fingers, computed torque control, and adaptive control based on estimated dynamics parameters.

Chapter 7 — *Multi-Fingered Haptic Interface*: This chapter focuses on the fundamentals of haptic interface control, and on hand manipulability-based control for multi-fingered haptic interfaces that equip haptic arms. Applications that use a multi-fingered haptic interface, including a hand-rehabilitation support system, a measurement system for breast engorgement, a surgery training system using plural devices, a breast palpation training system, and a skill-transfer system, are presented.

Chapter 8 — *Teleoperation of Robot Hands*: Teleoperation technology generally uses a master–slave system, where the 'master' is a haptic interface operated by a human, and the 'slave' is a remotely located robot. The slave then moves according to motions generated through the master system. Human biological signals have also been used as a type of master system for the command of robots. One typical application is the myoelectric controlled hand. This chapter presents the fundamentals of the master–slave system, teleoperation of robot hands through a multi-fingered haptic interface, and prophetic hand control by surface electromyogram (sEMG).

Answers to the Exercises are presented in the back-matter of the book.

Acknowledgments

I would like to offer special thanks to my colleagues for their collaboration. I am especially grateful to Dr. Tetsuya Mouri, Dr. Takahiro Endo, Dr. Satoshi Ueki, Dr. Satoshi Ito, Dr. Osama Halabi, Dr. Vytautas Daniulaitis, Dr. Masashi Hioki and Dr. Yasuhiko Ishigure. I would also like to express my appreciation to Ms. Miki Nomura for her work on translation and editing. Thanks to my students, anonymous reviewers of the book, the staff at KN International, Inc., and to the staff at World Scientific, especially Ms. Amanda Yun.

Finally, I wish to thank my wife Kyoko Kawasaki for her assistance, back-patting, and understanding.

Haruhisa Kawasaki.

Contents

List of Tables

List of Figures

Chapter 1

The Human Hand and the Robotic Hand

To enable a robotic hand to grasp a target object dexterously, like a human hand, structural features and control functions similar to those in human hands are necessary. Until now, a deep understanding of the human hand has been the province of medical professionals. Researchers in robotics have much to learn from them. This chapter describes the structure of the human hand, its grasping functions, and the structure of the humanoid robot hand called the Gifu Hand. Furthermore, the configuration of a multi-fingered haptic interface called the Haptic Interface Robot (HIRO), a new application for the robotic hand, will be discussed.

1.1. Joints of the Human Hand

Humans use hands to grasp, squeeze, and manipulate objects dexterously. As shown in Fig. 1.1, the **human hand** consists of a thumb, four fingers (index, middle, ring, and little finger) and a palm. There are 27 bones in a single hand, as well as muscles, tendons and other associated structures. This section focuses on the bones and joints.

From each fingertip to the base of the finger, there is a distal phalanx, middle phalanx, and proximal phalanx. The metacarpal bone lies between the base of the fingers and the carpal region. Between each phalanx bone lies joints called (starting from the fingertip) the distal interphalangeal (DIP) joint, the proximal interphalangeal (PIP) joint, metacarpophalangeal (MP) joint, and the carpometacarpal (CM) joint.

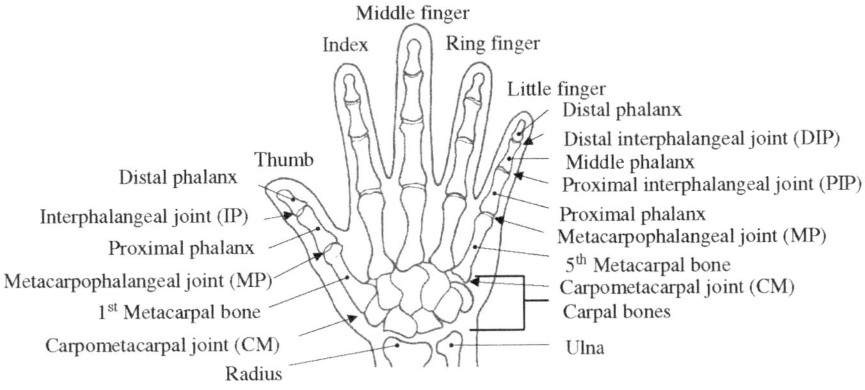

Fig. 1.1. Bones and joints in the human hand.

The MP joint has two degrees of freedom (DOFs), adduction/abduction and flexion/extension, along two axes that are approximately orthogonal to each other. The DIP and PIP joints have only flexion and extension with a single DOF. In the case of normal movement, the DIP joint moves in conjunction with the PIP joint; therefore the two joints are considered as a single entity with one DOF. The CM joint of the fingers is considered negligible due to its small operating range. As a result, robotic fingers can be modeled with just three DOFs.

On the other hand, the thumb has only a distal phalanx and proximal phalanx before the metacarpal bone. The joints between the bones are called (starting from the fingertip) the interphalangeal (IP) joint, MP joint, and CM joint. The CM joint has two DOFs; flexion/extension and adduction/abduction. This enables the thumb to oppose each finger — which is called thumb opposition. The thumb MP joint also has two DOFs in flexion/extension and adduction/abduction; however, it usually is modeled as one DOF of flexion/extension due to the limited operating range of adduction/abduction. The flexion/extension of a single DOF is possible for the thumb IP joint. Thus, although the human thumb has five DOFs from the point of view of kinematics, modeling four DOFs may be suitable for the purpose of simplifying a robotic hand. Table 1.1 shows the movable range of the human hand. The radial abduction

Table 1.1 Movable range of the human hand.

Name	Motion direction	Movable angle (degree)	Reference drawing
Thumb	Radial abduction Ulnar adduction	60 0	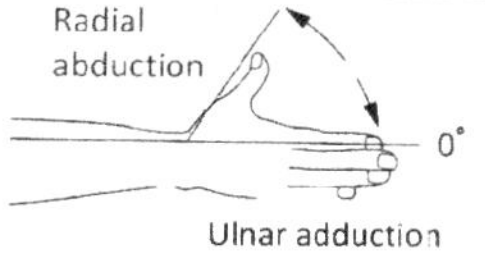
	Palmar abduction Palmar adduction	90 0	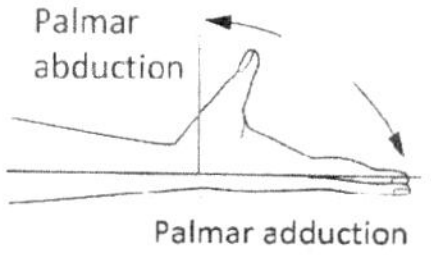
	MCP flexion MCP extension	60 10	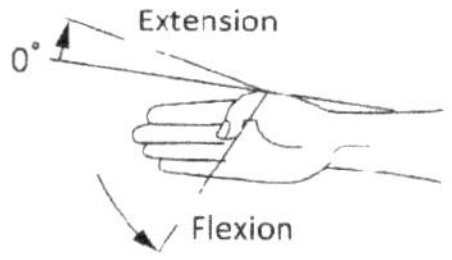
	IP flexion IP extension	80 10	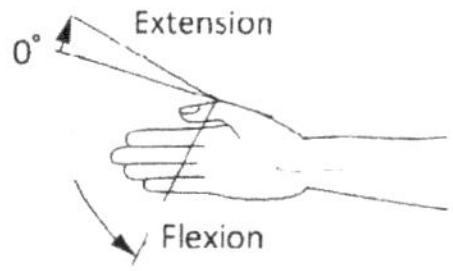
Fingers	Abduction Adduction		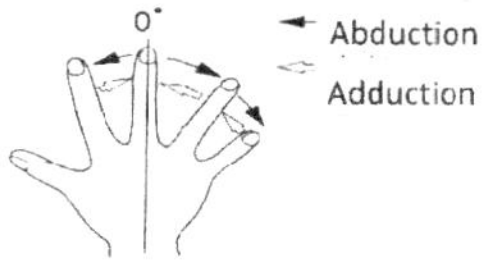
	MCP flexion MP extension	90 45	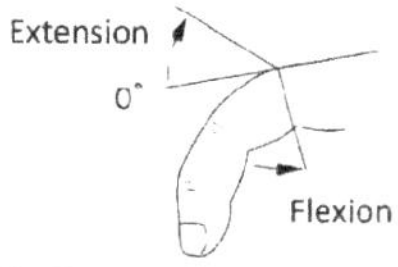
	PIP flexion PIP extension	100 0	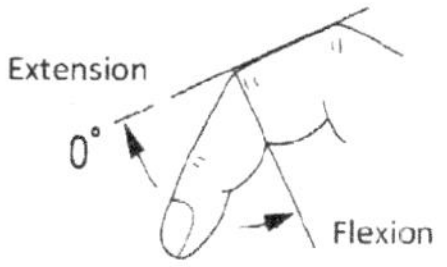
	DIP flexion DIP extension	80 0	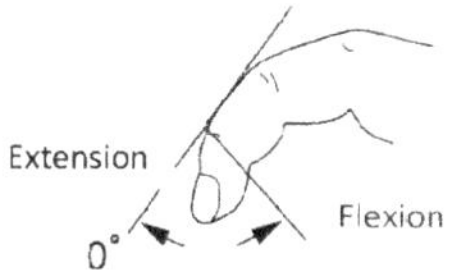

Source: Formulation by the Japanese Othopaedic Association and the Japanese Association of Rehabilitaion Medicine, 1995.

and ulnar adduction of the thumb are motions in the volar aspect, while palmar abduction and adduction are motions in a plane that is orthogonal to the volar aspect.

There are eight carpal bones in the carpal region of the palm. These constitute the middle carpal joint, although the operating range of this joint can be negligible. The back of the hand can be deployed on a plane when the palm is expanded, and becomes arched when the hand is strongly squeezed. This is called the transverse arch of the hand. It is caused by the mobility of the CM joint that consists of metacarpal and carpal bones, and varies with the various operations of the hand. Though the palm of the human hand has multiple DOFs in reality, it can nevertheless be said that there is approximately one DOF in the transverse arch of the palm.

Taking the information noted above into consideration, the robotic hand will be modeled with around 16–21 DOFs in total. Figure 1.2 is a model diagram of the hand when the finger is set to three DOFs, the thumb to four DOFs, and the palm to zero DOF.

1.2. Hand Features

Many mammals, including whales, bats, and dogs, retain the body structure of the reptilian ancestors of mammals. Hands consisting of five fingers, including the thumb with its distinctive shape, have been specialized to adapt to the environment. Whales have finned bodies

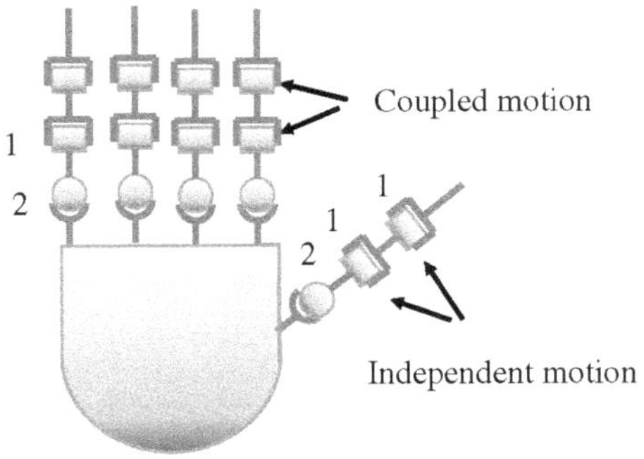

Fig. 1.2. Human hand model.

in order to live in the sea, while bats are shaped like cloaks so as facilitate flight, and dogs are shaped so as to move easily on land on four legs. All are specialized to their environments. While primates also have five fingers, many have shorter thumbs compared to humans — reflecting how life in the trees (e.g., hanging from branches) is easily accomplished with 2–4 fingers, making the thumb less important. For this purpose, the need for thumb–finger opposition is rare, so a primate's thumb is not suitable for grasping and manipulating small objects.

The human hand became generalized rather than specialized to a particular environment. Its manipulation capability was improved by the acquisition of an opposable thumb, creating a hand suitable for manipulating various objects. In addition, instead of simply bending in parallel directions, the four fingers can also move toward the thumb.

Regarding the motion of the hand, Napier (1956) pointed out that grasps can be divided into two broad categories: **power grasp** and **precision grasp**. Gripping a hammer is an example of a power grasp. The fingers clamp down on an object with the thumb creating counter pressure. In contrast, a precision grasp is exemplified by gripping a small ball; multiple fingertips and the thumb grasp the target object. Cutkosky (1989) proposed a grasp taxonomy containing 16 hierarchical structures based on the example of the grasping motion of workers in a machine factory. Kamakura *et al.* (1998) provided a more detailed analysis than Cutkosky's (1989). They included non-grasp grips used in daily activities from the point of view of hand function training.

Detailed classifications for grasping are not necessary from the point of view of robotic hand control. Whether there is a physical constraint of **contact** or **no contact** between the hand and the object is of greater importance. Figure 1.3 shows a simplified classification of hand functions, broadly classified into contact or non-contact with the object, based on Cutkosky's (1989) system. In the grasp range shown, dexterity requirements increase from left to right, while power requirements increase from right to left.

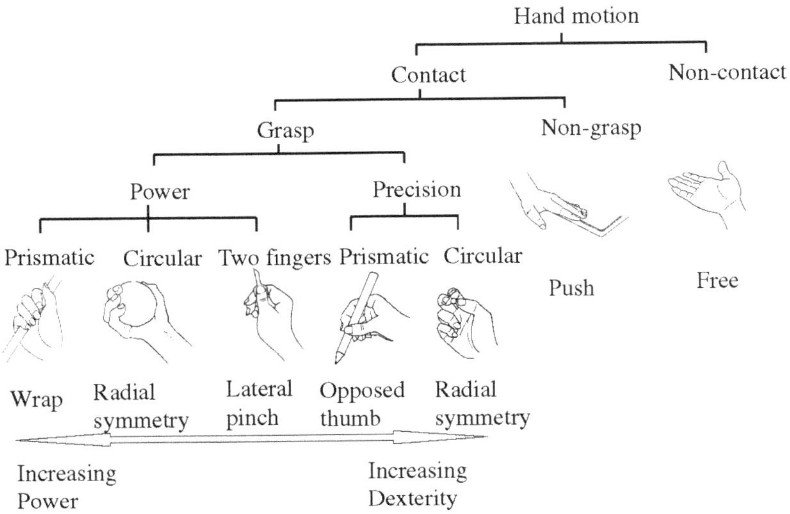

Fig. 1.3. Classification of hand functions (Modified from Cutkosky, 1989).

1.3. Tactile Sense

Human skin contains numerous cells known as **tactile receptors** (Grunwald, 2008). Skin is divided into two areas: hairy and glabrous, as shown in Fig. 1.4. The palm and finger pads are examples of glabrous skin. Glabrous skin is thick, able to retain a palm print and fingerprint, and plays an important role in obtaining information from the outside world through touch. The hairy areas of the skin are non-glabrous, and account for most of the skin. There are many hairy areas on the back of the hand. As seen in Fig. 1.4, from the outside in, both glabrous and hairy areas are divided into three layers: the epidermis, the dermis, and the subcutaneous. The epidermis consists of a papillary layer and stratum corneum that is layer of dead cells. Its thickness at the palmus manus is 0.7 mm. The dermis is a fibrous connective tissue with a thickness of 0.3–2.4 mm.

Receptor cells distributed on the inner part of the skin experience skin sensations such as touch, pain, and heat. The receptors involved in tactile sensing have either a specific structure or a free nerve terminal without a specific structure that spreads at the tip of the nerve fibers. These are called **mechanoreceptors**, because they

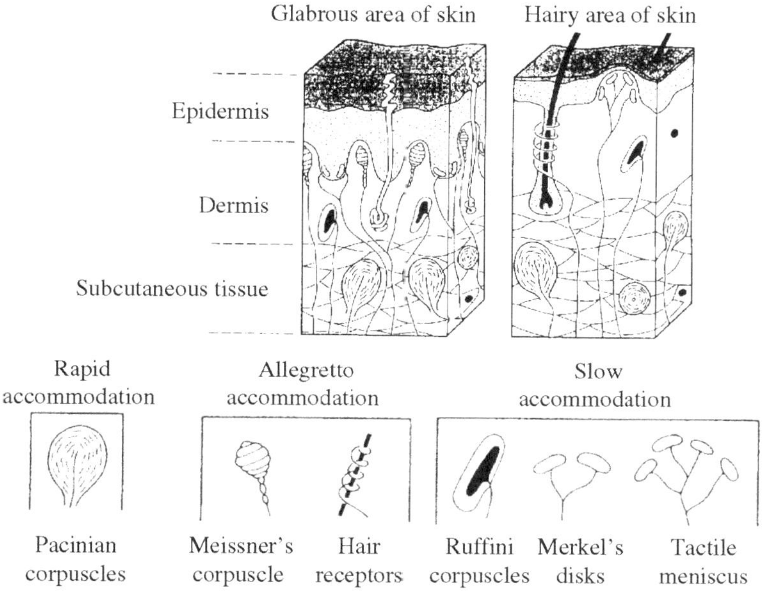

Fig. 1.4. Hand skin.

react to mechanical stimulation applied to the skin. The mechanoreceptors contain Pacinian corpuscles, Meissner's corpuscles, hair follicle receptors, Ruffini endings, Merkel's disks, tactile meniscus, and so on. These are further categorized based on their rates of adaptation into four types: fast adapting type I (FA I), which adapt to velocity; fast adapting type II (FA II), which adapt to acceleration; slow adapting type I (SA I), which respond to velocity and displacement; and slow adapting type II (SA II), which respond mainly to stimuli displacement.

Meissner's corpuscles are distributed evenly in the papilla of the dermis, and their structure is encapsulated by nerve endings and surrounded by nerve tissue. Corpuscles are 80–150 μm in length and 20–40 μm in diameter. The action potential of the nerve is immediately brought about by morphological deformation of the corpuscles, which quickly reduces and eventually subsides completely. The corpuscles return to their original form when there is no further stimulation. In their position at the front of the dermis, they are suitable for detecting the speed of skin displacement by tactile stimulation. Meissner's corpuscle can detect crude vibrations of less than 40 Hz, and so are classified as FA I.

Vater–Pacini corpuscles are oval-shaped 1 mm bodies distributed in the subcutaneous tissue; the capsule bodies consist entirely of a few tens of layers of nerve endings. Due to their layered structure, the corpuscles are selectively capable of conveying changes of mechanical stimulation (differential information) and detecting the acceleration of skin displacement. As a result, they respond to high-speed pressure changes and vibration, and their threshold value is lowest with a repeated stimulation of 200 Hz. Vater–Pacini corpuscles are easily excited on contact with the skin because they have remarkable sensitivity. They are classified as FA II.

Merkel's disks are located in the deepest part of the epidermis and formed from Merkel tactile cell and nerve endings existing at the boundary of the dermis. Their adaptation is slow showing a response that is proportional to the magnitude of the sustained skin displacement. Merkel's disks are classified as SA I.

Ruffini endings are nerve endings 0.5–2 mm in length, wrapped in a package located in the lower dermis and subcutaneous tissue. Like Merkel's disks, Ruffini endings are slow receptors with a response proportional to the magnitude of the sustainable skin displacement. Unlike Merkel's disks, they exist in the dermis layer, and have a tendency to be easily pulled by skin and respond somewhat to further displacement. For these reasons, they are classified as SA II.

The density of FA I and SA I mechanoreceptors (shown in Fig. 1.5; Johansson and Vallbo, 1983) is about 140 pieces/cm^2 at the fingertips. It decreases as distance from the fingertips increases, and is around 24 pieces/cm^2 in other areas of the fingers and the palm. In the case of FA II and SA II, all parts of the hand have the same low density of about 21 pieces/cm^2. It is said that there are about 17,000 pieces of mechanoreceptors in the human hand: 43% are FA I, 13% FA II, 25% SA I, and 19% SA II. There is a close relationship between cognitive ability and tactile density of the mechanoreceptors. Humans take advantage of the information from a number of receptors in order to manipulate objects dexterously. Human skin acts

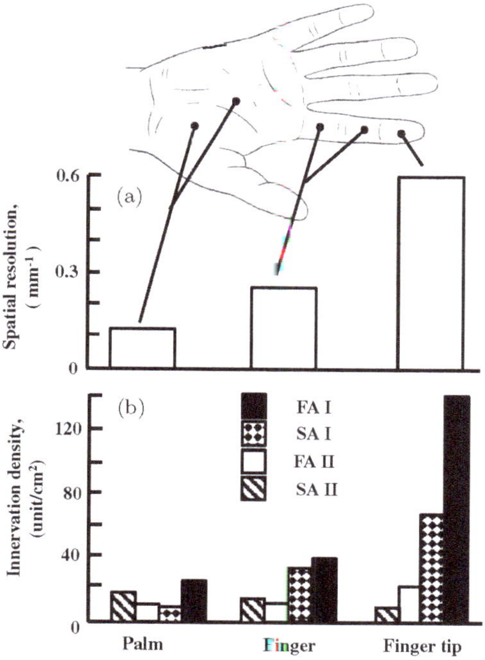

Fig. 1.5. Spatial resolution capacity in psychophysical tests compared to density of tactile units in three glabrous skin regions (Adapted from Johansson and Vallbo, 1983).

as the protective barrier between the internal body systems and the outside world. Its ability to perceive touch sensations gives human brains a wealth of information about the environment around them, such as temperature, pain, and pressure. Without the sense of touch, it would be very hard to get around in the world. Although we still don't know all the ins and outs of how the skin perceives touch, what we do know is fascinating.

1.4. The Robot Hand

1.4.1. *History of the Robotic Hand*

Future humanoid robots will execute various complicated tasks based on instructions received from the human users. These robots will be equipped with anthropomorphic **robotic hands** much like the human hand. The robots will eventually supplant human labor in the execution of intricate and dangerous tasks in areas such as manufacturing, and space and deep ocean exploration.

Many multi-fingered robot hands have been developed (e.g., the Stanford–JPL hand by Salisbury *et al.* (1982); the Utah/MIT hand by Jacobsen *et al.* (1984); and the Anthrobotic hand by Kyriakopoulos *et al.* (1997). These robot hands are driven by actuators located remotely from the robotic hand frame and are connected by tendon cables. The elasticity of the tendon cables can cause inaccurate joint angle control, and the long wiring of such cables may obstruct robot's motion, when the hand is attached to the tip of the robotic arm. These hands have been problematic commercial products, particularly in terms of maintenance, due to their mechanical complexity.

To solve these problems, robotic hands in which the actuators are built directly into the hand have been developed [e.g., the Omni hand by Rosheim (1994); the NTU hand by Lin and Huang (1996); and the DLR hand by Liu *et al.* (1999)]. However, these hands present a challenge in that their movement is unlike that of the human hand because the number of fingers in the robotic hand — and the number of joints in the finger — are insufficient. Many reports on the use of tactile sensors (e.g., Fearing, 1990; Howe, 1994; Johnston *et al.*, 1996) have been presented, all of which attempted to realize adequate object manipulation involving contact with the fingers and palm of a robotic hand. The Gifu Hand II (Kawasaki *et al.*, 2002) was developed to provide a standard robot hand for use in the study of grasping and dexterous manipulation. It was the first robotic hand equipped with six-axes force sensors at the fingertips and was covered by distributed tactile sensors.

1.4.2. *Design of the Gifu Hand*

The authors of this study developed the Gifu Hand I (Kawasaki and Komatsu, 1999), the Gifu Hand II (Kawasaki *et al.*, 2002), the Gifu Hand III (Mouri *et al*, 2002), as well as the kinematic humanoid (KH) hand type S (Mouri and Kawasaki, 2007). Figure 1.6 shows the Gifu Hand III and the KH hand type S. The aim of these developments

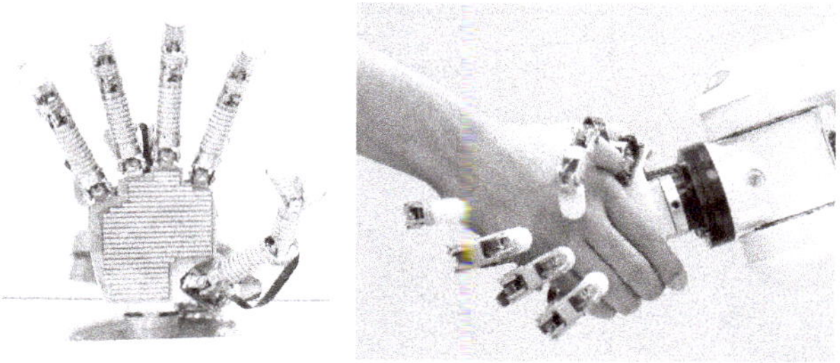

(a) Gifu Hand III (b) KH H and type S

Fig. 1.6. Anthropomorphic robotic hand.

was to build a standard robot hand that could be used to study grasping and dexterous manipulation. The design concepts for these robot hands are as follows:

1.4.2.1. *Size*

For skillful manipulation, it was desirable for the robotic hand to resemble the human hand in size. The robot hands were designed similar to a relatively large human hand, with a thumb and four fingers.

1.4.2.2. *Finger DOFs*

There is a mechanical difference between the thumb and the other fingers in the Gifu Hand III. The fourth joint of the thumb is driven by a servomotor and a gear train lodged in the second segment of the thumb; the fourth joint of other fingers, meanwhile, is driven through a planar four-bar linkage mechanism in the second segment of the finger. Thus, the thumb has one more DOF than the other fingers. As a result, the Gifu hand has a total of twenty joints and sixteen DOFs, which is similar to the number of joints and DOFs of the human hand. In the human hand, the motion of the fourth joint in all fingers except the thumb engages with that of the third joint. Figure 1.7 shows the relation between the third and fourth joint angles of the forefinger in the human hand (Kawasaki and Komatsu, 1999). As seen in the figure, the fourth joint angle engages with the third almost linearly. In the Gifu Hand III, the fourth joint is driven

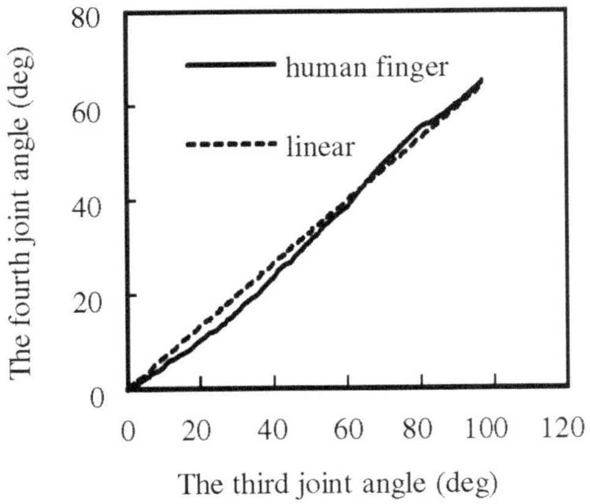

Fig. 1.7. Relation between the third and fourth joint angles of the forefinger in the human hand (Adapted from Kawasaki and Komatsu, 1999).

by a planar four-bar linkage mechanism, as shown in Fig. 1.8 (a); here, a_i is the length of the finger's second segment, r_{i3} and r_{i4} are the link radii of the third and fourth joints, q_{i3_0} and q_{i4_0} are the initial angles of the planar four-bar linkage mechanism, l_i is the length of the connection link, and the suffix i refers to the i-th finger. The relation between the third and fourth joint angles is made almost linear by setting proper link parameters, as shown in Fig. 1.8 (b).

1.4.2.3. *Configuration of human hand and opposability of the thumb*

In an illustration of three configurations of a human hand by Isobe (1985) (see Fig. 1.9), the following concepts were presented: (i) when all fingers and the thumb are retroflexed, the fingertips are located on the circumference of a circle whose radius is the length of the middle finger; (ii) the vertical axis of the palm corresponds to the median line of the middle finger, and (iii) each finger line at the flexion concentrates to the tubercle of the scaphoid bone. A line from the first joint of the index finger to that of the little finger crosses the vertical axis at 75 degrees.

Usually, the thumb of a human hand moves in opposition to the fingers, and this opposability is the source of the dexterity of the

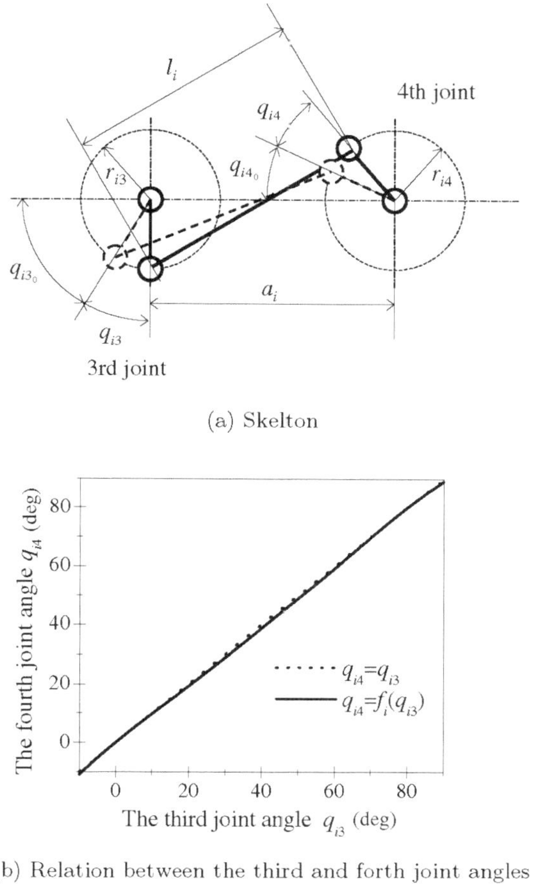

Fig. 1.8. Planar four-bar linkage mechanism (Adapted from Kawasaki and Komatsu, 1999).

human hand in object manipulation. The palm of the robot hand was designed to have high opposability of the thumb in consideration of concepts (i) and (ii).

A performance index of the thumb's opposability is given as follows (Mouri *et al.*, 2002):

$$J = \sum_{i=1}^{k} \frac{w_i v_i}{d_i^3}, \tag{1.1}$$

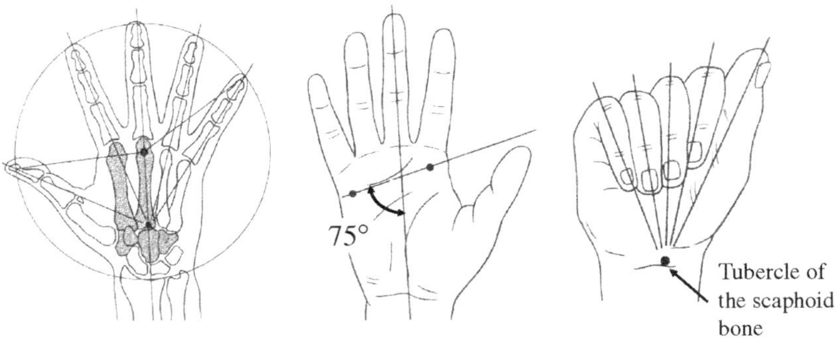

(a) Finger alignment (b) Axes of hand (c) Finger lines at flexion

Fig. 1.9. Human hand position (Adapted from Isobe, 1985).

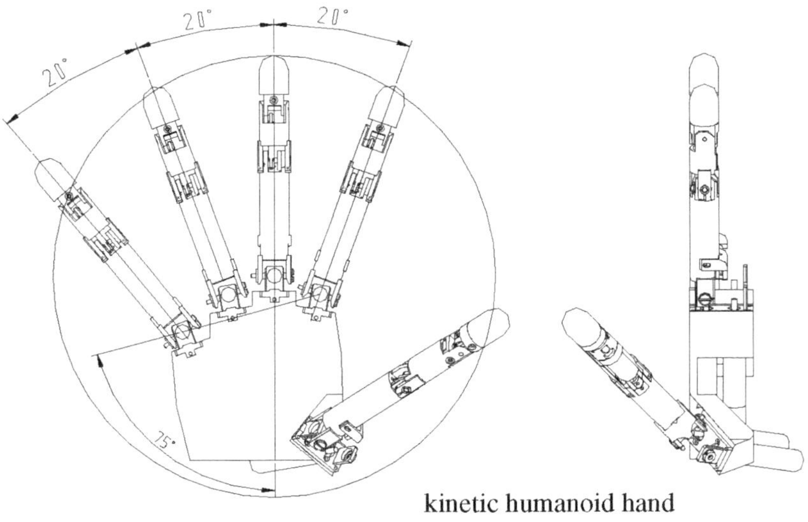

Fig. 1.10. Configuration of thumb and fingers.

where v_i denotes the volume of the intersection between the mobility space of the thumb and that of the i-th finger, k denotes the finger number (never the thumb), d_i is the maximum length of the fingers, and w_i is a weighting coefficient. A configuration of the fingers and thumb based on the features of the KH hand is shown in Fig. 1.10. Except for the little finger, the fingertips of the thumb and fingers

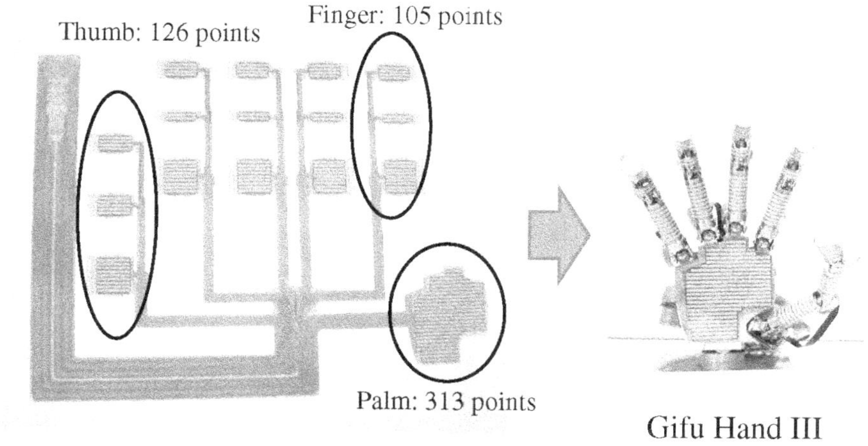

Fig. 1.11. Gifu hand III and distributed tactile sensors.

are located on the circumference of a circle whose radius is the length of the middle finger.

1.4.2.4. *Force sensor and distributed tactile sensors*

The robot hand grasps and manipulates objects dexterously with the help of force sensors and tactile sensors in the fingers. Each finger was designed to be equipped with a six-axes force sensor for compliant pinching. The robot hand's design included the mounting of distributed tactile sensors, as shown in Fig. 1.11. The number of detection points on the palm, thumb, and each finger is 313, 126, and 105, respectively, for a total of 895 detection points. The tactile sensors are composed of grid-pattern electrodes and use conductive ink in which the electric resistance changes in proportion to the pressure on the top and bottom of a thin film.

The extended output patterns of the tactile sensors, measured while the Gifu hand is holding a soft spherical object 95 mm in diameter and a hard cylindrical object 66 mm in diameter, are shown in Fig. 1.12. Each vertical line in the figure represents the output of a tactile sensor with the unit kN/m^2. The points on the horizontal plane are the measurement points of the tactile sensors, which are opened to show the measurement data of the force. It is clear that

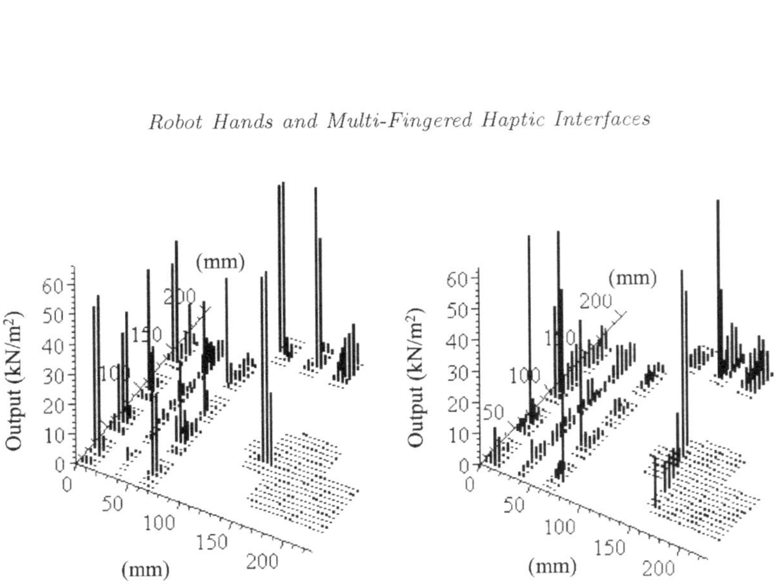

(a) Spherical object (b) Cylindrical object

Fig. 1.12. Output patterns of the tactile sensors by the Gifu hand II (Adapted from Kawasaki *et al.*, 2002).

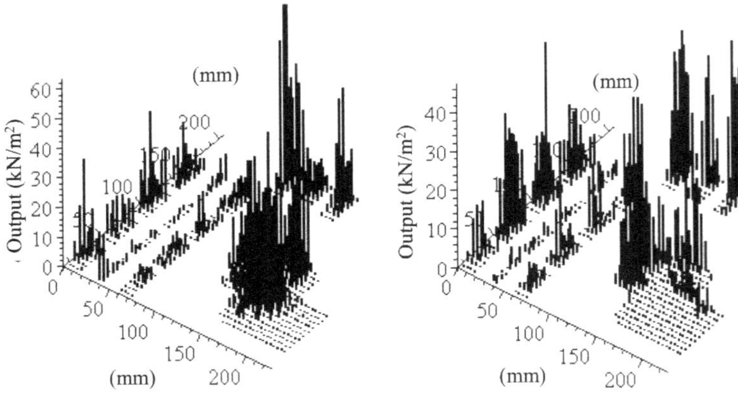

(a) Spherical object (b) Cylindrical object

Fig. 1.13. Output patterns of tactile sensors by the human hand (Adapted from Kawasaki *et al.*, 2002).

the output pattern depends on the shape of the object. Figure 1.13 compares these with the output patterns of the tactile sensors when the human hand grasps the same objects using the globe attached to the tactile sensors. In this case, a part of the tactile sensor was cut to fit the size of the human hand. Being covered with soft derma, the output pattern by the human hand differs from that by the robotic hand, in part. However, these show that the Gifu hand with the distributed tactile sensors has a high potential for dexterous grasping and manipulation.

1.5. Multi-Fingered Haptic Interfaces

Most human interfaces in virtual environments use information from visual and auditory sensations. With the addition of tactile sensations to these human interfaces, a human operator can handle objects with a realistic tactile sense and perform complex tasks in a virtual-reality environment. **Multi-fingered haptic interfaces** that allow multipoint contact between users and a virtual environment have greater potential for various applications than single-point haptic interfaces. The multipoint interaction allows a user to perform natural actions such as grasping, manipulation, and exploration of virtual objects. Such interaction will dramatically increase the believability of the haptic experience.

Over the last two decades, haptic interfaces that present force and tactile feelings to the fingertips of humans have been investigated extensively. Some of these interfaces (Walairacht *et al.,* 2001; Adachi *et al.,* 2002; Montoy *et al.,* 2008) provide three-dimensional (3D) force feelings to the human fingertip, and a few haptic interfaces cover the workspace of the human arm. Such interfaces are classified into two types: wearable-type interfaces (f.e., CyberGrasp by CyberGlove systems) which are mounted firmly on the human hand, and grounded-type interfaces, which are located in positions that match and are opposed to the human hand. The wearable-type haptic interface has a large workspace, but with this interface, it is difficult to present three-directional forces or to simulate the weight of virtual objects on the fingertips, because the hand mechanism is mounted on the back of a human hand and the exerted force is only one-directional. In contrast, the grounded-type haptic interface generally has a fairly small workspace compared with the wearable-type haptic interface, but a grounded-type haptic interface consisting of an arm and fingertips could be used in a large workspace. A haptic interface must be safe, function in a wide operation space, and represent the weight of virtual objects, as well as the force at the contact points. In addition, it should not produce an oppressive feeling when attached to a person, and it should not represent its own weight.

1.5.1. *Mechanical Structure of the Haptic Interface Robot (HIRO) Hand*

Multi-fingered haptic interfaces joined to the arm can provide a wide operation space. However, most of these interface systems are mounted on the back of the human hand (see, e.g., the CyberForce). Fixing the haptic interface to the hand can give the operator an oppressive feeling, since the interface must be firmly bound. To reduce this oppressive feeling and increase safety, we developed multi-fingered haptic interface robots that are located opposite to the human hand, including the HIRO (Kawasaki *et al.*, 2003), HIRO II^+ (Kawasaki and Mouri, 2007), and HIRO III (Endo *et al.*, 2011) (Fig. 1.14). The HIRO III can present 3D forces to the operator's five fingertips. The specifications of the HIRO III are listed in Table 1.2. The HIRO III is commercially available under the name Hand HIRO.

The HIRO consists of an interface arm, a haptic hand with five haptic fingers, and a controller; it can present a three-directional force at each of the five fingertips. The user places the HIRO opposite to the hand, and the haptic fingertips are connected to the human fingertips with finger holders and passive spherical magnet joints.

Fig. 1.14. Multi-fingered haptic interface robot, HIRO III.

Table 1.2 Specifications of HIRO III.

Hand	Degrees of freedom	15 DOFs
	Weight	0.78 kg
Finger	Degrees of freedom	3 DOFs
	Weight	0.12 kg
	Maximum output force	Over 3.6 N
	Workspace	705 cm^3 (Thumb) 587 cm^3 (Other)
Arm	Degrees of freedom	6 DOFs
	Weight	3.0 kg
	Maximum output force	Over 56 N
	Workspace	0.09 m^3

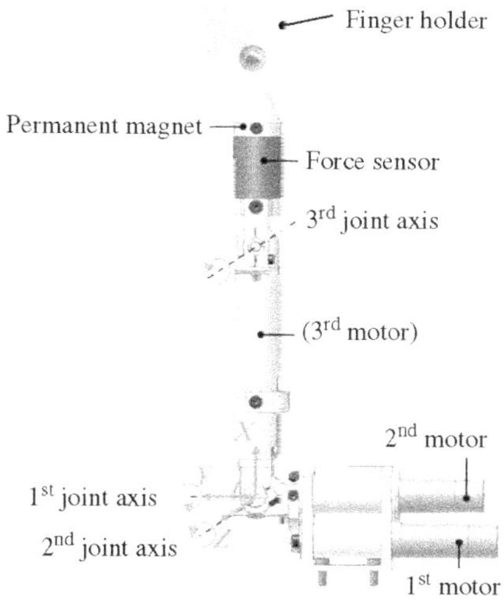

Fig. 1.15. Mechanical structure of the haptic finger.

When the operator moves his/her hand, HIRO follows the motion of the operator's fingers and represents the force's feelings. Thus, the operator is unlikely to experience much feeling of oppression, because the connection between the human hand and haptic interface is limited to the fingertips.

The mechanical structure of the haptic finger of HIRO III is shown in Fig. 1.15. It has three joints, allowing for three DOFs. The first

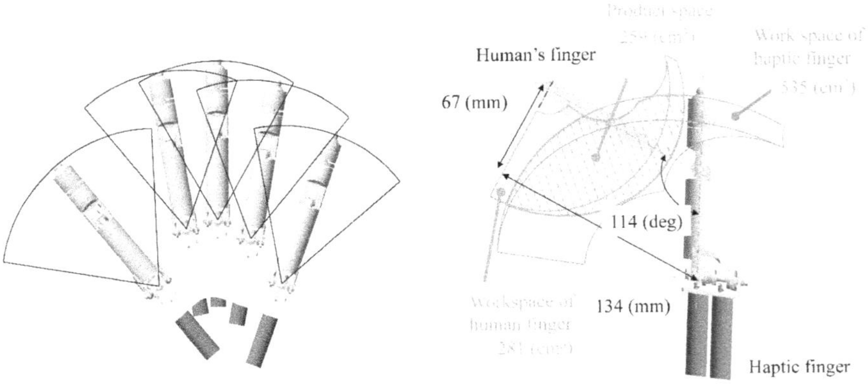

(a) Work space of haptic hand (b) Product space of index

Fig. 1.16. Product space of haptic finger and human index finger by HIRO II.

joint, relative to the hand base, allows abduction/adduction. The second and third joints allow flexion/extension. All joints are driven by DC servomotors with gear transmissions and rotary encoders. The workspace of the haptic finger is illustrated in Fig. 1.16 (b). The thumb is designed to work in a wide space, because the workspace of the human thumb is larger than that of the other fingers. The finger layout of the haptic hand is designed to maximize the volume formed by the intersecting workspaces of the haptic and human fingers, as shown in Figs. 1.16 (a) and (b). The haptic index finger was designed based on the statistical data for Japanese males. The position of the fingers on the haptic hand was designed based on considerations of this geometrical relation.

To read the loading of the fingers, a three-axes force sensor was installed in the second link of each haptic finger. To manipulate the haptic interface, the user has to wear a finger holder on his/her fingertips, as shown in Fig. 1.15. This finger holder has an iron sphere, which forms a passive spherical joint, when attached to the permanent magnet at the force sensor tip. The passive spherical joint has two roles. The first is to adjust for differences between the orientation of the human and haptic fingers. Each human finger has six DOFs, while each haptic finger has only three DOFs. Hence, an additional passive three DOFs are needed. The second role of the spherical joint is to ensure that the operator can remove his fingers from the haptic interface, in case of malfunction. The suction force by the permanent magnet is 4.3 N. The maximum output force at a fingertip is 3.6 N. The maximum fingertip velocities, v_x, v_y, and v_z, are 0.43, 0.43, and 0.23 m/s, respectively.

The interface arm with six DOFs is designed to be compact and covers the workspace of an average human desk. The lengths of the upper arm and the forearm are 0.2 and 0.17 m, respectively. The arm joints are actuated by servomotors equipped with rotary encoders and gear transmissions. As a haptic interface, the combination of the arm and hand makes a wide workspace.

1.5.2. *Control System*

The control system of the HIRO III is shown in Fig. 1.17. To present the force at the five human fingertips, 15 DC motors and five force

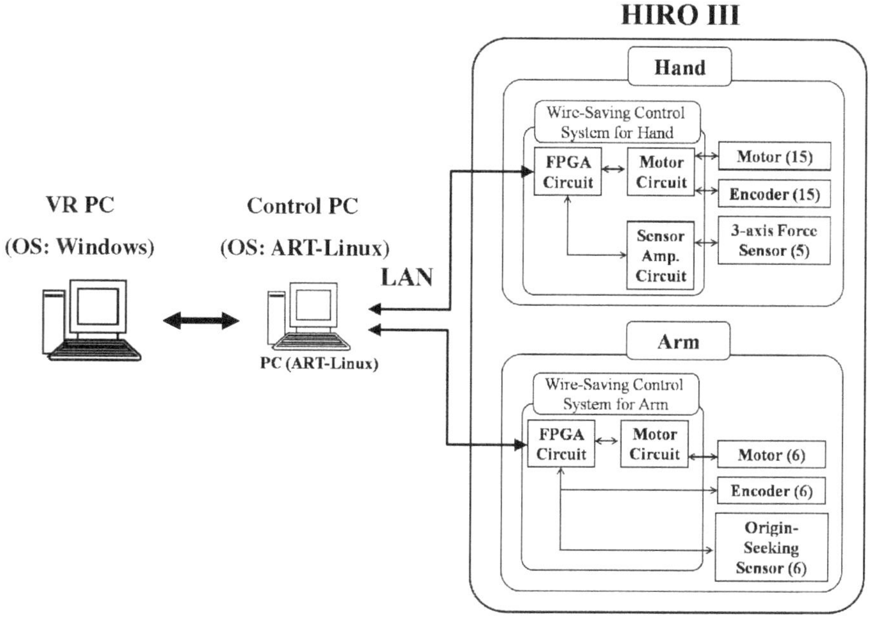

Fig. 1.17. Control system of HIRO III.

sensors are installed in the haptic hand. As a result, the communication cable between the control PC and the haptic hand consists of 60 wires for the 15 encoders, 30 wires for the 15 motors, and 40 wires for the five force sensors. Similarly, the communication cable between the control PC and the haptic arm consists of 102 wires for the six encoders, motors, and origin-seeking sensors. Therefore, there are a total of 202 wires between the haptic interface and the control PC. These wires greatly obstruct the smooth movement of the haptic interface. Thus, a wire-saving control system was developed for the haptic hand consisting of the interface Field-Programmable Gate Array (FPGA) circuit, the force sensor amplifier circuit, and the motor amplifier circuit. These circuits are mounted in the haptic hand frame. In the FPGA, Very High Speed Integrated Circuits (VHSIC) Hardware Description Language (VHDL) was used to create a counter driver to count the up/down pulses from the motor amplifier circuit, a Pulse Width Modulation (PWM) driver to provide the PWM outputs of 15 motors, and a sensor driver to read the digital values of the five force sensors and provide the zero adjustment value for them. The FPGA sends a PWM signal to the motor amplifier circuit when it receives the duty ratio signal from the control PC.

A three-axes force sensor is installed in the second link of each haptic finger and is connected to the sensor amplifier circuit (Endo *et al.*, 2008). Using the strain gauge in the pressure port of the sensor element, the force sensor measures the force from the deformation of the metal beam. There are several reasons for using the strain gauge: (1) it allows the sensor to have better linearity; (2) effects of temperature change are small; and (3) it makes the force sensor suitable for measurement of dynamic phenomena because the amount of displacement is extremely small.

1.5.3. *Human Perception Using the HIRO*

It is useful to measure human perceptual ability when using a haptic interface. In particular, it is important to know how accurately humans can perceive fingertip force through the haptic interface. The

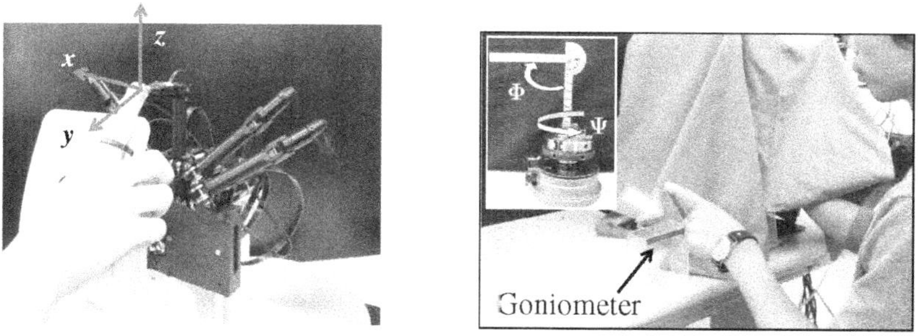

(a) Coordinate systems of finger (b) Experimental setup

Fig. 1.18. Measurement environment.

results of investigating the human perception of fingertip force form the foundation of HIRO applications.

1.5.3.1. *Force perception*

Human perceptual ability with regard to the direction of force has been evaluated experimentally using HIRO III, as shown in Fig. 1.18 (Endo *et al.*, 2010). Ten people (age range 21–24; nine males and one female) participated in the experiment. All the participants were right handed. Participants connected their index finger to the HIRO III at a bar, as shown in Fig. 1.18(a), and HIRO III presented the force to the participants. While measurements were being taken, the hands of both HIRO III and the participants were covered with a cloth, so that visual information was not available, as shown in Fig. 1.18(b). The participants responded to the direction of the presented force using a measuring instrument called a goniometer. The standard stimulus force was 2.2 N, and it was presented until the participant responded. Figs. 1.19(a) and (b) show the direction of the force presented in the horizontal ($\boldsymbol{xy}$) plane and vertical ($\boldsymbol{yz}$) planes, respectively. Figs. 1.19(c) and (d) show the measurement results of the average angular error between the presented force direction and the response force direction, at each presented angle in the horizontal and vertical planes. Human perception of force direction had an average error of 0.23 rad in the horizontal, and 0.25 rad in the vertical directions. In other words, there was not a big difference in the perception of force in the two planes. This also showed that angular error was lowest directly forward, and increased largely according to angle distance from the forward direction. This phenomenon was likely due to the haptic interface using a finger holder. Human perceptual ability with regard to magnitude of force was also evaluated experimentally using HIRO III in Endo and Kawasaki (2013).

1.5.3.2. *Identification of friction moment direction (FMD)*

It is necessary to display both the friction force and the friction moment acting on the human fingertips to enhance the reality of object manipulation in a virtual–reality environment. For example, when a person places a book on a shelf by grasping it near its corner or pours water from a bottle into a cup by grasping the cup handle with two fingertips, he or she feels the friction moment at the fingertips more strongly than the friction. A new finger holder has been developed to display the friction moment. It is almost the same shape as the previous finger holder, as shown in Fig. 1.20(a) (Kawasaki *et al.*, 2011). A small brushless motor was built into an 8 mm diameter ball attached to the finger holder. Specifications of the motor were as follows: outside diameter 2.4 mm; overall length 10.8 mm; mass 0.245 g; gear ratio 337; and maximum torque 0.69 mNm at a rated voltage of 3 volts.

When wearing the newly developed finger holder, subjects' identification of FMD was examined by the simple experimental system shown in Fig. 1.20(b). The subjects put the index finger with the holder on a disk with the specified pushing force of 2 N. Subjects could observe the force through the load-measurement instrument. There were 10 subjects, aged 21–24. The percentage of correctly identified FMD at a normal contact force of 2 N is shown in Fig. 1.21. To measure the percentage of correctly identified FMD, the friction moments were displayed to the subject's fingertip, as step inputs at random levels between 0.1 and 0.9 mNm for each subject. The display time of the force moment was 3 s. Before the experiment, subjects were trained four times to familiarize them with the experimental apparatus. After the finger holder displayed a friction moment to their fingertips, subjects were asked whether they had felt a friction

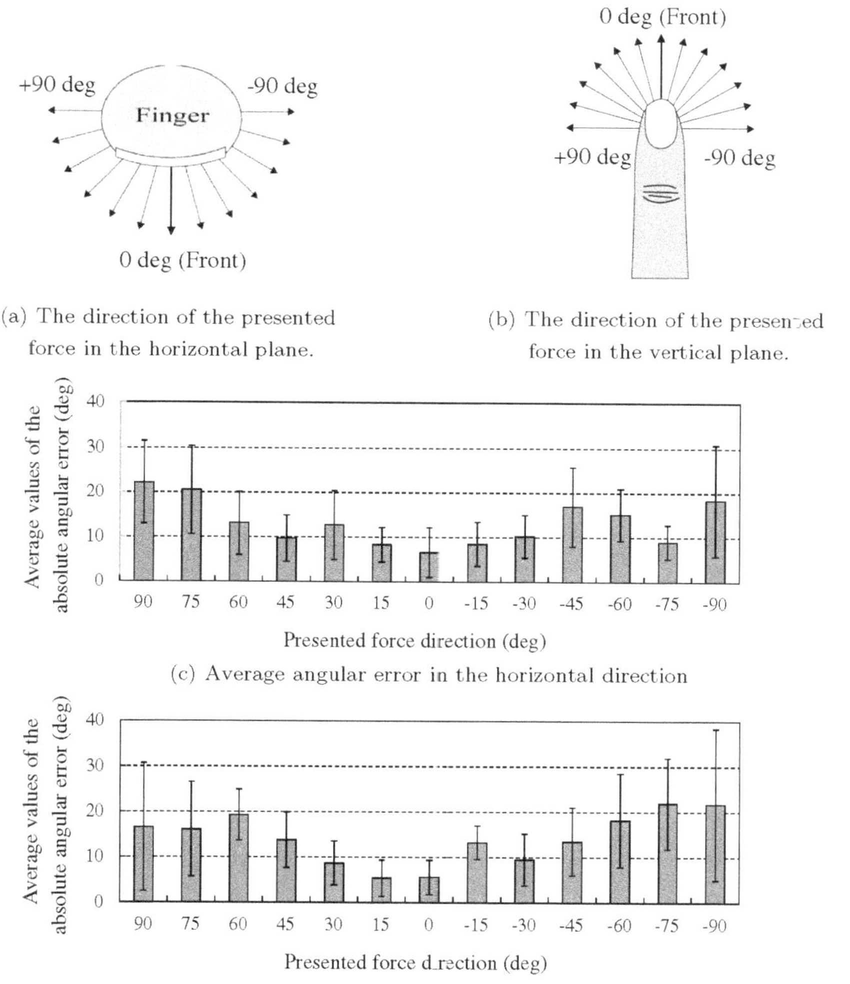

(a) The direction of the presented force in the horizontal plane.

(b) The direction of the presented force in the vertical plane.

(c) Average angular error in the horizontal direction

(d) Average angular error in the vertical direction

Fig. 1.19. Measurement results of the angular error at each presented force direction (Adapted from Endo and Kawasaki, 2013).

moment and to give its rotational direction if so; that is, to state whether the sensed moment (if any) had moved towards the right, towards the left, or was unclear. Subjects recognized FMD of more than 0.45 mNm with 80% accuracy This result was very similar to the one observed when a subject put his/her index finger directly

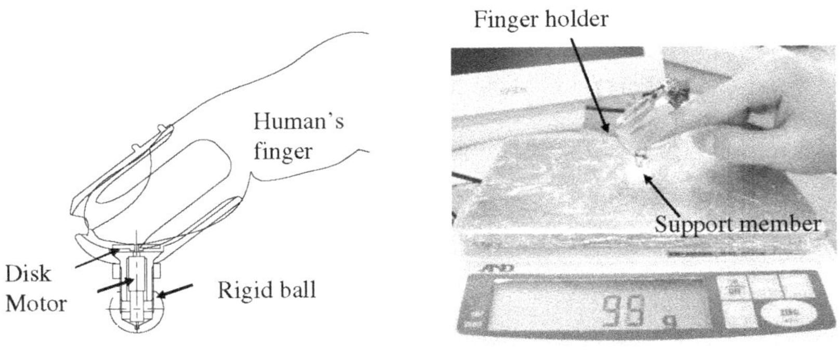

(a) Developed finger holder for displaying friction moment

(b) Measurement system

Fig. 1.20. Experimental system for perception of friction moment.

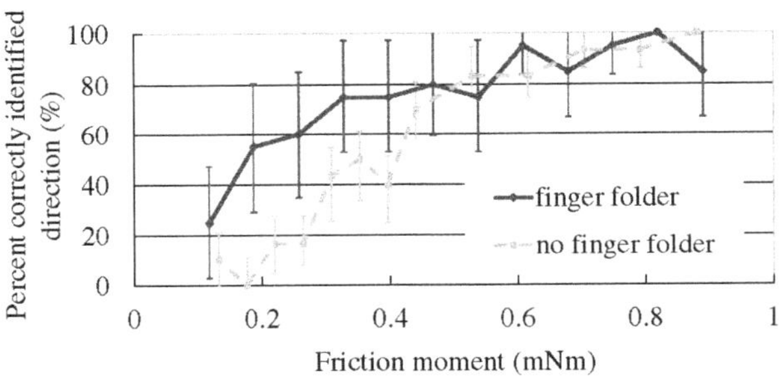

Fig. 1.21. Comparison of percent of correctly identified FMD with and without the finger holder (Adapted from Kawasaki *et al.*, 2011).

on a disk, driven by a DC motor, fixed on the load-measurement instrument without a finger holder. This finding indicated that the presence or absence of a finger holder had no effect on the identification of rotational direction of more than 0.45 mNm. The combination of the new finger holder and the computational method of friction moment enhanced the reality of object manipulation in a virtual–reality environment.

Exercise

1.1. The two-point threshold of tactile force is the smallest noticeable distance between two points, at which a sense of touch is detected on the skin, as the two legs of a drafting compass are closed. Examine this.

1.2. Penfield's studies of the control areas for motor and sensory control in the human cerebral cortex are well known (Penfield and Rasmussen, 1950). Examine this.

Chapter 2

Kinematics of Multi-Fingered Hands

When a robot hand grasps and manipulates an object, the hand and object interact at several contact points. Since contact conditions vary according to point contact, line contact, surface contact, and the existence or non-existence of friction at the contact point, the constraint conditions that the object receives are also diverse. This chapter presents basic **kinematics** in terms of the grasp of a multi-fingered robotic hand, including contact models, forms of grasp, constraint conditions, and manipulable grasp. Introduction of these topics are available in Yoshikawa, 1995, 1996a, 1996b and Zeng *et al.*, 1997.

2.1. Contact Statics

2.1.1. *Contact Models*

When a robot hand grasps and manipulates an object, contact models are important for the analysis of object constraint conditions and the possibilities of the grasp. They are also necessary for robotic hand control. When two objects come in contact with each other, there is potential for a point contact, a line contact, and a surface contact, as shown in Fig. 2.1. When a robot hand manipulates an object, the force and moment exerted by the hand on the object vary, depending on the existence or non-existence of friction at the contact surface. Let us define a vector $\boldsymbol{F}_C = [\boldsymbol{f}_C^T, \boldsymbol{n}_C^T]^T \in \mathbb{R}^6$, where the force $\boldsymbol{f}_C \in \mathbb{R}^3$ and the moment $\boldsymbol{n}_C \in \mathbb{R}^3$ combine at contact point C when exerted from the finger to the object. Note that a force and moment pair is called a **wrench**, a symbol $\mathbb{R}^n$ is n dimension real space, and a symbol $\in$ implies an element.

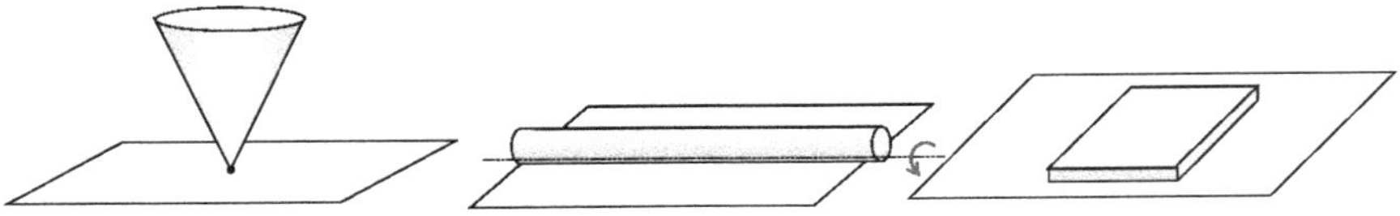

(a) Point contact (b) Line contact (c) Surface contact

Fig. 2.1. Type of contacts.

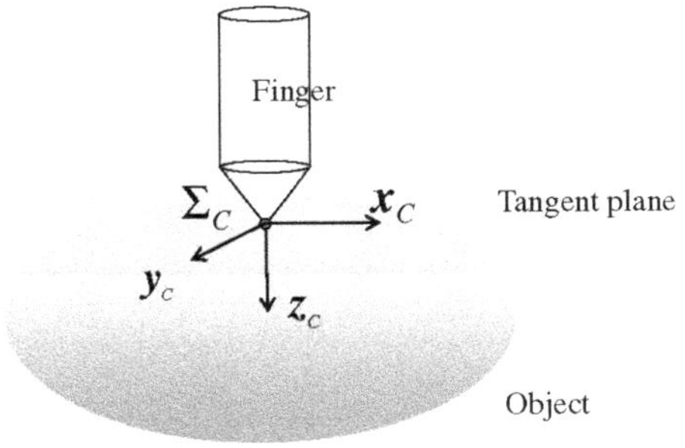

Fig. 2.2. Contact coordinate system.

For the analysis of the contact model by robotic fingers, a contact coordinate system Σ_C is set at the contact point of the object, as shown in Fig. 2.2. In the figure, the $\boldsymbol{z}_C$ axis of the contact coordinate system is a unit normal line toward the center of the object at the contact point, while the $\boldsymbol{x}_C$ and $\boldsymbol{y}_C$ axes are on the tangent plane. The contact with the object at the fingertip of the robotic hand can be considered a point contact when the body is rigid, and a surface contact when it is flexible. In addition, the existence or non-existence of friction at the contact surface produces differing transmissions of forces and moments. Moreover, there can be a ***fixed contact*** where the contact point between fingertip and object surface does not move and a ***slide contact*** where there is movement between the contact point of the fingertip and the object surface.

As shown in Fig. 2.3, we consider here a fixed contact, where the contact point does not move on the object surface. Depending on the existence or non-existence of friction, the contact may be classified as a **point contact without friction**, a **point contact with**

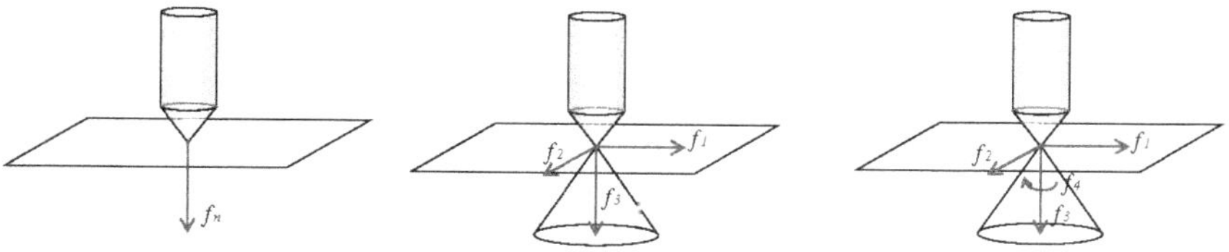

(a) Point contact without friction (b) Point contact with friction (c) Soft finger contact

Fig. 2.3. Contact model of fingers.

friction, or a **soft finger contact**. Hereafter, the left superscript of a vector represents the coordinate system in which the vector is expressed. The left superscript will be omitted if the coordinate system under consideration can be easily understood as the reference coordinate system Σ_R.

2.1.1.1. *Point contact model without friction*

A point contact without friction is obtained when there is no friction between the robot fingertip and the object, and the contact point can thus move freely on the contact surface. It is possible for the finger to apply force to the object in the normal direction, but it cannot exert force and moment except in the normal direction. The wrench exerted by the finger at the contact point in this situation is expressed with respect to the contact coordinate system and can be represented as,

$$ {}^{C}\boldsymbol{F}_C = \begin{bmatrix} 0 \\ 0 \\ 1 \\ 0 \\ 0 \\ 0 \end{bmatrix} f_n, \quad f_n \geq 0, \tag{2.1} $$

where f_n is the magnitude of the force exerted in the normal direction by the finger. The vector of the right side is called the **wrench basis** at the contact point without friction. In general, when a finger contacts an object, a contact can be a push on an object, but not a pull. This requires the condition of $f_n \geq 0$. This condition is called the **unilateral contact condition**. A point contact without friction almost never occurs in a practical situation, but it can serve as a useful model for contacts in which friction between the finger and the object is negligible.

2.1.1.2. *Point contact model with friction*

In the point contact model with friction, the friction resists the relative translational motion at the contact point between the fingertip and object, and the relative rotational motion around the contact point is free. The friction force created at the contact point is referred to as the **Coulomb friction model**. The frictional force is generated on the tangent surface and the magnitude of the frictional force is equivalent to the tangential force f_t on the contact point of the object surface. Without slipping on the contact point, the magnitude of the frictional force can be described as,

$$f_t \leq \mu f_n, \tag{2.2}$$

where $f_n = f_3, f_t = (f_1^2 + f_2^2)^{1/2}$, and $\mu > 0$ is the static friction coefficient. This equation represents any contact force that lies in a cone around the normal line of the tangent plane, as shown in Fig. 2.4. This cone is called the friction cone. One-half of the vertex angle of the cone with respect to the normal is given by,

$$\alpha = \tan^{-1} \mu. \tag{2.3}$$

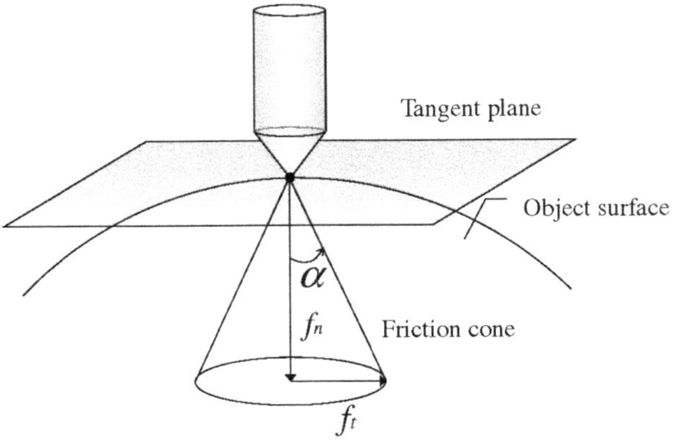

Fig. 2.4. Point contact.

Dynamic friction with sliding at the contact point occurs when the tangential force is added $f_t > \mu f_n$. The tangential force should be in a range that satisfies the conditions of Eq. (2.2) in order to keep a fixed contact. At this point, the applied wrench at the point contact with friction is represented by,

$$^C\boldsymbol{F}_C = \begin{bmatrix} 1 & 0 & 0 \\ 0 & 1 & 0 \\ 0 & 0 & 1 \\ 0 & 0 & 0 \\ 0 & 0 & 0 \\ 0 & 0 & 0 \end{bmatrix} \boldsymbol{f}_{FC}, \quad \boldsymbol{f}_{FC} \in FC, \tag{2.4}$$

where FC is a set of vectors within the friction cone given by,

$$FC = \{\boldsymbol{f} = [f_1, f_2, f_3]^T : \sqrt{f_1^2 + f_2^2} \leq \mu f_3, f_3 \geq 0\}.$$

The matrix on the right-hand side represents the wrench basis. In the same way that a frictionless point contact can push an object but not pull it, this means that the condition of $f_3 \geq 0$ is satisfied. In the point contact with friction model, it is possible for the finger to exert force on the object, but it cannot add to the moment. Table 2.1 shows the friction coefficients for most commonly used materials.

2.1.1.3. *Soft finger contact model*

Here we consider a contact model of a soft fingertip that is used as a surface contact instead of a point contact. In this model, the friction

Table 2.1 Friction coefficients.

Friction piece	Friction surface	Friction coefficient μ
Wood	Metal	0.6 (Dry) ~0.2 (Wet)
Rubber	Rubber	0.5
Leather	Metal	0.4~0.6
Teflon	Tefron (iron)	0.04
Carbon	Soft steel	0.21 (✻1)
Hard steel	Hard steel	0.35~0.4

Note: (✻1) shows the motion solid friction coefficient.
Reference: Mechanical Engineering Handbook, Published by the Japan Society of Mechanical Engineers.

moment works around the normal line at the contact point to resist the relative rotational motion in both the curved surface of the object and the fingertips, in addition to the friction force to resist relative translational motion. However, the moment is limited to maintain the static friction state and the wrench is represented as,

$$
{}^{C}\boldsymbol{F}_{C} = \begin{bmatrix} 1 & 0 & 0 & 0 \\ 0 & 1 & 0 & 0 \\ 0 & 0 & 1 & 0 \\ 0 & 0 & 0 & 0 \\ 0 & 0 & 0 & 0 \\ 0 & 0 & 0 & 1 \end{bmatrix} \boldsymbol{f}_{FC}, \quad \boldsymbol{f}_{FC} \in FC. \tag{2.5}
$$

The friction cone is given by,

$$
FC = \{\boldsymbol{f} = [f_1, \ldots, f_4]^T : \sqrt{f_1^2 + f_2^2} \le \mu f_3, f_3 \ge 0, |f_4| \le \gamma f_3\}.
$$

Here, $\gamma > 0$ is called the torsional friction coefficient. The matrix of the right hand side of Eq. (2.5) represents the wrench basis.

The three contact models discussed above are typical. The wrench applied to the object by the finger in the three contact models can be described as,

$$
{}^{C}\boldsymbol{F}_{C} = {}^{C}\boldsymbol{H}_{C}\boldsymbol{f}_{FC}, \quad \boldsymbol{f}_{FC} \in FC, \tag{2.6}
$$

where ${}^{C}\boldsymbol{H}_{C}$ is a matrix that represents the wrench basis expressed with respect to the contact coordinate system. The dimensions of the wrench basis represent the number of independent forces and the moment that can be applied at the contact point. Defining m as the dimension of the wrench basis, the friction cone (FC) has the following properties.

Property 2.1. (Friction Cone)

(i) *FC is a closed subset in space $\mathbb{R}^m$ with non-empty interior.*
(ii) *If $\boldsymbol{f}_{FC_1}, \boldsymbol{f}_{FC_2} \in FC$, then $\alpha\boldsymbol{f}_{FC_1} + \beta\boldsymbol{f}_{FC_2} \in FC$ for any given $\alpha, \beta > 0$.*

With the description noted above, the wrench is expressed with respect to the contact point coordinate system. To represent it with respect to the reference coordinate system, a rotation matrix from the contact coordinate system to the reference coordinate system is denoted as ${}^{C}\boldsymbol{R} \in \mathbb{R}^{3\times3}$. In addition, it is possible to define an extended rotation matrix whose diagonal sub matrix is the rotation matrix as given by,

$$\Re({}^{C}\boldsymbol{R}) = \begin{bmatrix} {}^{C}\boldsymbol{R} & 0 \\ 0 & {}^{C}\boldsymbol{R} \end{bmatrix}. \tag{2.7}$$

Obviously, $\Re({}^{C}\boldsymbol{R})$ is an orthogonal matrix, and $\Re({}^{C}\boldsymbol{R})^{-1} = \Re(\boldsymbol{R}_C)^T$. Then, the relationships of $\boldsymbol{F}_C = \Re(\boldsymbol{R}_C)^{C}\boldsymbol{F}_C$ and $\boldsymbol{H}_C = \Re(\boldsymbol{R}_C)^{C}\boldsymbol{H}_C$ derives $\boldsymbol{F}_C = \boldsymbol{H}_C\boldsymbol{f}_{FC}$.

2.1.2. *Kinematic Constraints Based on Contact Model*

The kinematic constraint occurs according to the contact condition between the finger and the object. Therefore, the relative linear velocity and relative angular velocity are defined as $\boldsymbol{v}_S = \boldsymbol{v}_{FC} - \boldsymbol{v}_C$, where $\boldsymbol{v}_{FC} \in \mathbb{R}^6$ is the linear and angular velocity at the finger contact point and $\boldsymbol{v}_C \in \mathbb{R}^6$ is the linear and angular velocity at the object contact point. At this time, the linear and angular velocity expressed with respect to the contact point coordinate system can be written by,

$${}^{C}\boldsymbol{v}_S = \Re({}^{C}\boldsymbol{R})\boldsymbol{v}_S. \tag{2.8}$$

First, consider the frictionless point contact model. In this case, the relative linear velocity in the normal direction always becomes zero at the contact point; however, the relative angular velocity and relative linear velocity of other directions may take any value. Therefore, the kinematics constraints between the finger and the object can be written as

$$[0\;0\;1\;0\;0\;0]\,{}^{C}\boldsymbol{v}_S = 0.$$

In terms of the point contact model with friction, the relative linear velocity between the fingertip and the object at the contact point is zero, and the relative angular velocity can take any value. Therefore, the kinematic constraints between the object and the finger can be written as

$$\begin{bmatrix} 1 & 0 & 0 & 0 & 0 & 0 \\ 0 & 1 & 0 & 0 & 0 & 0 \\ 0 & 0 & 1 & 0 & 0 & 0 \end{bmatrix} {}^C\boldsymbol{v}_S = 0.$$

In the soft finger contact model, the relative angular velocity around the normal line of the tangent plane and the relative linear velocity between the fingertip and the object at the contact point are zero. The relative angular velocity around other axes can take any value. Therefore, the kinematic constraints between the object and the finger can be written as

$$\begin{bmatrix} 1 & 0 & 0 & 0 & 0 & 0 \\ 0 & 1 & 0 & 0 & 0 & 0 \\ 0 & 0 & 1 & 0 & 0 & 0 \\ 0 & 0 & 0 & 0 & 0 & 1 \end{bmatrix} {}^C\boldsymbol{v}_S = 0.$$

Taken together, the kinematic constraints between the object and fingertip at the contact point can be described as

$${}^C\boldsymbol{H}_C^T {}^C\boldsymbol{v}_S = 0. \tag{2.9}$$

This equation shows a dual relationship with Eq. (2.6), demonstrating the fact that the relative motion is restricted to zero in the axis in which the force and moment are generated. The relationship of the equation can be shown using the reference coordinate system as follows

$$\boldsymbol{H}_C^T \boldsymbol{v}_S = 0. \tag{2.10}$$

2.2. Forms and Static Balance of Grasp

A multi-fingered hand manipulates an object with fingertip and/or finger pad contacts. Grasping an object can take the form of a **fingertip grasp** (picking up or holding with only the fingertips) or an **enveloping grasp** (in which the object is wrapped in the palm and finger pads). The enveloping grasp limits the degrees of freedom of object manipulation, but can maintain the grasp even if an external

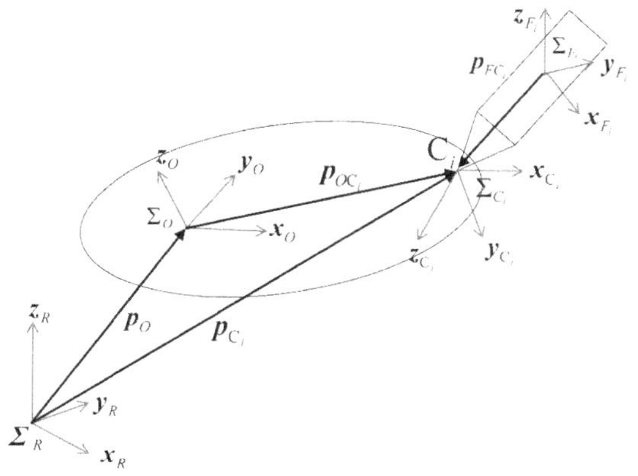

Fig. 2.5. Definition of the position vector and the coordinate.

force acts on the object. The **power grasp** is a type of enveloping grasp that restricts object motion completely. In the following, it is assumed that the hand has k number of fingers and the grasp is a fingertip grasp. An extension to the enveloping grasp is also spread out by a similar approach.

2.2.1. *Grasp Mapping*

Let us consider the force and moment acting on an object when it comes in contact with k fingers. As shown in Fig. 2.5, we set the object coordinate system Σ_O on the object, the contact coordinate system Σ_{C_i}on the i-th contact point C_i in the object surface, and the fingertip link coordinate system Σ_{F_i} at the i-th fingertip link. Note that $\boldsymbol{p}_O \in \mathbb{R}^3$ is a position vector from the origin of the reference coordinate system to the origin of the object coordinate system, $\boldsymbol{p}_{OC_i} \in \mathbb{R}^3$ is a position vector from the origin of the object coordinate system to point C_i, $\boldsymbol{p}_{FC_i} \in \mathbb{R}^3$ is a position vector from the origin of the i-th fingertip coordinate system to point C_i, $\boldsymbol{R}_O$ is a rotation matrix of the object coordinate system, $\boldsymbol{R}_{C_i} \in \mathbb{R}^{3\times 3}$ is a rotation matrix of the contact coordinate system, and $\boldsymbol{R}_{F_i} \in \mathbb{R}^{3\times 3}$ is a rotation matrix of the i-th fingertip link coordinate system, where these rotation matrices represent the orientation of the specified coordinate system and are expressed with respect to the reference coordinate system.

When the force $\boldsymbol{f}_{C_i} \in \mathbb{R}^3$ and the moment $\boldsymbol{n}_{C_i} \in \mathbb{R}^3$ act on the contact point C_i, then equivalent force $\boldsymbol{f}_{O_i}$ and moment $\boldsymbol{n}_{O_i}$ at the origin of the reference coordinate system are given by

$$\boldsymbol{f}_{O_i} = \boldsymbol{f}_{C_i}, \tag{2.11}$$

and

$$\boldsymbol{n}_{O_i} = \boldsymbol{p}_{OC_i} \times \boldsymbol{f}_{C_i} + \boldsymbol{n}_{C_i}. \tag{2.12}$$

where $\boldsymbol{p}_{OC_i} = \boldsymbol{R}_O{}^O\boldsymbol{p}_{OC_i}$. The cross product of an arbitrary vector $\boldsymbol{a}, \boldsymbol{b} \in \mathbb{R}^3$ is described as $\boldsymbol{a} \times \boldsymbol{b} = [\boldsymbol{a}\times]\boldsymbol{b}$, where $[\boldsymbol{a}\times]$ is a skew-symmetric matrix given by

$$[\boldsymbol{a}\times] = \begin{bmatrix} 0 & -a_z & a_y \\ a_z & 0 & -a_x \\ -a_y & a_x & 0 \end{bmatrix}. \tag{2.13}$$

By using this notion, the wrench at the origin of the object coordinate system $\boldsymbol{F}_{O_i} = [\boldsymbol{f}_{O_i}^T \ \boldsymbol{n}_{O_i}^T]^T$, which is equivalent to the wrench at the contact point C_i, $\boldsymbol{F}_{C_i} = [\boldsymbol{f}_{C_i}^T \ \boldsymbol{n}_{C_i}^T]^T$, are represented as

$$\boldsymbol{F}_{O_i} = \boldsymbol{W}_{OC_i}\boldsymbol{F}_{C_i} \tag{2.14}$$

and

$$\boldsymbol{W}_{OC_i} = \begin{bmatrix} \boldsymbol{I}_{3\times3} & 0 \\ [\boldsymbol{p}_{OC_i}\times] & \boldsymbol{I}_{3\times3} \end{bmatrix}, \tag{2.15}$$

where $\boldsymbol{I}_{i\times i}$ is an $i \times i$ unit matrix. As shown in Eq. (2.6), the object does not always receive all the force and moment of the fingertips. If the wrench basis matrix at the contact point C_i expressed with respect to the i-th contact coordinate system is ${}^{C_i}\boldsymbol{H}_{C_i}$, the friction cone of the i-th contact point is FC_i; the wrench generated at the origin of the object coordinate system is then

$$\boldsymbol{F}_{O_i} = \boldsymbol{W}_{OC_i}\boldsymbol{H}_{C_i}\boldsymbol{f}_{FC_i} = \boldsymbol{G}_{C_i}\boldsymbol{f}_{FC_i}, \quad \boldsymbol{f}_{FC_i} \in FC_i, \tag{2.16}$$

where $\boldsymbol{G}_{C_i} = \boldsymbol{W}_{OC_i}\boldsymbol{H}_{C_i}$.

As mentioned above, we have considered the contact point C_i. The wrench applied to the object from all contact points can be represented as the sum of the wrench, and therefore becomes:

$$\boldsymbol{F}_O = \sum_{i=1}^{k} \boldsymbol{F}_{O_i} = \boldsymbol{W}_{OC}\boldsymbol{F}_C = \boldsymbol{G}_C \boldsymbol{f}_{FC}, \tag{2.17}$$

where $\boldsymbol{W}_{OC} = [\boldsymbol{W}_{OC_1}, \ldots, \boldsymbol{W}_{OC_k}], \boldsymbol{F}_C = [\boldsymbol{F}_{C_1}^T, \ldots, \boldsymbol{F}_{C_k}^T]^T, \boldsymbol{f}_{FC} = [\boldsymbol{f}_{FC_1}^T, \ldots, \boldsymbol{f}_{FC_k}^T]^T, \boldsymbol{G}_C = \boldsymbol{W}_{OC}\boldsymbol{H}_C, \boldsymbol{H}_C = block\ diag[\boldsymbol{H}_{C_1}, \ldots, \boldsymbol{H}_{C_k}]$, $FC = FC_1 \times FC_2 \times \cdots \times FC_k \in \mathbb{R}^m, m = m_1 + m_2 + \cdots m_k$, and $\boldsymbol{G}_C : \mathbb{R}^m \to \mathbb{R}^p$ is a mapping of the object motion from m-dimensional space to p-dimensional space, which is called the **grasp mapping**. As a result, the grasp can be described by the friction cone FC and the grasp mapping. Let's look at the contact model described above.

2.2.1.1. *Grasp mapping of point contact without friction*

By substituting Eq. (2.1) with $\boldsymbol{F}_{C_i} = \Re(\boldsymbol{R}_{C_i})^{C_i}\boldsymbol{F}_{C_i}$, Eq. (2.17) can be represented by

$$\boldsymbol{F}_O = \sum_{i=1}^{k} \boldsymbol{W}_{OC_i}\Re(\boldsymbol{R}_{C_i}) \begin{bmatrix} 0 \\ 0 \\ 1 \\ 0 \\ 0 \\ 0 \end{bmatrix} f_{n_i} = \boldsymbol{G}_C \boldsymbol{f}_{FC}, \tag{2.18}$$

where

$$\boldsymbol{G}_C = \begin{bmatrix} \boldsymbol{z}_{C_1} & \cdots & \boldsymbol{z}_{C_k} \\ \boldsymbol{p}_{OC_1} \times \boldsymbol{z}_{C_1} & \cdots & \boldsymbol{p}_{OC_k} \times \boldsymbol{z}_{C_k} \end{bmatrix} \in \mathbb{R}^{6\times k},$$

$$\boldsymbol{f}_{FC} = [f_{n_1}, \ldots, f_{n_k}]^T \geq 0 \in \mathbb{R}^k,$$

and $\boldsymbol{z}_{C_i} \in \mathbb{R}^3$ is the third column vector of $\boldsymbol{R}_{C_i}$. For arbitrary vector $\boldsymbol{a}$, we use the notation $\boldsymbol{a} > 0$ if all the elements of the vector are positive, and $\boldsymbol{a} \geq 0$ if all the elements of the vector are not negative.

2.2.1.2. *Grasp mapping of point contacts with friction*

By substituting Eq. (2.4) with $\boldsymbol{F}_{C_i} = \Re(\boldsymbol{R}_{C_i})^{C_i}\boldsymbol{F}_{C_i}$, Eq. (2.17) can be represented by

$$\boldsymbol{F}_O = \sum_{i=1}^{k} \boldsymbol{W}_{OC_i}\Re(\boldsymbol{R}_{C_i}) \begin{bmatrix} 1 & 0 & 0 \\ 0 & 1 & 0 \\ 0 & 0 & 1 \\ 0 & 0 & 0 \\ 0 & 0 & 0 \\ 0 & 0 & 0 \end{bmatrix} \boldsymbol{f}_{FC_i} = \boldsymbol{G}_C \boldsymbol{f}_{FC}, \qquad (2.19)$$

where

$$\boldsymbol{G}_C = \begin{bmatrix} \boldsymbol{R}_{C_1} & \cdots & \boldsymbol{R}_{C_k} \\ [\boldsymbol{p}_{OC_1}\times]\boldsymbol{R}_{C_1} & \cdots & [\boldsymbol{p}_{OC_k}\times]\boldsymbol{R}_{C_k} \end{bmatrix} \in \mathbb{R}^{6\times 3k},$$

$$\boldsymbol{f}_{FC_i} = [f_1, f_2, f_3]^T,$$

and

$$\boldsymbol{f}_{FC} = [\boldsymbol{f}_{FC_1}^T, \ldots, \boldsymbol{f}_{FC_k}^T]^T \in \mathbb{R}^{3k}.$$

2.2.1.3. *Grasp mapping of soft finger*

By substituting Eq. (2.5) with $\boldsymbol{F}_{C_i} = \Re(\boldsymbol{R}_{C_i})^{C_i}\boldsymbol{F}_{C_i}$, Eq. (2.17) can be represented by

$$\boldsymbol{F}_O = \sum_{i=1}^{k} \boldsymbol{W}_{OC_i}\Re(\boldsymbol{R}_{C_i}) \begin{bmatrix} 1 & 0 & 0 & 0 \\ 0 & 1 & 0 & 0 \\ 0 & 0 & 1 & 0 \\ 0 & 0 & 0 & 0 \\ 0 & 0 & 0 & 0 \\ 0 & 0 & 0 & 1 \end{bmatrix} \boldsymbol{f}_{FC_i} = \boldsymbol{G}_C \boldsymbol{f}_{FC}, \qquad (2.20)$$

where

$$\boldsymbol{G}_C = \begin{bmatrix} \boldsymbol{R}_{C_1} & 0 & \cdots & \boldsymbol{R}_{C_k} & 0 \\ [\boldsymbol{p}_{OC_1}\times]\boldsymbol{R}_{C_1} & \boldsymbol{z}_{C_1} & \cdots & [\boldsymbol{p}_{OC_k}\times]\boldsymbol{R}_{C_k} & \boldsymbol{z}_{C_k} \end{bmatrix} \in \mathbb{R}^{6\times 4k},$$

$$\boldsymbol{f}_{FC_i} = [f_1, \ldots, f_4]^T,$$

and

$$\boldsymbol{f}_{FC} = [\boldsymbol{f}_{FC_1}^T, \ldots, \boldsymbol{f}_{FC_k}^T]^T \in \mathbb{R}^{4k}.$$

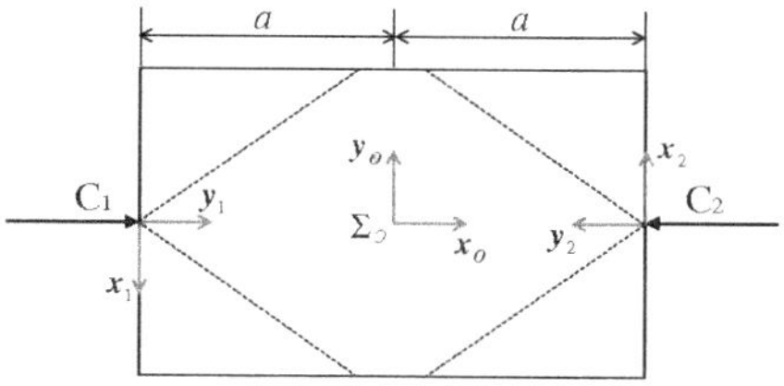

Fig. 2.6. Grasp in the plane.

Example 2.1. Let's consider the grasp at point contacts with friction on a rectangular object in a plane, as shown in Fig. 2.6. Here, the reference coordinate system is supposed to be in an overlapping position parallel to the object coordinate system.

In the planar motion, the position vector of the contact point C_i from the object coordinate origin is $\boldsymbol{p}_{OC_i} = [p_{xi}, p_{yi}]^T$. In addition, $\boldsymbol{W}_{OC_i}$ and $\Re(\boldsymbol{R}_{C_i})$ can be written as

$$\boldsymbol{W}_{OC_i} = \begin{bmatrix} \boldsymbol{I}_{2\times 2} & 0 \\ [-p_{yi}\ p_{xi}] & 1 \end{bmatrix}, \quad \Re(\boldsymbol{R}_{C_i}) = \begin{bmatrix} \boldsymbol{R}_{C_i} & 0 \\ 0 & 1 \end{bmatrix}. \tag{1}$$

In this case, the force $\boldsymbol{f}_{O_i} \in \mathbb{R}^2$ and the moment $\boldsymbol{n}_{O_i} \in \mathbb{R}^1$ at the origin of the object coordinate system, which are equivalent to the force $\boldsymbol{f}_{C_i} \in \mathbb{R}^2$ and the moment $\boldsymbol{n}_{C_i} \in \mathbb{R}^1$ at the origin of i-th finger contact coordinate system, can be represented by

$$\boldsymbol{F}_{O_i} = \boldsymbol{W}_{OC_i} \Re(\boldsymbol{R}_{C_i})^{C_i}\boldsymbol{F}_{C_i}, \tag{2}$$

where $\boldsymbol{F}_{C_i} = [\boldsymbol{f}_{C_i}^T, \boldsymbol{n}_{C_i}]^T$ and $\boldsymbol{F}_{O_i} = [\boldsymbol{f}_{O_i}^T, \boldsymbol{n}_{O_i}]^T$. From the relationship of the coordinate system shown in the figure, the position vector and the rotation matrix in a plane are represented by

$$\boldsymbol{R}_{C_1} = \begin{bmatrix} 0 & 1 \\ -1 & 0 \end{bmatrix}, \quad \boldsymbol{R}_{C_2} = \begin{bmatrix} 0 & -1 \\ 1 & 0 \end{bmatrix},$$

$$\boldsymbol{p}_{OC_1} = \begin{bmatrix} -a \\ 0 \end{bmatrix}, \quad \boldsymbol{p}_{OC_2} = \begin{bmatrix} a \\ 0 \end{bmatrix}. \tag{3}$$

The wrench basis at the i-th contact point is

$$^{C_i}\boldsymbol{H}_{C_i} = \begin{bmatrix} 1 & 0 \\ 0 & 1 \\ 0 & 0 \end{bmatrix}, \quad \text{for } i = 1, 2. \tag{4}$$

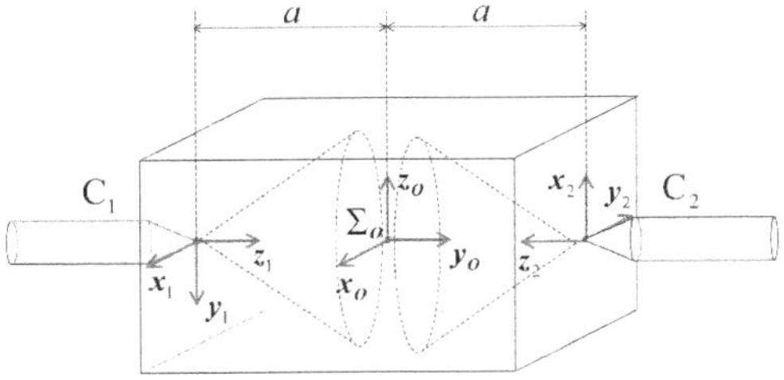

Fig. 2.7. Grasp by two soft fingers.

The friction cone is given by

$$FC_{OC_1} = \{\boldsymbol{f}_{FC} = [f_1, f_2]^T : |f_1| \le \mu f_2, f_2 \ge 0\}. \tag{5}$$

Using these relations, the grasp map can be represented as,

$$\boldsymbol{G}_C = [\,\boldsymbol{W}_{OC_1}\Re(\boldsymbol{R}_{C_1})^{C_1}\boldsymbol{H}_{C_1}\;\boldsymbol{W}_{OC_2}\Re(\boldsymbol{R}_{C_2})^{C_2}\boldsymbol{H}_{C_2}]$$

$$= \begin{bmatrix} 0 & 1 & 0 & -1 \\ -1 & 0 & 1 & 0 \\ a & 0 & a & 0 \end{bmatrix}. \tag{6}$$

Example 2.2. In this example, we find the grasp mapping, friction cone, and contact force when a rectangular solid is grasped by two soft fingers, as shown in Fig. 2.7.

According to Eq. (2.17), the equivalent wrench at the origin of the object coordinate system by the force $\boldsymbol{f}_{C_i} \in \mathbb{R}^4$ at the contact point C_i (for $i = 1, 2$), $\boldsymbol{F}_{O_i} = [\boldsymbol{f}_{O_i}^T, \boldsymbol{n}_{O_i}^T]^T$, can be written as

$$\boldsymbol{F}_O = \sum_{i=1}^{2} \boldsymbol{W}_{OC_i}\Re(\boldsymbol{R}_{C_i})^{C_i}\boldsymbol{H}_{C_i}\boldsymbol{f}_{FC_i} = \boldsymbol{G}_C\boldsymbol{f}_{FC}, \tag{1}$$

where the wrench basis ${}^{C_i}\boldsymbol{H}_{C_i}$ at the contact point C_i is, as referred to in Eq. (2.5),

$${}^{C_1}\boldsymbol{H}_{C_1} = {}^{C_2}\boldsymbol{H}_{C_2} = \begin{bmatrix} 1 & 0 & 0 & 0 \\ 0 & 1 & 0 & 0 \\ 0 & 0 & 1 & 0 \\ 0 & 0 & 0 & 0 \\ 0 & 0 & 0 & 0 \\ 0 & 0 & 0 & 1 \end{bmatrix}, \tag{2}$$

and the friction cone FC_i is

$$FC_1 = \left\{ \boldsymbol{f}_{FC_1} = [f_1, f_2, f_3, f_4]^T \in \mathbb{R}^4 : \sqrt{f_1^2 + f_2^2} \leq \mu f_3, |f_4| \leq \gamma f_3 \right\}, \tag{3}$$

$$FC_2 = \left\{ \boldsymbol{f}_{FC_2} = [f_1', f_2', f_3', f_4']^T \in \mathbb{R}^4 : \sqrt{f_1'^2 + f_2'^2} \leq \mu f_3', |f_4'| \leq \gamma f_3' \right\}, \tag{4}$$

where $\mu > 0$ is a static friction coefficient and $\gamma > 0$ is a torsional friction coefficient.

From the relationship of the coordinate systems shown in the figure, the rotation matrix and the position vector at the contact point C_i are given by

$$\boldsymbol{R}_{C_1} = \begin{bmatrix} 1 & 0 & 0 \\ 0 & 0 & 1 \\ 0 & -1 & 0 \end{bmatrix}, \quad \boldsymbol{R}_{C_2} = \begin{bmatrix} 0 & -1 & 0 \\ 0 & 0 & -1 \\ 1 & 0 & 0 \end{bmatrix},$$

$$\boldsymbol{p}_{OC_1} = \begin{bmatrix} 0 \\ -a \\ 0 \end{bmatrix}, \quad \boldsymbol{p}_{OC_2} = \begin{bmatrix} 0 \\ a \\ 0 \end{bmatrix}. \tag{5}$$

Therefore, the grasp map, according to Eq. (1), is represented as

$$\boldsymbol{G}_C = \begin{bmatrix} 1 & 0 & 0 & 0 & 0 & -1 & 0 & 0 \\ 0 & 0 & 1 & 0 & 0 & 0 & -1 & 0 \\ 0 & -1 & 0 & 0 & 1 & 0 & 0 & 0 \\ 0 & a & 0 & 0 & a & 0 & 0 & 0 \\ 0 & 0 & 0 & 1 & 0 & 0 & 0 & -1 \\ a & 0 & 0 & 0 & 0 & a & 0 & 0 \end{bmatrix}. \tag{6}$$

The force, $\boldsymbol{F}_O$, applied to the object by the contact force $\boldsymbol{f}_{FC} = [\boldsymbol{f}_{FC_1}^T, \boldsymbol{f}_{FC_2}^T]^T$ is represented by

$$\begin{aligned} \boldsymbol{F}_O &= \boldsymbol{G}_C \boldsymbol{f}_{FC} \\ &= \left[f_1 - f_2', f_3 - f_3', -f_2 + f_1', af_2 + af_1', f_4 - f_4', af_1 + af_2' \right]^T. \end{aligned} \tag{7}$$

2.2.2. *Active Constraint and Passive Constraint*

Multi-fingered robotic hands for grasping and manipulating objects or jigging to fix an object are a means of applying some kinds of mechanical or kinematic constraints, which we call the **constraint mechanism**. This section discusses the point contact model, with and without friction, applied with one-sided constraint for a rigid body object by a multi-fingered hand.

Passive closure is a constraint that maintains the position and orientation of an object without changing the driving force of the constraining limb; it generates a reaction force that balances the external force by the constraint mechanism. Passive constraints include **passive force closure**, in which it is necessary to apply a preliminary force by gravity and/or joint driving force, and **passive form closure**, which does not require a preliminary force. This type of passive constraint may be simply called **form closure**.

When the constraint mechanism can apply arbitrary force and moment to the object, the constraint is called an **active closure**. An active constraint that occurred based on the force is called an **active force closure** or a **force closure**.

Figure 2.8 shows simple examples of these concepts. Figure 2.8(a) shows a passive form closure in which an object is constrained with two fingers with no joint; Fig. 2.8(b) represents a passive force closure where one finger is constrained by the environment and the other can move freely in the horizontal direction; Fig. 2.8(c) demonstrates

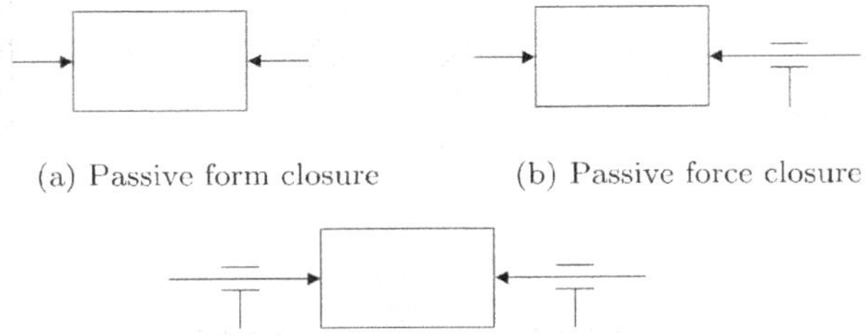

(a) Passive form closure (b) Passive force closure

(c) Active closure

Fig. 2.8. Constraint of one-dimensional motion.

active closure where each finger has a degree of freedom (DOF). In this figure, both fingers move freely in the horizontal direction and arbitrary force can be applied. Furthermore, an operator would be able to move the object with active closure, but not with passive force closure. In general, it is necessary to control the right and left joint drive force by measuring the force in the case of active closure. It would be possible for active closure to work similarly to a passive force closure if the right and left joint drive forces are controlled such that one is set as a fixed contact point and force is added from one to the other.

2.2.3. *Form Closure*

To fix an object is a basic operation of handling. **Form closure** describes a state in which object movement does not occur due to geometric constraint without assuming any friction. Figure 2.9 shows examples of form closures. Figure 2.9(a) shows that an object cannot be set in motion in any direction where there is contact at four points on a rectangular object in a two-dimensional (2D) plane. In the planar motion, the form closure is realized by four frictionless point contacts, which is a minimum number; Fig. 2.9(b) is an example of how an object can be constrained with six degrees of freedom by seven frictionless point contacts, which is the minimum number in three-dimensional (3D) space.

Let's consider the relational expressions of form closure. Assume that an object has a small displacement of translation and rotation

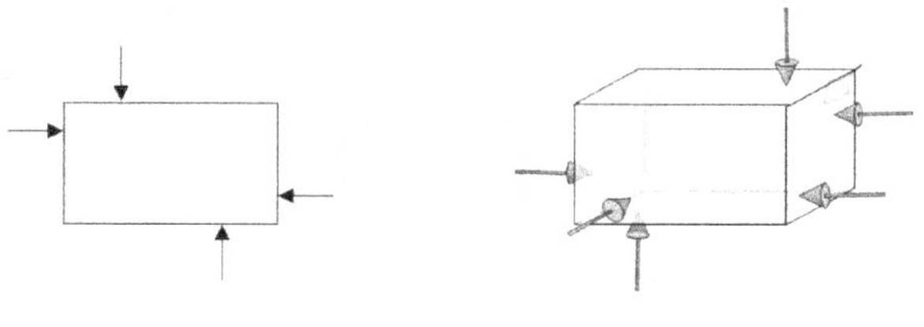

(a) Form closure in the plane (b) Form closure in 3D space

Fig. 2.9. Examples of form closure.

as $\Delta \boldsymbol{r}_O \in \mathbb{R}^6$. The translational displacement of the object at the i-th contact point, $\Delta \boldsymbol{p}_i \in \mathbb{R}^3$, is written as,

$$\Delta \boldsymbol{p}_i = [\boldsymbol{I}_{3\times 3} - [\boldsymbol{p}_{OC_i}\times]]\Delta \boldsymbol{r}_O. \tag{2.21}$$

Because it is a one-sided constraint, the direction component, $\boldsymbol{z}_{C_i}$ of $\Delta \boldsymbol{p}_i$, is represented by using a formula of the vector triple product for arbitrary vectors $\boldsymbol{a}, \boldsymbol{b}, \boldsymbol{c} \in \mathbb{R}^3$, use $\boldsymbol{a}^T(\boldsymbol{b} \times \boldsymbol{c}) = \boldsymbol{b}^T(\boldsymbol{c} \times \boldsymbol{a}) = \boldsymbol{c}^T(\boldsymbol{a} \times \boldsymbol{b})$ as

$$\boldsymbol{z}_{C_i}^T \Delta \boldsymbol{p}_i = \left[\boldsymbol{z}_{C_i}^T \quad (\boldsymbol{p}_{OC_i} \times \boldsymbol{z}_{C_i})^T\right] \Delta \boldsymbol{r}_o \geq 0. \tag{2.22}$$

Therefore, a mixture of these formulas, $\Delta \boldsymbol{P} = [\boldsymbol{z}_{C_1}^T \Delta \boldsymbol{p}_1, \cdots, \boldsymbol{z}_{C_k}^T \Delta \boldsymbol{p}_k]^T$, satisfies the following formula using Eq. (2.18):

$$\Delta \boldsymbol{P} = \begin{bmatrix} \boldsymbol{z}_{C_1} & \cdots & \boldsymbol{z}_{C_k} \\ \boldsymbol{p}_{OC_1} \times \boldsymbol{z}_{C_1} & \cdots & \boldsymbol{p}_{OC_k} \times \boldsymbol{z}_{C_k} \end{bmatrix}^T \Delta \boldsymbol{r}_O = \boldsymbol{G}_C^T \Delta \boldsymbol{r}_O \geq 0. \tag{2.23}$$

When no motion has occurred, it can be $\Delta \boldsymbol{P} = 0$. Therefore, the necessary and sufficient condition of the form closure is that the solution of $\Delta \boldsymbol{r}_O$ that satisfies the condition $\Delta \boldsymbol{P} = 0$ can only be $\Delta \boldsymbol{r}_O = 0$.

Thus far, we have discussed the necessary and sufficient conditions from the perspective of understanding the relationship of object displacement. From the viewpoint of power, the perspective of force can be described as follows. As shown in Eq. (2.18), the force applied to the object by k fingers in the frictionless contact model can be represented as $\boldsymbol{F}_O = \boldsymbol{G}_C \boldsymbol{f}_{FC}$. From this perspective, the necessary and sufficient condition that the object is kept in the static state for any external force $\boldsymbol{F}_e \in \mathbb{R}^6$ is that there exists a solution of Eq. (2.24) in which $\boldsymbol{f}_{FC} \geq 0$.

$$\boldsymbol{G}_C \boldsymbol{f}_{FC} = -\boldsymbol{F}_e. \tag{2.24}$$

In addition, in the case of a change in the sign of $\boldsymbol{F}_e$, a solution of Eq. (2.24) with $\boldsymbol{f}_{FC} \geq 0$ should also exist. Therefore, conditions for form closure can be described as follows.

Property 2.2. (Form Closure) *When there is a constraint by frictionless point contacts, the necessary and sufficient conditions of* *form closure for an object in 3D space are: (i) rank* $\boldsymbol{G}_C = 6$*, and (ii) that there exists a solution of* $\boldsymbol{G}_C \boldsymbol{f}_{FC} = 0$ *with* $\boldsymbol{f}_{FC} > 0$.

If $k = 6$, even if condition (i) is satisfied, (ii) will not be, because the solution of $\boldsymbol{G}_C \boldsymbol{f}_{FC} = 0$ is $\boldsymbol{f}_{FC} = 0$ only. Therefore, k should be ≥ 7. As shown in Fig. 2.9(b), seven is the minimum number of frictionless contact points for perfect constraint of object movement in 3D space.

2.2.4. *Force Closure*

If a grasp can resist any applied wrench, that grasp is called a force closure.

Definition 2.1. (Force closure)

A grasp is a **force closure** if, given any external wrench $\boldsymbol{F}_e \in \mathbb{R}^6$ applied to the object, there exists contact force $\boldsymbol{f}_{FC} \in FC$ such that

$$\boldsymbol{G}_C \boldsymbol{f}_{FC} = -\boldsymbol{F}_e. \tag{2.25}$$

The difference between form closure and force closure is that in the former, point contacts without friction and condition $\boldsymbol{f}_{FC} \geq 0$ should be satisfied, whereas in the latter point contacts are with friction, including soft finger contact. Among the elements of $\boldsymbol{f}_{FC}$, the unilateral contact condition exists along the normal direction at each contact point. Now, we introduce the definition of surjective mapping and internal forces. Here, for an arbitrary matrix $\boldsymbol{A}$, the range of the matrix is described as Im $(\boldsymbol{A})$, the kernel of the matrix is described as Ker $(\boldsymbol{A})$ as shown in Fig. 2.10 and the interior of the friction cone is described as int (FC).

Definition 2.2. (Surjective mapping)

If a range space of arbitrary mapping $\boldsymbol{G} \in \mathbb{R}^{p \times m}$ is equal to $\mathbb{R}^p$, it is called a **surjective mapping** and described as $Im(\boldsymbol{G}) = \mathbb{R}^p$.

Definition 2.3. (Internal force)

If $\boldsymbol{f}_N \in Ker(\boldsymbol{G}_C) \cap FC$, then $\boldsymbol{f}_N$ is called an **internal force**, that is, $\boldsymbol{G}_C \boldsymbol{f}_N = 0$. If $\boldsymbol{f}_N \in Ker(\boldsymbol{G}_C) \cap int(FC)$, then $\boldsymbol{f}_N$ is called the **strictly internal force**.

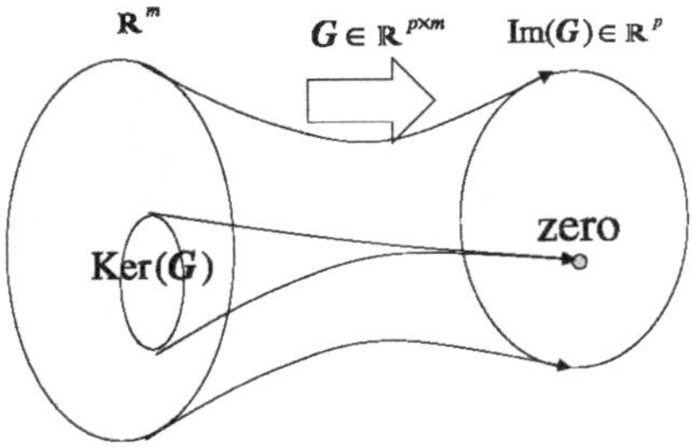

Fig. 2.10. Matrix map.

Although the internal force does not generate object movement, it is necessary for a stable grasp. The conditions of force closure can be described as follows.

Theorem 2.1. (Necessary and sufficient conditions for force closure)

A grasp is a force closure if and only if the grasp mapping $\boldsymbol{G}_C$ *is surjective and a strict internal force* $\boldsymbol{f}_N \in Ker(\boldsymbol{G}_C) \cap int(FC)$ *with non-zero condition (this is called the* ***existence conditions of non-zero internal force****) exits.*

Proof. (Sufficiency) There exists a contact force to realize a given wrench because the mapping $\boldsymbol{G}_C$ is surjective. For arbitrary selected $\boldsymbol{F}_O \in \mathbb{R}^6$, there exists a force $\boldsymbol{f}$ such that $\boldsymbol{G}_C\boldsymbol{f} = \boldsymbol{F}_O$, and an internal force $\boldsymbol{f}_N$ such that $\boldsymbol{G}_C\boldsymbol{f}_N = 0$. Let us show that there exists an α such that $\boldsymbol{f} + \alpha\boldsymbol{f}_N \in \text{int}(FC)$. Notice that $\lim_{\alpha\to\infty}(\boldsymbol{f} + \alpha\boldsymbol{f}_N)/\alpha = \boldsymbol{f}_N \in \text{int}(FC)$. Therefore, there exists a sufficiently large α' such that $(\boldsymbol{f} + \alpha'\boldsymbol{f}_N)/\alpha' \in \text{int}(FC) \subset FC$. From the properties of the friction cone, it follows that $\boldsymbol{f}_{FC} = \boldsymbol{f} + \alpha'\boldsymbol{f}_N \in \text{int}(FC)$, and $\boldsymbol{G}_C\boldsymbol{f} = \boldsymbol{G}_C\boldsymbol{f}_{FC} = \boldsymbol{F}_O$.

(Necessity) Suppose that a grasp is a force closure. Choose $\boldsymbol{f}_1 \in FC$ such that $\boldsymbol{G}_C\boldsymbol{f}_1 = \boldsymbol{F}_O \neq 0$, and $\boldsymbol{f}_2 \in FC$ such that $\boldsymbol{G}_C\boldsymbol{f}_2 = -\boldsymbol{F}_O$, and define $\boldsymbol{f}_N = \boldsymbol{f}_1 + \boldsymbol{f}_2$. Then we see that $\boldsymbol{G}_C\boldsymbol{f}_N = \boldsymbol{F}_O - \boldsymbol{F}_O = 0$ and $\boldsymbol{f}_N \in FC$ from the properties of a cone. Therefore, there exists internal forces.

In the case of a 2D plane, the minimum number of contact points for force closure is two, because of the unilateral constraint. The

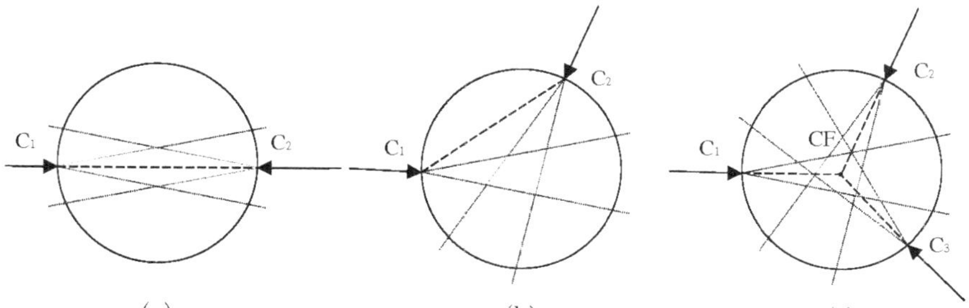

(a) (b) (c)

Fig. 2.11. Force closure in the 2D space.

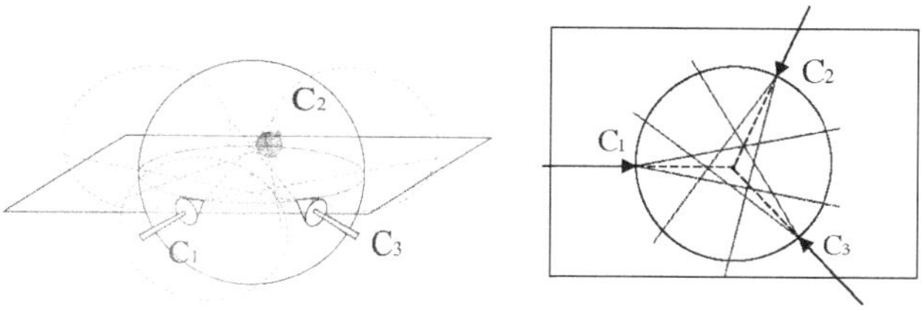

(a) Contact at three points (b) Grasp plane

Fig. 2.12. Force closure in 3D space.

existing condition of an internal force is that a straight line connecting two contact points be surrounded by two friction cones. In Fig. 2.11(a), demonstrates a force closure, and (b) demonstrates a situation that would not be a force closure. When one point contact is added to the case illustrated in Fig. 2.11(b) as in Fig. 2.11(c), it becomes a force closure because of the existence of non-zero internal forces such that one arbitrary point, CF, can be considered a balanced point of normal forces within the common area of the three friction cones.

In the case of a 3D space, the minimum number of contact points for force closure is three, because of the unilateral constraint, as shown in Fig. 2.12. The existence of internal forces is judged based on whether there exists a common area on the surface of the three friction cones that passes through three contact points, as shown in Fig. 2.12(b).

In the frictionless point contact model, if a geometric constraint of an object is form closure, a constraint by a robotic hand under the same contact conditions is a force closure, and the opposite is also true.

2.2.5. *Force Distribution*

A fingertip force that balances an external force given by Eq. (2.25) in the case of a force closure can be obtained as

$$\boldsymbol{f}_{FC} = -\boldsymbol{G}_C^{+}\boldsymbol{F}_e + (\boldsymbol{I} - \boldsymbol{G}_C^{+}\boldsymbol{G}_C)\boldsymbol{k}, \tag{2.26}$$

where $\boldsymbol{G}_C^{+}$ is a pseudo-inverse matrix of $\boldsymbol{G}_C$ and is given by

$$\boldsymbol{G}_C^{+} = \boldsymbol{G}_C^T(\boldsymbol{G}_C\boldsymbol{G}_C^T)^{-1},$$

and $\boldsymbol{k}$ is an arbitrary vector. There are countless solutions. However, the $\boldsymbol{k}$ must be selected such that each fingertip force exists in the friction cone at the contact point. This indeterminate problem should be solved using evaluation functions. Examples of the evaluation functions are: to minimize the norm of the internal forces, to minimize the norm of $\boldsymbol{f}_{FC} \in FC$, to minimize the maximum value of the elements of $\boldsymbol{f}_{FC} \in FC$, and so on.

Example 2.3. Let us consider the two-finger frictionless point contact in Example 2.1 as a force closure in the plane. We will also determine the fingertip force that minimizes the norm of $\boldsymbol{f}_{FC}$ in the balance of the external force when gravity $\boldsymbol{F}_e = [0, -mg, 0]^T$ is applied, where m is the mass of the object, and g is the gravitational acceleration.

First, see the force closure. According to $\boldsymbol{G}_C$ given by Eq. (5) in Example 2.1, $\boldsymbol{G}_C^{+}$ is obtained as

$$\boldsymbol{G}_C^{+} = \begin{bmatrix} 0 & -\dfrac{1}{2} & \dfrac{1}{2a} \\ \dfrac{1}{2} & 0 & 0 \\ 0 & \dfrac{1}{2} & \dfrac{1}{2a} \\ -\dfrac{1}{2} & 0 & 0 \end{bmatrix}. \tag{1}$$

By substituting the external force and the arbitrary vector $\boldsymbol{k} = [k_1\ k_2\ k_3\ k_4]^T$ into Eq. (2.26), the constraint force is obtained as

$$\boldsymbol{f}_{FC} = -\begin{bmatrix} \frac{1}{2}mg \\ 0 \\ -\frac{1}{2}mg \\ 0 \end{bmatrix} + \begin{bmatrix} 0 \\ \frac{1}{2}k_2 + \frac{1}{2}k_4 \\ 0 \\ \frac{1}{2}k_2 + \frac{1}{2}k_4 \end{bmatrix} = \begin{bmatrix} \frac{1}{2}mg \\ \frac{1}{2}k_2 + \frac{1}{2}k_4 \\ -\frac{1}{2}mg \\ \frac{1}{2}k_2 + \frac{1}{2}k_4 \end{bmatrix}. \tag{2}$$

Since the friction cones at the right and left contact points are given by

$$FC = \left\{\boldsymbol{f}_{FC} \in \mathbb{R}^2 : |f_1| \leq \mu f_2, f_2 \geq 0\right\}, \tag{3}$$

the necessary condition to be force closure is represented as

$$k_2 + k_4 \geq \frac{1}{\mu}mg. \tag{4}$$

Obviously, this is a force closure because there exists a vector $\boldsymbol{k}$ that satisfies Eq. (4).

Next, the norm of $\boldsymbol{f}_{FC}$ is represented as

$$\|\boldsymbol{f}_{FC}\| = \sqrt{\frac{1}{2}((mg)^2 + (k_2 + k_4)^2)}, \tag{5}$$

and it should be minimized when $k_2 + k_4 = mg/\mu$. Then

$$\boldsymbol{f}_{FC} = \frac{mg}{2}\begin{bmatrix} 1 \\ 1/\mu \\ -1 \\ 1/\mu \end{bmatrix}, \quad \|\boldsymbol{f}_{FC}\| = \sqrt{\frac{1}{2}\left(1 + \frac{1}{\mu}\right)}mg. \tag{6}$$

This shows that a large grasp force is required to counteract the gravitational force if the friction coefficient is small.

Example 2.4. In this example, we find the internal force of Example 2.1 from the relationship of $\boldsymbol{G}_C\boldsymbol{f}_N = 0$. The internal force $\boldsymbol{f}_N$ satisfies

$\boldsymbol{f}_N \in Ker(\boldsymbol{G}_C) \cap FC$. Let us denote $\boldsymbol{f}_N = [f_{N1}\ f_{N2}\ f_{N3}\ f_{N4}]^T$, then Example 2.1 (5) derives

$$\boldsymbol{G}_C \boldsymbol{f}_N = \begin{bmatrix} f_{N2} - f_{N4} \\ -f_{N1} + f_{N3} \\ af_{N1} + af_{N3} \end{bmatrix}. \tag{1}$$

The condition of $\boldsymbol{G}_C \boldsymbol{f}_N = \boldsymbol{0}$ derives $f_{N1} = f_{N3}, f_{N2} = f_{N4}$ and $a(f_{N1} + f_{N3}) = 0$. Because a is not 0 and $\boldsymbol{f}_N$ must exist within the friction cone, the following should be satisfied.

$$FC = \left\{ \boldsymbol{f}_N = \begin{bmatrix} f_{N1} \\ f_{N2} \\ f_{N3} \\ f_{N4} \end{bmatrix} : |f_{N1}| \leq \mu f_{N2}, f_{N2} \geq 0, |f_{N3}| \leq \mu f_{N4}, f_{N4} \geq 0 \right\}. \tag{2}$$

Therefore,

$$f_{N1} = f_{N3} = 0, \quad f_{N2} = f_{N4} \geq 0.$$

Finally, $\boldsymbol{f}_N$ is represented as

$$\boldsymbol{f}_N = \begin{bmatrix} 0 \\ f_n \\ 0 \\ f_n \end{bmatrix}. \tag{3}$$

Here, $f_n = f_{N2} = f_{N4} \geq 0$. Using this equation and Eq. (4) in Example 2.3, the grasp force is obtained as

$$\boldsymbol{f}_{FC} = \frac{mg}{2} \begin{bmatrix} 1 \\ 1/\mu \\ -1 \\ 1/\mu \end{bmatrix} + \begin{bmatrix} 0 \\ f_n \\ 0 \\ f_n \end{bmatrix}. \tag{4}$$

2.3. Non-Fixed Contact

We have been considering fixed contacts where the object contact point does not move relative to the finger contact point when the

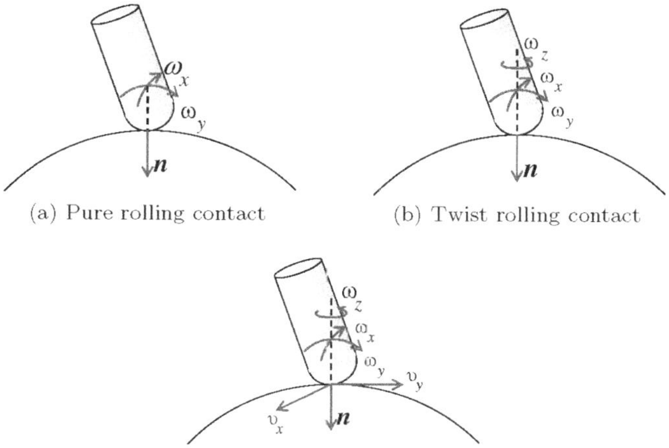

(a) Pure rolling contact (b) Twist rolling contact

(c) Slide rolling contact

Fig. 2.13. Relative motion at contact point.

object and the fingertips are in contact. However, in an actual contact, relative motion such as sliding or rolling of the fingertips on the object surface may occur. This section explains the relative motion of the object and the finger(s) when the surfaces of the finger and the object are smooth. There is at least one constraint in the relative motions that maintain contact, with a maximum of five degrees of freedom. We call this the **contact degrees of freedom**. Considering the existence of friction on a contact surface, there are three types of contact under practical conditions, as shown in Fig. 2.13. We call them the **non-fixed contacts**.

2.3.1. *Pure Rolling Contact Model*

In a model where the friction force and friction moment around the normal to the object surface are at work on the contact point, the contact becomes a rolling contact on both curved surfaces. When no rotation can be seen around the normal to the contact as shown in Fig. 2.13(a), it is called a **pure rolling contact model**. The DOF of the relative motion for this model is two. The wrench basis at the contact point is the same as that of the soft finger contact model with the fixed contact.

2.3.2. *Twist Rolling Contact Model*

A model where the friction moment around the normal to the object surface does not act due to small friction and a rolling around the normal to the object surface occurs, as shown in Fig. 2.13(b), is called a **twist rolling contact model**. The contact DOF of this model is three. The wrench basis at the contact point is the same as that of the point contact model with friction at the fixed contact.

2.3.3. *Slide Rolling Contact Model*

A model where the friction force and moment do not act at the contact point and the contact point slides on both curved surfaces, as shown in Fig. 2.13(c), is called a **slide rolling contact model**. The contact DOF of this model is five. The wrench basis at the contact point is the same as that of the frictionless point contact model at the fixed contact.

In addition, a case where the friction force does not act but the friction moment does is not considered an actual contact problem of the object and the finger; it is therefore excluded from the classification.

2.4. Constraint Conditions

2.4.1. *Contact Model and Constraint Condition*

We will now consider the position constraint conditions of each contact model (Yoshikawa, 1995,1996a, 1996b; Cole *et al.*, 1989). Let us denote a position of an arbitrary point referred to the object coordinate system Σ_O as ${}^{O}\boldsymbol{p} \in \mathbb{R}^3$, and a position of a arbitrary point referred to the i-th fingertip link coordinate system Σ_{F_i} as ${}^{F_i}\boldsymbol{p} \in \mathbb{R}^3$. It is assumed that shapes of the fingertip surface and the object surface can be written as

$$S_O({}^{O}\boldsymbol{p}) = 0, S_{F_i}({}^{F_i}\boldsymbol{p}) = 0, \tag{2.27}$$

where ${}^{O}\boldsymbol{p}$ for $S_O({}^{O}\boldsymbol{p}) < 0$ is an inner point of the object, and ${}^{O}\boldsymbol{p}$ for $S_O({}^{O}\boldsymbol{p}) > 0$ is an outer point of the object. This description is also used in the case of $S_{F_i}({}^{F_i}\boldsymbol{p})$.

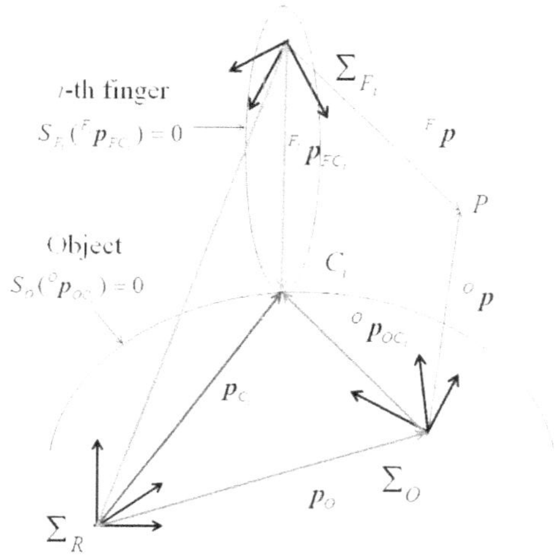

Fig. 2.14. Contact and constraint.

First, let's discuss the necessary and sufficient conditions for the point contact. If ${}^{O}\boldsymbol{p}_{OC_i}$ and ${}^{F_i}\boldsymbol{p}_{FC_i}$ are the same to the contact point $\boldsymbol{p}_{C_i}$ on each curved surface, as shown in Fig. 2.14,

$$S_O({}^{O}\boldsymbol{p}_{OC_i}) = 0, S_{F_i}({}^{F_i}\boldsymbol{p}_{FC_i}) = 0, \tag{2.28}$$

$$\boldsymbol{p}_{C_i} = \boldsymbol{p}_O + \boldsymbol{R}_O {}^{O}\boldsymbol{p}_{OC_i} = \boldsymbol{p}_{F_i} + \boldsymbol{R}_{F_i} {}^{F_i}\boldsymbol{p}_{FC_i}. \tag{2.29}$$

On the other hand, points except for the contact point $\boldsymbol{p}_{C_i}$ do not interfere with each other, so they satisfy the following condition:

$$\max_{\boldsymbol{p} \neq \boldsymbol{p}_{C_i}} \{S_O({}^{O}\boldsymbol{p}), S_{F_i}({}^{F_i}\boldsymbol{p})\} > 0. \tag{2.30}$$

The three equations above are the conditions for the point contact. They are consistent with constraints in the case of a contact point where slipping occurs in all directions. We call this a slide contact, which does not include a rolling contact.

Second, consider a fixed-point contact. In this case, in addition to the conditions from Eq. (2.28) to Eq. (2.30), the contact point does not alter when viewed from the object coordinate system and the i-th fingertip link coordinate system. Therefore the following conditions are applied

$${}^{O}\boldsymbol{p}_{OC_i} = \text{constant}, \quad {}^{F_i}\boldsymbol{p}_{FC_i} = \text{constant}. \tag{2.31}$$

Third, consider the case of rolling contacts. Besides the conditions from Eq. (2.28) to Eq. (2.30), the velocity of the contact point on the object plane, ${}^{O}\dot{\boldsymbol{p}}_{OC_i}$, and the velocity of the contact point on the fingertip link coordinates, ${}^{F_i}\dot{\boldsymbol{p}}_{FC_i}$, are equal when viewed from the reference coordinates. Therefore the following condition is added:

$$\boldsymbol{R}_O{}^{O}\dot{\boldsymbol{p}}_{OC_i} = \boldsymbol{R}_{F_i}{}^{F_i}\dot{\boldsymbol{p}}_{FC_i}. \tag{2.32}$$

The conditions of rolling contact listed above can be applied to any pure and twist rolling contacts. In case of the pure rolling contact, a condition is added that the rolling velocity around the normal contact line be zero.

In terms of constraint conditions, constraints by contact model can be arranged as follows:

Slide contact model: Eqs. (2.28)–(2.30)
Fixed contact model: Eqs. (2.28)–(2.31)
Rolling contact model: Eqs. (2.28)–(2.30) and (2.32)

Furthermore, constraints for slide contacts and fixed contacts are holonomic constraints, while those of rolling contacts are non-holonomic constraints. Refer Chapter 3 for more details about holonomic and non-holonomic constraints.

2.4.2. *Velocity Constraints*

General relations between the object contact point velocity and the fingertip contact point velocity can be obtained by differentiating Eq. (2.29); that is,

$$\begin{aligned}\dot{\boldsymbol{p}}_{C_i} &= \dot{\boldsymbol{p}}_O + \boldsymbol{\omega}_O \times \boldsymbol{R}_O{}^{O}\boldsymbol{p}_{OC_i} + \boldsymbol{R}_O{}^{O}\dot{\boldsymbol{p}}_{OC_i} \\ &= \dot{\boldsymbol{p}}_{F_i} + \boldsymbol{\omega}_{F_i} \times \boldsymbol{R}_{F_i}{}^{F_i}\boldsymbol{p}_{FC_i} + \boldsymbol{R}_{F_i}{}^{F_i}\dot{\boldsymbol{p}}_{FC_i}\end{aligned} \tag{2.33}$$

where $\boldsymbol{\omega}_O \in \mathbb{R}^3$ is the object angular velocity, and $\boldsymbol{\omega}_{F_i} \in \mathbb{R}^3$ is the i-th fingertip angular velocity. Let us denote the linear velocity and angular velocity of the object as $\boldsymbol{v}_O = [\dot{\boldsymbol{p}}_O^T, \boldsymbol{\omega}_O^T]^T$ and that of the origin of i-th fingertip link coordinate system as $\boldsymbol{v}_{F_i} = [\dot{\boldsymbol{p}}_{F_i}^T, \boldsymbol{\omega}_{F_i}^T]^T$. The above equation is then represented by

$$\dot{\boldsymbol{p}}_{C_i} = \boldsymbol{D}_{OC_i}^T \boldsymbol{v}_O + \boldsymbol{R}_O{}^{O}\dot{\boldsymbol{p}}_{OC_i} = \boldsymbol{D}_{FC_i}^T \boldsymbol{v}_{F_i} + \boldsymbol{R}_{F_i}{}^{F_i}\dot{\boldsymbol{p}}_{FC_i}, \tag{2.34}$$

where

$$D_{OC_i} = \begin{bmatrix} I_{3\times 3} \\ [p_{OC_i}\times] \end{bmatrix}, \tag{2.35-1}$$

$$D_{FC_i} = \begin{bmatrix} I_{3\times 3} \\ [p_{FC_i}\times] \end{bmatrix}. \tag{2.35-2}$$

The constraint condition of each contact model is added to Eq. (2.34). Since the fixed contact is ${}^{O}\dot{p}_{OC_i} = {}^{F_i}\dot{p}_{OC_i} = 0$, therefore,

$$D_{OC_i}^T v_O = D_{FC_i}^T v_{F_i}. \tag{2.36}$$

In addition, because there is a relationship with Eq. (2.32) in the case of the rolling contact, the above equation can be established as well. However, ${}^{O}p_{OC_i}$ and ${}^{F_i}p_{FC_i}$ are constant in the fixed contact, but not in the rolling contact.

Equation (2.34) is established in the case of the slide contact because no condition is applied. At this time, let us define the relative translation velocity between the contact point of the object surface and that of the fingertips as a slide velocity $\dot{p}_{S_i} \in \mathbb{R}^3$. It is then represented by

$$\dot{p}_{S_i} = R_{F_i}{}^{F_i}\dot{p}_{FC_i} - R_O{}^{O}\dot{p}_{OC_i}. \tag{2.37}$$

When the fingertip contact points have no motion in the slide contact, that is, ${}^{F_i}\dot{p}_{FC_i} = 0$, the contact point velocity is $\dot{p}_{C_i} = D_{FC_i} v_{F_i}$, and the slide velocity is $\dot{p}_{S_i} = -R_O{}^{O}\dot{p}_{OC_i}$.

2.5. Manipulable Grasp of Hands

Consider the condition where a multi-fingered hand can manipulate an object freely.

2.5.1. *Kinematics and Statics of Fingers*

Assume that the DOF of the i-th finger joint of the robot hand is n_i. Then, determine the relationship between the force and moment acting on the fingertip link coordinate system and the contact point coordinate system (see Fig. 2.15).

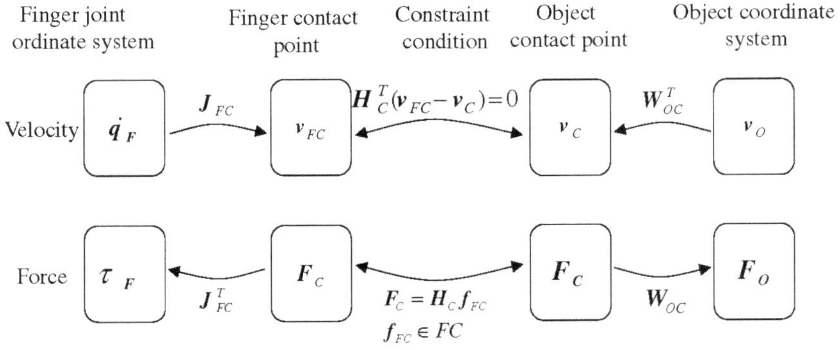

Fig. 2.15. Diagram of force and velocity.

The relationship between finger joint velocity $\dot{\boldsymbol{q}}_{F_i} \in \mathbb{R}^{n_i}$ and the velocity of the fingertip link coordinate system origin $\boldsymbol{v}_{F_i} \in \mathbb{R}^6$ is the same as the relationship between joint velocity and end-effect velocity of a robotic arm, therefore,

$$\boldsymbol{v}_{F_i} = \boldsymbol{J}_{F_i}\dot{\boldsymbol{q}}_{F_i}, \tag{2.38}$$

where $\boldsymbol{J}_{F_i} \in \mathbb{R}^{6\times n_i}$ is a kinematic Jacobian of the i-th finger. Also, the relationship between the joint torque of the i-th finger $\boldsymbol{\tau}_{F_i} \in \mathbb{R}^{n_i}$ and the wrench acting on the origin of the fingertip link coordinate system $\boldsymbol{F}_{F_i} \in \mathbb{R}^6$ is the same as the relationship between the joint torque and wrench in the robotic arm, so

$$\boldsymbol{\tau}_{F_i} = \boldsymbol{J}_{F_i}^T \boldsymbol{F}_{F_i}. \tag{2.39}$$

This relationship can be applied to all contact points, therefore,

$$\boldsymbol{v}_F = \boldsymbol{J}_F\dot{\boldsymbol{q}}_F, \tag{2.40}$$

$$\boldsymbol{\tau}_F = \boldsymbol{J}_F^T\boldsymbol{F}_F, \tag{2.41}$$

where, $\boldsymbol{\tau}_F = [\boldsymbol{\tau}_{F_1}^T, \ldots, \boldsymbol{\tau}_{F_k}^T]^T, \boldsymbol{v}_F = [\boldsymbol{v}_{F_1}^T, \ldots, \boldsymbol{v}_{F_k}^T]^T, \boldsymbol{q}_F = [\boldsymbol{q}_{F_1}^T, \ldots, \boldsymbol{q}_{F_k}^T]^T$, and $\boldsymbol{J}_F = block\ diag[\boldsymbol{J}_{F_1}, \ldots, \boldsymbol{J}_{F_k}]$.

In the case of a fixed contact, as seen in Eq. (2.10), the relative velocity between the finger contact point and the object contact point is $\boldsymbol{H}_{C_i}^T\boldsymbol{v}_{S_i} = \boldsymbol{H}_{C_i}^T(\boldsymbol{v}_{FC_i} - \boldsymbol{v}_{C_i}) = 0$, and we have the relationship of

$\boldsymbol{v}_{FC_i} = \boldsymbol{W}_{FC_i}^T \boldsymbol{v}_{F_i}, \boldsymbol{v}_{C_i} = \boldsymbol{W}_{OC_i}^T \boldsymbol{v}_O$. Therefore,

$$\boldsymbol{G}_{C_i}^T \boldsymbol{v}_O = \boldsymbol{H}_{C_i}^T \boldsymbol{J}_{FC_i} \dot{\boldsymbol{q}}_{F_i}. \tag{2.42}$$

where $\boldsymbol{G}_{C_i} = \boldsymbol{W}_{C_i} \boldsymbol{H}_{C_i}$. Here, let us denote the position vector from the i-th fingertip coordinate origin to the contact point C_i as $\boldsymbol{p}_{FC_i}$, then,

$$\boldsymbol{J}_{FC_i} = \boldsymbol{W}_{FC_i}^T \boldsymbol{J}_{F_i}, \boldsymbol{J}_{C_i} = \boldsymbol{H}_{C_i}^T \boldsymbol{J}_{FC_i}, \boldsymbol{G}_{C_i} = \boldsymbol{W}_{C_i} \boldsymbol{H}_{OC_i}, \tag{2.43}$$

where

$$\boldsymbol{W}_{FC_i} = \begin{bmatrix} \boldsymbol{I}_{3\times 3} & 0 \\ [\boldsymbol{p}_{FC_i} \times] & \boldsymbol{I}_{3\times 3} \end{bmatrix}. \tag{2.44}$$

This equation shows a constraint between the joint velocity and the object velocity. Eq. (2.36) demonstrates the constraint of translational velocity at the contact point; however, Eq. (2.42) is a constraint that includes the angular velocity at the contact point. Assume all contact points have such constraint and we get

$$\boldsymbol{G}_C^T \boldsymbol{v}_O = \boldsymbol{H}_C^T \boldsymbol{J}_{FC} \dot{\boldsymbol{q}}_F = \boldsymbol{J}_C \dot{\boldsymbol{q}}_F, \tag{2.45}$$

where $\boldsymbol{J}_{FC}, \boldsymbol{J}_C$ are block diagonal matrices whose i-th sub block matrices are $\boldsymbol{J}_{FC_i}, \boldsymbol{J}_{C_i}$, respectively.

In addition, there is a relationship with Eq. (2.34) in the case of the rolling contact model at the fingertip. Moreover, the wrench basis of the pure and twist rolling contact model is the same as that of the soft finger contact model and point contact with friction model in the fixed contact. Therefore there should exist the same constraints as in Eq. (2.45).

2.5.2. *Manipulable Grasp*

The grasp by a multi-fingered hand can be described by the hand **Jacobian** $\boldsymbol{J}_{FC}$, the grasp mapping $\boldsymbol{G}$, and the friction cone FC. This section discusses the concept of manipulable grasp, which is about the operation of an object by a multi-fingered hand.

Definition 2.4. (Manipulable grasp)

A multi-fingered grasp is manipulable if $\dot{\boldsymbol{q}}_F$, which satisfies Eq. (2.45) for arbitrary object motion $\boldsymbol{v}_O$, exists.

Property 2.3. (Manipulable grasp)

The necessary and sufficient condition for a manipulable grasp is $\mathrm{Im}(\boldsymbol{G}_C^T) \subseteq Im(\boldsymbol{J}_C)$.

It is easy to prove this property. There are multiple $\dot{\boldsymbol{q}}_F$ for object motion $\boldsymbol{v}_O$ in the manipulability grasp, because $\boldsymbol{J}_C$ does not need to be injective. When focusing on one finger, this occurs when $n > m$, where m is the DOF of object movement and n is the dimension of $\boldsymbol{q}_F$. On the other hand, finger joint velocity satisfying $\dot{\boldsymbol{q}}_F \in Ker(\boldsymbol{J}_C)$ does not influence object motion at all. Such $\dot{\boldsymbol{q}}_F$ is called an internal motion of the multi-fingered hand. When the internal motion is added to an arbitrary finger joint velocity, it does not affect the object motion. This means that each finger mechanism may not be redundant, but together they are redundant as a hand.

Now, we will discuss the relationship of the force by using Eq. (2.45) when internal motion exists. As can be seen in Fig. 2.16, there are four different cases, because force closure and manipulable grasp are different concepts. Figure 2.16(a) shows a case of multi-joint fingers on both the right and left side, and there is friction at each contact point. The grasp is manipulable and a force closure. In this case, an arbitrary contact force can be generated by the joint torques. However, joint torque corresponding to the contact force is not unique if $\boldsymbol{J}_C$ is not a square matrix. Figure 2.16(b) presents a

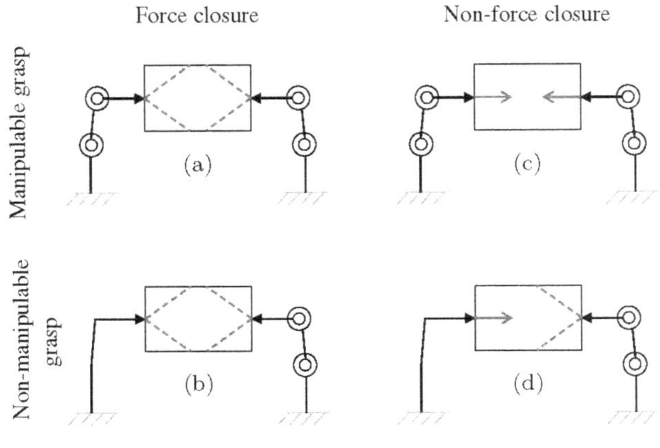

Fig. 2.16. Categories of grasp.

case with one multi-joint finger and a second finger with no joint. Here again, there is friction at each contact point. In this case, the grasp is not manipulable but it is a force closure, and so it is not possible to generate arbitrary contact velocity. Figure 2.16(c) shows a case of multi-joint fingers on both the right and left, but without friction at each contact point. In this case, the grasp is manipulable but not a force closure. It is not possible to balance the force if an external force acts on the object in the vertical direction. If there is no external force, an arbitrary object velocity is generated in a plane by the finger joint velocities. Finally, Fig. 2.16(d) is a case with one multi-joint finger with friction at the contact point and a second finger with no joint, and there is no friction at the contact point. In this case, the grasp is neither manipulable nor a force closure.

Example 2.5. Let's examine the manipulability of grasp when two fingers configured with translational joints grasp a rectangular object at point contacts with friction, as shown in Fig. 2.17.

The necessary and sufficient condition of the manipulable grasp is $\mathrm{Im}(\boldsymbol{G}_C^T) \subseteq \mathrm{Im}(\boldsymbol{H}_C^T \boldsymbol{J}_{FC})$. As in the case of Example 2.1, rotation matrices representing the orientation of each contact point coordinate system and the position vectors of contact points with respect to the reference coordinate system are represented as

$$\boldsymbol{R}_{C_1} = \begin{bmatrix} 0 & 1 \\ -1 & 0 \end{bmatrix}, \boldsymbol{R}_{C_2} = \begin{bmatrix} 0 & -1 \\ 1 & 0 \end{bmatrix}, \boldsymbol{p}_{C_1} = \begin{bmatrix} -a \\ d_{11} \end{bmatrix}, \boldsymbol{p}_{C_2} = \begin{bmatrix} a \\ d_{21} \end{bmatrix}. \quad (1)$$

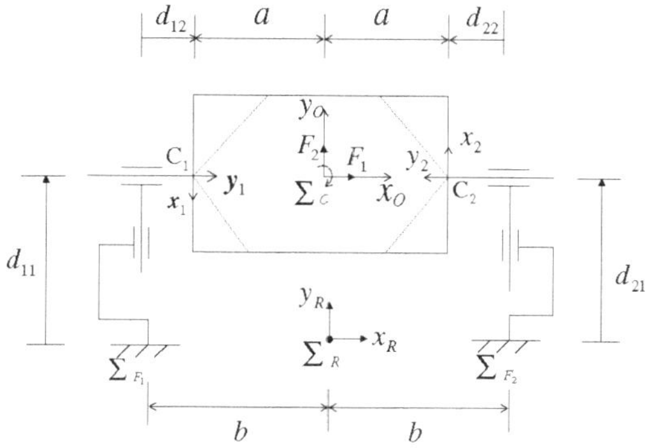

Fig. 2.17. Grasping an object with two jointed fingers with linear motion in the plane.

Equation (4) in Example 2.1 and

$$\boldsymbol{H}_{C_i} = \begin{bmatrix} \boldsymbol{R}_{C_i} & 0 \\ 0 & 1 \end{bmatrix} {}^{C_i}\boldsymbol{H}_{C_i}$$

derives

$$\boldsymbol{H}_{C_1} = \begin{bmatrix} 0 & 1 \\ -1 & 0 \\ 0 & 0 \end{bmatrix}, \quad \boldsymbol{H}_{C_2} = \begin{bmatrix} 0 & -1 \\ 1 & 0 \\ 0 & 0 \end{bmatrix}. \tag{2}$$

Hence, we obtain

$$\boldsymbol{H}_C = \begin{bmatrix} 0 & 1 & 0 & 0 \\ -1 & 0 & 0 & 0 \\ 0 & 0 & 0 & 0 \\ 0 & 0 & 0 & -1 \\ 0 & 0 & 1 & 0 \\ 0 & 0 & 0 & 0 \end{bmatrix}. \tag{3}$$

Next, Eq. (1) in Example 2.1 derives

$$\boldsymbol{W}_{OC_1} = \begin{bmatrix} 1 & 0 & 0 \\ 0 & 1 & 0 \\ -d_{11} & -a & 1 \end{bmatrix}, \quad \boldsymbol{W}_{OC_2} = \begin{bmatrix} 1 & 0 & 0 \\ 0 & 1 & 0 \\ -d_{21} & a & 1 \end{bmatrix} \tag{4}$$

Hence, from $\boldsymbol{G}_{C_i} = \boldsymbol{W}_{OC_i}\boldsymbol{H}_{C_i}$, we obtain

$$\boldsymbol{G}_{C_1} = \begin{bmatrix} 0 & 1 \\ -1 & 0 \\ a & -d_{11} \end{bmatrix}, \quad \boldsymbol{G}_{C_2} = \begin{bmatrix} 0 & -1 \\ 1 & 0 \\ a & d_{21} \end{bmatrix} \tag{5}$$

These equations derive

$$\boldsymbol{G}_C = [\boldsymbol{G}_{C_1}\boldsymbol{G}_{C_2}] = \begin{bmatrix} 0 & 1 & 0 & -1 \\ -1 & 0 & 1 & 0 \\ a & -d_{11} & a & d_{21} \end{bmatrix}. \tag{6}$$

It is evident that $rank\ (\boldsymbol{G}_C^T) = 3$. Position and orientation of the i-th fingertip referred to the reference coordinate $\boldsymbol{r}_i$ and joint valuable of

i–th finger $\boldsymbol{q}_i$ are represented as

$$\boldsymbol{r}_1 = \begin{bmatrix} d_{12} - b \\ d_{11} \\ -\pi/2 \end{bmatrix}, \quad \boldsymbol{r}_2 = \begin{bmatrix} -d_{22} + b \\ d_{21} \\ \pi/2 \end{bmatrix}, \quad \boldsymbol{q}_1 = \begin{bmatrix} d_{11} \\ d_{12} \end{bmatrix}, \quad \boldsymbol{q}_2 = \begin{bmatrix} d_{21} \\ d_{22} \end{bmatrix}. \tag{7}$$

Theses relations derive

$$\boldsymbol{J}_{FC_1} = \frac{\partial \boldsymbol{r}_1}{\partial \boldsymbol{q}_1} = \begin{bmatrix} 0 & 1 \\ 1 & 0 \\ 0 & 0 \end{bmatrix}, \quad \boldsymbol{J}_{FC_2} = \frac{\partial \boldsymbol{r}_2}{\partial \boldsymbol{q}_2} = \begin{bmatrix} 0 & -1 \\ 1 & 0 \\ 0 & 0 \end{bmatrix}. \tag{8}$$

From these relations, $\boldsymbol{J}_{FC}$ is represented as

$$\boldsymbol{J}_{FC} = \begin{bmatrix} 0 & 1 & 0 & 0 \\ 1 & 0 & 0 & 0 \\ 0 & 0 & 0 & 0 \\ 0 & 0 & 0 & -1 \\ 0 & 0 & 1 & 0 \\ 0 & 0 & 0 & 0 \end{bmatrix}. \tag{9}$$

Hence,

$$\boldsymbol{J}_C = \boldsymbol{H}_C^T \boldsymbol{J}_{FC} = \begin{bmatrix} -1 & 0 & 0 & 0 \\ 0 & 1 & 0 & 0 \\ 0 & 0 & 1 & 0 \\ 0 & 0 & 0 & 1 \end{bmatrix}. \tag{10}$$

This shows that the grasp is manipulable because of $rank(\boldsymbol{J}_C) = 4$, which means $\mathrm{Im}(\boldsymbol{G}_C^T) \subset \mathrm{Im}(\boldsymbol{J}_C)$.

2.6. Motion Representation of Rolling Contact by Curved Surface Coordinate

2.6.1. *Curved Surface Coordinate*

Now, consider a smooth object. By using the mapping $\boldsymbol{c} : \mathbb{R}^2 \to \mathbb{R}^3$ from 2D to 3D, as shown in Fig. 2.18, a point on the 2D surface

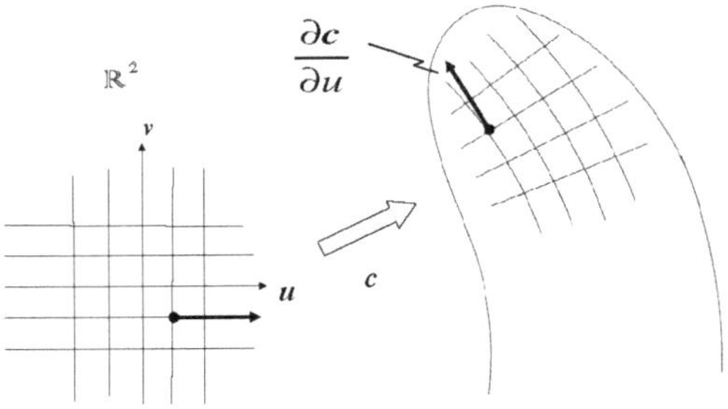

Fig. 2.18. Mapping $\boldsymbol{c}$.

$\boldsymbol{\alpha} = (u, v)^T$ is mapped to a point P on the object's 3D surface $\boldsymbol{p} = (x, y, z)^T$ as

$$\boldsymbol{p} = \boldsymbol{c}(\boldsymbol{\alpha}), \tag{2.46}$$

where it is assumed that the mapping $\boldsymbol{c}$ is a diffeomorphic mapping and that the coordinate system on 2D surfaces are orthogonal for the sake of simplicity. In this case, a tangent plane of the object is a space spanned by two vectors $\boldsymbol{c}_u = \partial \boldsymbol{c}/\partial u$ and $\boldsymbol{c}_v = \partial \boldsymbol{c}/\partial v$. The contact coordinate system $\Sigma_C = [\boldsymbol{x}, \boldsymbol{y}, \boldsymbol{z}]$ that consists of the two axes on the tangent plane and the outward normal orthogonal line is represented by

$$[\boldsymbol{x}, \boldsymbol{y}, \boldsymbol{z}] = \left[\frac{\boldsymbol{c}_u}{\|\boldsymbol{c}_u\|}, \frac{\boldsymbol{c}_v}{\|\boldsymbol{c}_v\|}, \frac{\boldsymbol{c}_u \times \boldsymbol{c}_v}{\|\boldsymbol{c}_u \times \boldsymbol{c}_v\|}\right]. \tag{2.47}$$

This is called a **normalized Gauss frame**. Here, the metric tensor $\boldsymbol{M}_c$ on the curved surface is given as

$$\boldsymbol{M}_p = \begin{bmatrix} \|\boldsymbol{c}_u\| & 0 \\ 0 & \|\boldsymbol{c}_v\| \end{bmatrix}. \tag{2.48}$$

Also, the **curvature tensor** $\boldsymbol{K}_p$ on the curved surface is given as

$$\boldsymbol{K}_p = \begin{bmatrix} \boldsymbol{x}^T \\ \boldsymbol{y}^T \end{bmatrix} \left[\frac{\boldsymbol{z}_u}{\|\boldsymbol{c}_u\|}, \frac{\boldsymbol{z}_v}{\|\boldsymbol{c}_v\|}\right], \tag{2.49}$$

where $\boldsymbol{z}_u = \partial \boldsymbol{z}/\partial u$ and $\boldsymbol{z}_v = \partial \boldsymbol{z}/\partial v$. This curvature tensor is considered as a magnitude of the fluctuation of the unit normal to the

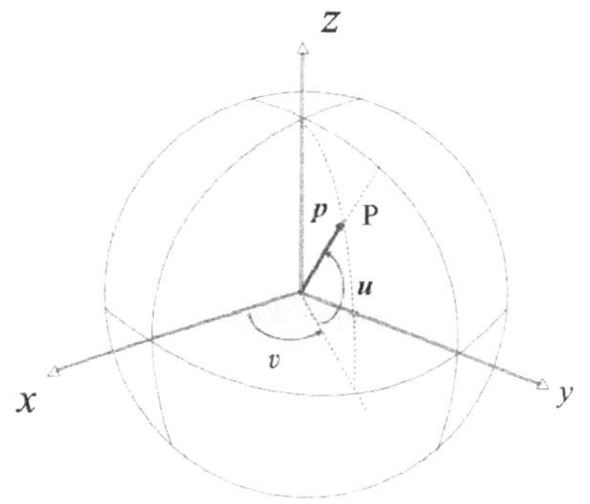

Fig. 2.19. Spherical coordinate.

tangent plane. Twist on the curved surface is defined as the change rates of curvature along the curved surface and is represented by

$$\boldsymbol{T}_p = \boldsymbol{y}^T \left[\frac{\boldsymbol{x}_u}{\|\boldsymbol{c}_u\|}, \frac{\boldsymbol{x}_v}{\|\boldsymbol{c}_v\|} \right], \tag{2.50}$$

Where $\boldsymbol{x}_u = \partial \boldsymbol{x}/\partial u$ and $\boldsymbol{x}_v = \partial \boldsymbol{x}/\partial v$. These $(\boldsymbol{M}_p, \boldsymbol{K}_p, \boldsymbol{T}_p)$ are collectively referred to as the geometric parameters of the surface.

When the object surface is a plane, then $\boldsymbol{z}_u = \boldsymbol{z}_v = \boldsymbol{x}_u = \boldsymbol{x}_v = 0$, and the following are obtained: $\boldsymbol{K}_p = 0$ and $\boldsymbol{T}_p = 0$.

Example 2.6. Let us set a spherical coordinate system $[u, v]$ for a spherical surface with radius ρ, as shown in Fig. 2.19. Find the geometric parameters $(\boldsymbol{M}_p, \boldsymbol{K}_p, \boldsymbol{T}_p)$.

A point P on the sphere surface is represented by

$$\boldsymbol{p} = \boldsymbol{c}(\boldsymbol{\alpha}) = \begin{bmatrix} \rho \cos u \cos v \\ \rho \cos u \sin v \\ \rho \sin u \end{bmatrix}, \tag{1}$$

where $-\pi/2 < u < \pi/2$ and $-\pi < v < \pi$. The partial differential of $\boldsymbol{c}$ regarding u and v are given by

$$\boldsymbol{c}_u = \begin{bmatrix} -\rho \sin u \cos v \\ -\rho \sin u \sin v \\ \rho \cos u \end{bmatrix}, \quad \boldsymbol{c}_v = \begin{bmatrix} -\rho \cos u \sin v \\ \rho \cos u \cos v \\ 0 \end{bmatrix}. \tag{2}$$

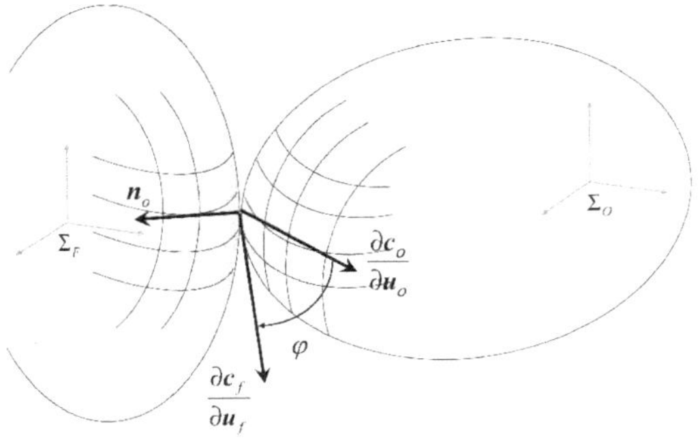

Fig. 2.20. Contact coordinate and twist angle.

Clearly, $\boldsymbol{c}_u$ and $\boldsymbol{c}_v$ are orthogonal. Hence,

$$\boldsymbol{M}_p = \begin{bmatrix} \rho & 0 \\ 0 & \rho\cos u \end{bmatrix}, \boldsymbol{K}_p = \begin{bmatrix} 1/\rho & 0 \\ 0 & 1/\rho \end{bmatrix}, \quad \boldsymbol{T}_p = \begin{bmatrix} 0, -\frac{\tan u}{\rho} \end{bmatrix}. \quad (3)$$

2.6.2. *Contacts between Object and Finger*

Consider the motion of a rigid finger and a rigid object, both with smooth curved surfaces, while a point of contact is maintained between the finger and the object, as shown in Fig. 2.20. By using the mapping $\boldsymbol{c}_f$: $\mathbb{R}^2 \to \mathbb{R}^3$ and $\boldsymbol{c}_o$: $\mathbb{R}^2 \to \mathbb{R}^3$ to represent the surface shapes of the object and the finger, and 2D orthogonal surface coordinates $\boldsymbol{\alpha}_f = [u_f, v_f]^T$ and $\boldsymbol{\alpha}_o = [u_o, v_o]^T$, a point of the fingertip surface expressed by the finger coordinates and a point of the object surface expressed by the object coordinates are represented as

$$^F\boldsymbol{p}_{FC} = \boldsymbol{c}_f(\boldsymbol{\alpha}_f), {}^O\boldsymbol{p}_{OC} = \boldsymbol{c}_o(\boldsymbol{\alpha}_o). \quad (2.51)$$

The deviation of relative attitude between the contact coordinate system on the finger side $\Sigma_{FC} = [\boldsymbol{x}_{FC}, \boldsymbol{y}_{FC}, \boldsymbol{z}_{FC}]$ and that on the object side $\Sigma_{OC} = [\boldsymbol{x}_{OC}, \boldsymbol{y}_{OC}, \boldsymbol{z}_{OC}]$ is represented with the angle φ from $\boldsymbol{x}_{FC}$ to $\boldsymbol{x}_{OC}$, which is called the angle of contact. Note that the $\boldsymbol{z}$-axis in each coordinate should be taken in the outward normal direction. Let us denote the relative translational velocity and the relative angular velocity of the finger contact point when referred with the object contact point as $^{FC}\dot{\boldsymbol{p}}_s = [v_x, v_y, v_z]^T$ and

${}^{FC}\boldsymbol{\omega}_s = [\omega_x, \omega_y, \omega_z]^T$, respectively. Then

$$\boldsymbol{q} = \begin{bmatrix} \boldsymbol{\alpha}_f \\ \boldsymbol{\alpha}_o \\ \phi \end{bmatrix} \in \mathbb{R}^5 \tag{2.52}$$

is called the **contact coordinates** on the two object surfaces. The geometrical parameters for each contact coordinate of the object side and the finger side are represented as $(\boldsymbol{M}_f, \boldsymbol{K}_f, \boldsymbol{T}_f)$ and $(\boldsymbol{M}_o, \boldsymbol{K}_o, \boldsymbol{T}_o)$, respectively. At the contact point, the following theorem is verified (Montana, 1988).

Theorem 2.2. (Equation of motion at the contact point) *The motion of contact coordinates are described as*

$$\dot{\boldsymbol{q}} = \begin{bmatrix} \boldsymbol{M}_f^{-1}(\boldsymbol{K}_f + \tilde{\boldsymbol{K}}_o)^{-1}\left(\begin{bmatrix} -\omega_y \\ \omega_x \end{bmatrix} - \tilde{\boldsymbol{K}}_o \begin{bmatrix} v_x \\ v_y \end{bmatrix}\right) \\ \boldsymbol{M}_o^{-1}\boldsymbol{R}_\varphi(\boldsymbol{K}_f + \tilde{\boldsymbol{K}}_o)^{-1}\left(\begin{bmatrix} -\omega_y \\ \omega_x \end{bmatrix} + \boldsymbol{K}_f \begin{bmatrix} v_x \\ v_y \end{bmatrix}\right) \\ \omega_z + \boldsymbol{T}_f\boldsymbol{M}_f\dot{\boldsymbol{\alpha}}_f + \boldsymbol{T}_o\boldsymbol{M}_o\dot{\boldsymbol{\alpha}}_o \end{bmatrix} \tag{2.53}$$

$$v_z = 0$$

where $\boldsymbol{R}_\varphi \in \mathbb{R}^{2\times 2}$ *is a rotation matrix that represents the orientation of the finger contact coordinate system* Σ_{FC} *referred with the object contact point coordinate system* Σ_{OC}*, and given by*

$$\boldsymbol{R}_\varphi = \begin{bmatrix} \cos\varphi & -\sin\varphi \\ -\sin\varphi & -\cos\varphi \end{bmatrix}, \tag{2.54}$$

and $\tilde{\boldsymbol{K}}_o = \boldsymbol{R}_\varphi \boldsymbol{K}_o \boldsymbol{R}_\varphi$.

For the proof of the theorem, see Example 2.5. Now, apply this theorem to the pure rolling contact. In the pure rolling contact,

$$\dot{\boldsymbol{p}}_s = [v_x, v_y, v_z]^T = 0, \quad \omega_z = 0. \tag{2.55}$$

Therefore, Eq. (2.53) is written as

$$\boldsymbol{M}_f\dot{\boldsymbol{\alpha}}_f - \boldsymbol{R}_\varphi\boldsymbol{M}_o\dot{\boldsymbol{\alpha}}_0 = 0$$

$$\boldsymbol{T}_f\boldsymbol{M}_f\dot{\boldsymbol{\alpha}}_f + \boldsymbol{T}_o\boldsymbol{M}_o\dot{\boldsymbol{\alpha}}_o - \dot{\varphi} = 0$$

These equations derive

$$A_q \dot{q} = 0, \tag{2.56}$$

where

$$A_q = \begin{bmatrix} M_f & -R_\varphi M_o & 0 \\ T_f M_f & T_o M_o & -1 \end{bmatrix}. \tag{2.57}$$

Equation (2.56) represents the velocity constraint described by the surface coordinates for pure rolling contacts.

Example 2.7. Consider the situation where the motion of a spherical-shaped fingertip with a radius of ρ has a pure rolling contact on a plane. As shown in Example 2.6, when we set the spherical coordinates at the contact point of the fingertip, the contact position is represented as

$$p = c_f(\alpha_f) = \begin{bmatrix} \rho \cos u_f \cos v_f \\ \rho \cos u_f \sin v_v \\ \rho \sin u_f \end{bmatrix}. \tag{1}$$

On the other hand, when we set a plane coordinate system at the contact point on the plane, the contact position is represented as

$$p = c_o(\alpha_o) = \begin{bmatrix} u_o \\ v_o \\ 0 \end{bmatrix}. \tag{2}$$

Then, the following are derived:

$$M_f = \begin{bmatrix} \rho & 0 \\ 0 & \rho \cos u_f \end{bmatrix}, \quad K_f = \begin{bmatrix} 1/\rho & 0 \\ 0 & 1/\rho \end{bmatrix}, \quad T_f = \begin{bmatrix} 0, -\frac{\tan u_f}{\rho} \end{bmatrix}, \tag{3}$$

$$M_o = \begin{bmatrix} 1 & 0 \\ 0 & 1 \end{bmatrix}, \quad K_o = \begin{bmatrix} 0 & 0 \\ 0 & 0 \end{bmatrix}, T_o = [0, 0]. \tag{4}$$

On the condition of pure rolling contact, we get

$$^{FC}v_s = 0, \omega_z = 0.$$

According to Eq. (2.53), the motion of contact coordinate systems is represented as

$$\dot{\boldsymbol{q}} = \begin{bmatrix} 0 \\ \sec u_f \\ -\rho \sin \varphi \\ -\rho \cos \varphi \\ -\tan u_f \end{bmatrix} \omega_x + \begin{bmatrix} -1 \\ 0 \\ -\rho \cos \varphi \\ \rho \sin \varphi \\ 0 \end{bmatrix} \omega_y \tag{5}$$

Exercises

2.1. At the contacts between two rigid objects, there exists a line contact and a surface contact, as shown in Fig. 2.21. Find the matrix that represents these wrench bases.

2.2. As seen in Fig. 2.22, find the grasp map, the friction cone, and the contact force when grasping a rectangular solid at the soft finger contact C_1, and the frictionless point contact C_2.

2.3. Find $\boldsymbol{G}$, $\boldsymbol{W}_C$, $\boldsymbol{J}_{FC}$ in the case of two fingers with two rotational joints and a translational joint grasping a rectangular solid, as shown in Fig. 2.23. Also, determine whether it is a force closure and if it is manipulable. Note that the contact model is a soft finger contact. In the figure, the θ_1 and θ_2 are 0.

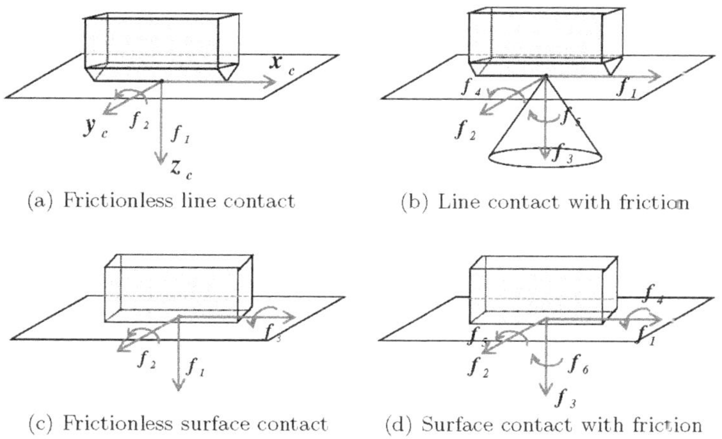

(a) Frictionless line contact (b) Line contact with friction

(c) Frictionless surface contact (d) Surface contact with friction

Fig. 2.21. Contact between objects.

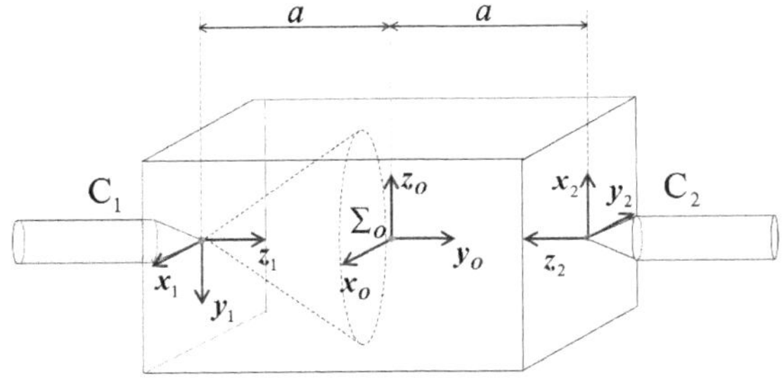

Fig. 2.22. Grasp by two fingers.

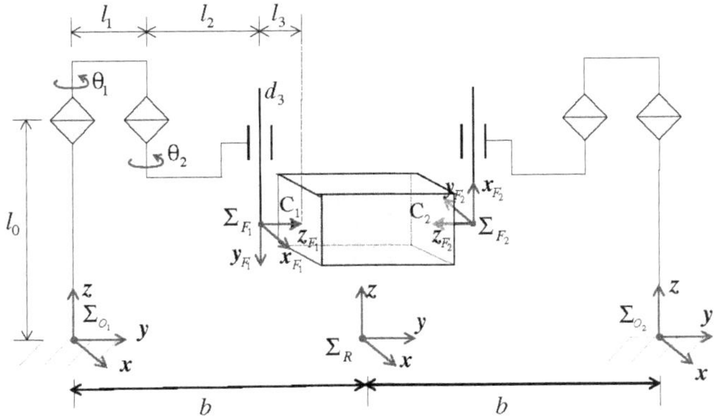

Fig. 2.23. Grasping an object with a horizontal, articulated robot.

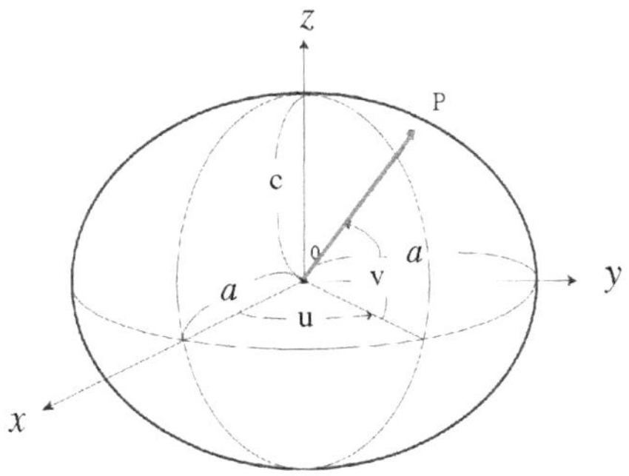

Fig. 2.24. Elliptic coordinates.

2.4. Find the geometric parameters of the ellipsoid as shown in Fig. 2.24.

$$\left(\frac{x}{a}\right)^2 + \left(\frac{y}{a}\right)^2 + \left(\frac{z}{c}\right)^2 = 1$$

2.5. Prove Theorem 2.2.

Chapter 3

Kinematic Constraint and Controllability

When the robot manipulates an object, the kinematics of fingertips are constrained by the contact with the object, as presented in Chapter 2. The kinematics of an automobile are constrained because the front and rear wheels are only allowed to roll and spin, but not slide sideways. These motions are under the kinematic constraints. The kinematic constraints are generally classified into two types: holonomic constraints and non-holonomic constraints. The holonomic constraints are expressed by algebraic equations in generalized coordinates. The other constraints are non-holonomic constraints, which are typically expressed by differential equations in generalized coordinates. We know that independently driven two-wheel mobile robots can take any position and orientation on the plane in time. This is due to the effect of non-holonomic constraints.

In this chapter, the concepts of holonomic and non-holonomic constraints, the effect of non-holonomic constraints, and the basic controllability analysis for nonlinear control systems are presented. More details on these topics are available in references of. Neimark and Fufaev (1972), Murray *et al.* (1993), Nakamura (1993a-1993d, 1994), Mason (1998), Soltakhanov *et al.*, (2009), Bloch (2010) and Kwatny and Blankenship (2000).

3.1. Holonomic and Non-Holonomic Constraints

When a robot hand manipulates an object, there exist **kinematic constraints** between the object and environment and between the object and fingers in general. The generalized coordinates $\boldsymbol{q} = [q_1, q_2, \ldots, q_n]^T$ are used to describe the kinematics of the mechanical

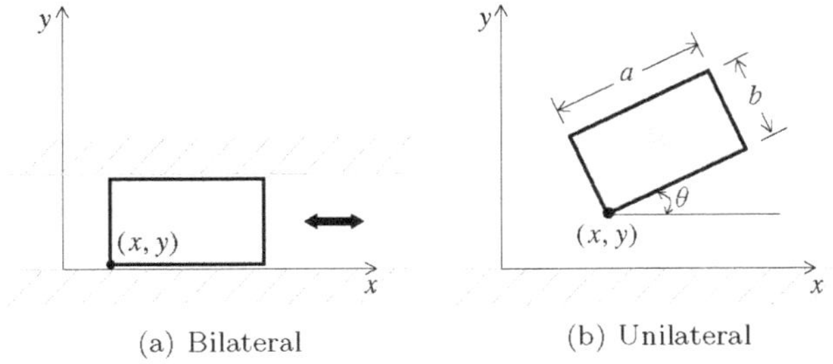

Fig. 3.1. Bilateral and unilateral constraints.

system. There are two types of kinematic constraints in the physical classification: **bilateral constraints**, which are represented by two-sided equality constraints as typified by $\varphi = 0$, and **unilateral constraints**, which are represented by one-sided inequality constraints as typified by $\varphi \geq 0$. Consider the examples of kinematic constraints shown in Fig. 3.1, where a rectangular object with size $a \times b$ moves at the corner of the lower left point in the xy plane. Set the generalized coordinates in the plane as $\boldsymbol{q} = [x, y, \theta]^T$, where $[x, y]$ are the coordinates of the left lower apex, and θ is the angle between the x axis and the bottom line of the rectangular object. Figure 3.1(a) shows a bilateral constraint when the rectangular object moves in a conduit placed parallel to the x axis and is constrained by both sides. The constraint is represented as

$$y = 0, \quad \theta = 0.$$

Figure 3.1(b) shows a unilateral constraint when the x axis is a wall and the constant is only one side. Then, the four apexes are subject to constraints represented as

$$y \geq 0, y + a\sin(\theta) \geq 0, y + a\sin(\theta) + b\cos(\theta) \geq 0, y + b\cos(\theta) \geq 0.$$

In another classification, the kinematic constraints are classified as **holonomic constraints**, which are represented by the velocity variable-independent equation $\varphi(\boldsymbol{q}, t) = 0$, and **non-holonomic constraints**, which cannot be represented by the equation $\varphi(\boldsymbol{q}, t) = 0$ but can be represented by an equation including the velocity

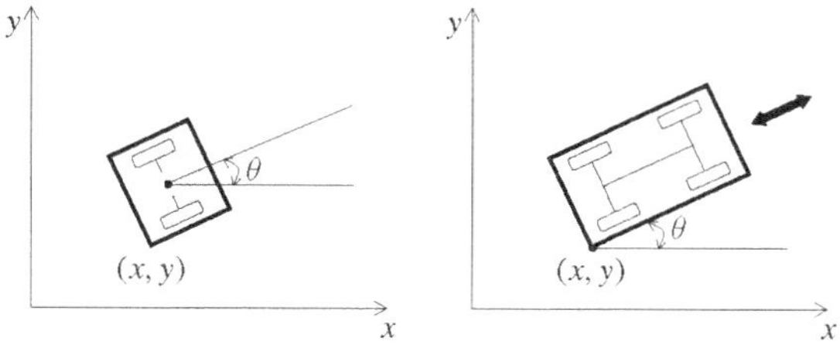

(a) Non-holonomic (b) Holonomic

Fig. 3.2. Non-holonomic and holonomic constraints.

variable: $\varphi(\boldsymbol{q}, \dot{\boldsymbol{q}}, t) = 0$. The latter constraints are also called **non-integral constraints** because the integration of the constraint equation has no physical meaning.

The examples of these kinematic constraints are shown in Fig. 3.2. Figure 3.2(a) shows a two-wheel vehicle whose wheels are located to the right and left of the body center and can be driven independently. It can move in the direction $[x, y]$ and rotate with angle θ around the body center. At this time, it is not possible for the vehicle to move in a direction $[\sin\theta, -\cos\theta]^T$ perpendicular to the traveling direction $[\dot{x}, \dot{y}]^T$. Therefore, the kinematic constraint is represented by

$$\dot{x}\sin\theta - \dot{y}\cos\theta = 0.$$

Since this constraint cannot be represented only by the generalized coordinate $\boldsymbol{q}$, this is a non-holonomic constraint. In general, non-holonomic constraints include constraints which are represented by inequality. However, in this book they are limited to constraints represented by equalities including velocity variables. Figure 3.2(b) shows a four-wheel vehicle that has no handle and only moves back and forth while maintaining the current posture. This is an example of a holonomic constraint. The vehicle cannot move in a direction perpendicular to the traveling direction, and therefore the kinematic constraint is represented as

$$\dot{x}\sin\theta - \dot{y}\cos\theta = 0, \quad \dot{\theta} = 0.$$

In this example, the constraint contains velocity variables, but it can be rewritten by integrating the above equations as follows:

$$(x - x_0)\sin\theta - (y - y_0)\cos\theta = 0, \quad \theta - \theta_0 = 0,$$

where $[x_0, y_0, \theta_0]^T$ is the initial state. This constraint does not include the velocity variables. In this way, even if a constraint includes velocity variables, it is not a non-holonomic constraint if it can be rewritten by a constraint which does not include the velocity variables.

Let us denote $\boldsymbol{q}$ as the **generalized coordinates** of the system. In general, if a constraint restricts the motion of the system on a hypersurface $M \subset \mathbb{R}^n$ expressing the generalized coordinates $\boldsymbol{q}$, it is said to be a holonomic constraint. The hypersurface is called a **manifold**. Assume that the local algebraic constraint is expressed by

$$\varphi_i(\boldsymbol{q}) = 0, \quad \text{for } i = 1, \ldots, k. \tag{3.1}$$

These constraints are summarized as $\boldsymbol{\varphi}(= [\varphi_1, \ldots, \varphi_k]^T \in \mathbb{R}^k)$, which is a vector-valued function. A set of $\boldsymbol{q}$ satisfying Eq. (3.1) is expressed by $M = \{\boldsymbol{q} : \boldsymbol{\varphi}(\boldsymbol{q}) = \boldsymbol{0}\}$, which is a manifold satisfying constraint. If all row vectors of $\boldsymbol{\varphi}$ are independent at each point $\boldsymbol{q}$, then the dimension of the manifold M is $n-k$. If the gradient of the constraint with respect to $\boldsymbol{q}$,

$$\frac{\partial \boldsymbol{\varphi}}{\partial \boldsymbol{q}} = \begin{bmatrix} \frac{\partial \varphi_1}{\partial q_1} & \cdots & \frac{\partial \varphi_1}{\partial q_n} \\ \vdots & \ddots & \vdots \\ \frac{\partial \varphi_k}{\partial q_1} & \cdots & \frac{\partial \varphi_k}{\partial q_n} \end{bmatrix}, \tag{3.2}$$

has full row rank, then the constraint force that is generated by the holonomic constraint Eq. (3.1) is expressed by

$$\boldsymbol{f} = \left(\frac{\partial \boldsymbol{\varphi}}{\partial \boldsymbol{q}}\right)^T \boldsymbol{\lambda}, \tag{3.3}$$

where $\boldsymbol{\lambda} \in \mathbb{R}^k$ is a Lagrange multiplier that represents the scale factor of the column vector of the gradient. This constraint force is the normal force applied against the constraint surface and is always orthogonal to the velocity as shown in the following relation:

$$\boldsymbol{f}^T \dot{\boldsymbol{q}} = \boldsymbol{\lambda}^T \frac{\partial \boldsymbol{\varphi}}{\partial \boldsymbol{q}} \dot{\boldsymbol{q}} = \boldsymbol{\lambda}^T \frac{d}{dt}(\boldsymbol{\varphi}(\boldsymbol{q})) = 0. \tag{3.4}$$

If the non-holonomic constraint is expressed by an equation $\varphi(\boldsymbol{q}, \dot{\boldsymbol{q}}) = 0$ that does not include the time t explicitly, then it is said to be **seleronomic**. If the non-holonomic constraint is expressed by an equation $\varphi(\boldsymbol{q}, \dot{\boldsymbol{q}}, t) = 0$ that includes the time explicitly, then it is said to be **rheonomic**.

The constraints for mechanical systems such as the motion of mobile robots, the rolling contact by a robotic hand, and the motion of space robots based on the momentum conservation law contain only $\boldsymbol{q}$ and $\dot{\boldsymbol{q}}$, but do not contain the time explicitly. Therefore, it is assumed hereafter that the constraints do not include the time explicitly.

3.2. Properties of Non-Holonomic Constraints

Consider a velocity constraint of the form

$$\boldsymbol{A}(\boldsymbol{q})\dot{\boldsymbol{q}} = 0, \tag{3.5}$$

where $\boldsymbol{A} \in \mathbb{R}^{k\times n}$ represents a set of k velocity constraints. A constraint of this form is called a **Pfaffian constraint**. It is assumed that the constraint has full row rank at each point $\boldsymbol{q}$ satisfying Eq. (3.5). The constraint is said to be **integrable** if there exists a vector-valued function φ such that

$$\varphi(\boldsymbol{q}) = 0 \Leftrightarrow \boldsymbol{A}(\boldsymbol{q})\dot{\boldsymbol{q}} = 0, \tag{3.6}$$

where A $\Leftrightarrow$ B means A is true if and only if B is true. An integrable Pfaffian constraint is equivalent to a holonomic constraint. A Pfaffian constraint which is not integrable is a non-holonomic constraint and there is no manifold M satisfying $\varphi(\boldsymbol{q}) = 0$.

Consider an integrable Pfaffian constraint. Let the velocity $\dot{\boldsymbol{q}}$ satisfying the constraint (3.5) be expressed by $\boldsymbol{p}_i(\boldsymbol{q})(i = 1, 2, \dots, n-k)$ as the function of $\boldsymbol{q}$. The $n-k$ number of $\boldsymbol{p}_i$ makes a set of basis vectors for the null space of $\boldsymbol{A}(\boldsymbol{q})$, which represents the entire feasible $\boldsymbol{p}_i$ satisfying the constraint. The derivative of the holonomic constraint given by Eq. (3.1) is represented by

$$\frac{\partial \varphi}{\partial \boldsymbol{q}}\dot{\boldsymbol{q}} = 0.$$

This leads to

$$\boldsymbol{A}(\boldsymbol{q}) = \frac{\partial \boldsymbol{\varphi}}{\partial \boldsymbol{q}}. \tag{3.7}$$

From this relationship, $\boldsymbol{p}_i$ represents a tangent vector of the manifold M satisfying the constraint and is called a **vector field**. The $n-k$ dimensional set of $\boldsymbol{p}_i$ $(i = 1, 2, \ldots, n-k)$ equals a tangent hyperplane of the manifold. We denote the tangent hyperplane as TM. This indicates that the locally possible motion under the holonomic constraint exists in the tangent hyperplane TM. On the other hand, in the non-holonomic constraint, the $n-k$ dimensional subspace is still the tangent hyperplane TM, but there is no manifold satisfied by Eq. (3.1) and the locally possible motion is not limited in the tangent hyperplane.

Let us consider these facts using a series of motions with four time units given by

$$\dot{\boldsymbol{q}}(t) = \begin{cases} \boldsymbol{p}_1(\boldsymbol{q}(t)) & (0 \le t < T) \\ \boldsymbol{p}_2(\boldsymbol{q}(t)) & (T \le t < 2T) \\ -\boldsymbol{p}_1(\boldsymbol{q}(t)) & (2T \le t < 3T) \\ -\boldsymbol{p}_2(\boldsymbol{q}(t)) & (3T \le t < 4T) \end{cases}. \tag{3.8}$$

The results of this motion can be determined as follows. The $\boldsymbol{q}(T)$ is calculated by a Taylor series expansion up to second order. First,

$$\dot{\boldsymbol{q}}(0) = \boldsymbol{p}_1(\boldsymbol{q}(0))$$

$$\ddot{\boldsymbol{q}}(0) = \frac{\partial \boldsymbol{p}_1}{\partial \boldsymbol{q}}(\boldsymbol{q}(0))\dot{\boldsymbol{q}}(0) = \frac{\partial \boldsymbol{p}_1}{\partial \boldsymbol{q}}(\boldsymbol{q}(0))\boldsymbol{p}_1(\boldsymbol{q}(0)).$$

Therefore,

$$\begin{aligned} \boldsymbol{q}(T) &= \boldsymbol{q}(0) + T\dot{\boldsymbol{q}}(0) + \frac{1}{2}T^2\ddot{\boldsymbol{q}}(0) + O(T^3) \\ &= \boldsymbol{q}(0) + T\boldsymbol{p}_1(\boldsymbol{q}(0)) + \frac{1}{2}T^2\frac{\partial \boldsymbol{p}_1}{\partial \boldsymbol{q}}(\boldsymbol{q}(0))\boldsymbol{p}_1(\boldsymbol{q}(0)) + O(T^3), \end{aligned}$$

where $O(T^k)$ represents all terms of a Taylor series expansion of more than k-th order. Next, from the relationship of

$$\dot{\boldsymbol{q}}(T) = \boldsymbol{p}_2(\boldsymbol{q}(T)) = \boldsymbol{p}_2(\boldsymbol{q}(0)) + T\frac{\partial \boldsymbol{p}_2}{\partial \boldsymbol{q}}(\boldsymbol{q}(0))\boldsymbol{p}_1(\boldsymbol{q}(0)) + O(T^2)$$

$$\ddot{\boldsymbol{q}}(T) = \frac{\partial \boldsymbol{p}_2}{\partial \boldsymbol{q}}(\boldsymbol{q}(T))\boldsymbol{p}_2(\boldsymbol{q}(T)) = \frac{\partial \boldsymbol{p}_2}{\partial \boldsymbol{q}}(\boldsymbol{q}(0))\boldsymbol{p}_2(\boldsymbol{q}(0)) + O(T),$$

the state $\boldsymbol{q}$ at $t = 2T$ is calculated by

$$\begin{aligned}\boldsymbol{q}(2T) &= \boldsymbol{q}(T) + T\dot{\boldsymbol{q}}(T) + \frac{1}{2}T^2\ddot{\boldsymbol{q}}(T) + O(T^3)\\ &= \boldsymbol{q}(0) + T(\boldsymbol{p}_1(\boldsymbol{q}(0)) + \boldsymbol{p}_2(\boldsymbol{q}(0))) + \frac{1}{2}T^2\left(\frac{\partial \boldsymbol{p}_1}{\partial \boldsymbol{q}}(\boldsymbol{q}(0))\boldsymbol{p}_1(\boldsymbol{q}(0))\right.\\ &\quad \left. + 2\frac{\partial \boldsymbol{p}_2}{\partial \boldsymbol{q}}(\boldsymbol{q}(0))\boldsymbol{p}_1(\boldsymbol{q}(0)) + \frac{\partial \boldsymbol{p}_2}{\partial \boldsymbol{q}}(\boldsymbol{q}(0))\boldsymbol{p}_2(\boldsymbol{q}(0))\right) + O(T^3).\end{aligned}$$

Similarly, $\dot{\boldsymbol{q}}$ and $\dot{\boldsymbol{q}}$ at $t = 2T$ are

$$\begin{aligned}\dot{\boldsymbol{q}}(2T) &= -\boldsymbol{p}_1(\boldsymbol{q}(2T))\\ &= -\boldsymbol{p}_1(\boldsymbol{q}(T)) - T\frac{\partial \boldsymbol{p}_1}{\partial \boldsymbol{q}}(\boldsymbol{q}(T))\boldsymbol{p}_2(\boldsymbol{q}(T)) + O(T^2)\\ &= -\boldsymbol{p}_1(\boldsymbol{q}(0))\\ &\quad -T\left(\frac{\partial \boldsymbol{p}_1}{\partial \boldsymbol{q}}(\boldsymbol{q}(0))\boldsymbol{p}_1(\boldsymbol{q}(0)) + \frac{\partial \boldsymbol{p}_1}{\partial \boldsymbol{q}}(\boldsymbol{q}(0))\boldsymbol{p}_2(\boldsymbol{q}(0))\right) + O(T^2)\\ \ddot{\boldsymbol{q}}(2T) &= -\frac{\partial \boldsymbol{p}_1}{\partial \boldsymbol{q}}(\boldsymbol{q}(2T))(-\boldsymbol{p}_1(\boldsymbol{q}(2T)))\\ &= \frac{\partial \boldsymbol{p}_1}{\partial \boldsymbol{q}}(\boldsymbol{q}(0))\boldsymbol{p}_1(\boldsymbol{q}(0)) + O(T).\end{aligned}$$

From these relations, the state $\boldsymbol{q}$ at $t = 3T$ is calculated by

$$\begin{aligned}\boldsymbol{q}(3T) &= \boldsymbol{q}(2T) + T\dot{\boldsymbol{q}}(2T) + \frac{1}{2}T^2\ddot{\boldsymbol{q}}(2T) + O(T^3)\\ &= \boldsymbol{q}(0) + T\boldsymbol{p}_2(\boldsymbol{q}(0))\\ &\quad + T^2\left(\frac{\partial \boldsymbol{p}_2}{\partial \boldsymbol{q}}(\boldsymbol{q}(0))\boldsymbol{p}_1(\boldsymbol{q}(0)) - \frac{\partial \boldsymbol{p}_1}{\partial \boldsymbol{q}}(\boldsymbol{q}(0))\boldsymbol{p}_2(\boldsymbol{q}(0))\right.\\ &\quad \left. + \frac{1}{2}\frac{\partial \boldsymbol{p}_2}{\partial \boldsymbol{q}}(\boldsymbol{q}(0))\boldsymbol{p}_2(\boldsymbol{q}(0))\right) + O(T^3).\end{aligned}$$

In addition, $\dot{\boldsymbol{q}}$ and $\dot{\boldsymbol{q}}$ at $t = 3T$ are

$$\begin{aligned}
\dot{\boldsymbol{q}}(3T) &= -\boldsymbol{p}_2(\boldsymbol{q}(3T)) \\
&= -\boldsymbol{p}_2(\boldsymbol{q}(2T)) - T\frac{\partial \boldsymbol{p}_2}{\partial \boldsymbol{q}}(\boldsymbol{q}(2T))(-\boldsymbol{p}_1(\boldsymbol{q}(2T))) + O(T^2) \\
&= -\boldsymbol{p}_2(\boldsymbol{q}(T)) \\
&\quad - T\left(\frac{\partial \boldsymbol{p}_2}{\partial \boldsymbol{q}}(\boldsymbol{q}(T))\boldsymbol{p}_2(\boldsymbol{q}(T) - \frac{\partial \boldsymbol{p}_2}{\partial \boldsymbol{q}}(\boldsymbol{q}(T))\boldsymbol{p}_1(\boldsymbol{q}(T))\right) + O(T^2) \\
&= -\boldsymbol{p}_2(\boldsymbol{q}(0)) - T\frac{\partial \boldsymbol{p}_2}{\partial \boldsymbol{q}}(\boldsymbol{q}(0))\boldsymbol{p}_2(\boldsymbol{q}(0)) + O(T^2) \\
\ddot{\boldsymbol{q}}(3T) &= -\frac{\partial \boldsymbol{p}_2}{\partial \boldsymbol{q}}(\boldsymbol{q}(3T))(-\boldsymbol{p}_2(\boldsymbol{q}(3T))) \\
&= \frac{\partial \boldsymbol{p}_2}{\partial \boldsymbol{q}}(\boldsymbol{q}(0))\boldsymbol{p}_2(\boldsymbol{q}(0)) + O(T).
\end{aligned}$$

Using the above relations, the state $\boldsymbol{q}$ at $t = 4T$ is calculated by

$$\begin{aligned}
\boldsymbol{q}(4T) &= \boldsymbol{q}(3T) + T\dot{\boldsymbol{q}}(3T) + \frac{1}{2}T^2\ddot{\boldsymbol{q}}(3T) + O(T^3) \\
&= \boldsymbol{q}(0) + T^2\left(\frac{\partial \boldsymbol{p}_2}{\partial \boldsymbol{q}}(\boldsymbol{q}(0))\boldsymbol{p}_1(\boldsymbol{q}(0) - \frac{\partial \boldsymbol{p}_1}{\partial \boldsymbol{q}}(\boldsymbol{q}(0))\boldsymbol{p}_2(\boldsymbol{q}(0))\right) \\
&\quad + O(T^3) \qquad (3.9)
\end{aligned}$$

This equation shows that $\boldsymbol{q}$ at $t = 4T$ returns to $\boldsymbol{q}$ at $t = 0$ if T is sufficiently small and $\boldsymbol{q}(4T)$ can be approximated by using the first order term of T. This also applies to both the holonomic and the non-holonomic constraints.

The difference between the holonomic and non-holonomic constraints is seen by the second order term of T in Eq. (3.9). Now, let this term be denoted by

$$[\boldsymbol{p}_1, \boldsymbol{p}_2] = \frac{\partial \boldsymbol{p}_2}{\partial \boldsymbol{q}}\boldsymbol{p}_1 - \frac{\partial \boldsymbol{p}_1}{\partial \boldsymbol{q}}\boldsymbol{p}_2. \qquad (3.10)$$

The notation $[\boldsymbol{p}_1, \boldsymbol{p}_2]$ is called the **Lie bracket** of two vector fields $\boldsymbol{p}_1$ and $\boldsymbol{p}_2$. It can be shown below that if the constraint is holonomic, then the vector $[\boldsymbol{p}_1, \boldsymbol{p}_2]$ exists in the tangent hyperplane TM of the manifold satisfying the constraint.

Assume that the constraint is holonomic and there exists the relation in Eq. (3.6). By multiplying $\boldsymbol{A}(\boldsymbol{q})$ to $[\boldsymbol{p}_1, \boldsymbol{p}_2]$ from the left side and substituting Eq. (3.6), we get

$$\boldsymbol{A}(\boldsymbol{q})[\boldsymbol{p}_1, \boldsymbol{p}_2] = \frac{\partial \boldsymbol{\varphi}}{\partial \boldsymbol{q}} \frac{\partial \boldsymbol{p}_2}{\partial \boldsymbol{q}} \boldsymbol{p}_1 - \frac{\partial \boldsymbol{\varphi}}{\partial \boldsymbol{q}} \frac{\partial \boldsymbol{p}_1}{\partial \boldsymbol{q}} \boldsymbol{p}_2. \tag{3.11}$$

Since $\boldsymbol{p}_i (i = 1, 2) \in TM$,

$$\boldsymbol{A}(\boldsymbol{q}) \boldsymbol{p}_i(\boldsymbol{q}) = \frac{\partial \boldsymbol{\varphi}}{\partial \boldsymbol{q}} \boldsymbol{p}_i(\boldsymbol{q}) = 0. \tag{3.12}$$

The partial derivative of the above equation with respect to $\boldsymbol{q}$ is represented by

$$\frac{\partial}{\partial \boldsymbol{q}} \left[\frac{\partial \boldsymbol{\varphi}}{\partial \boldsymbol{q}} \boldsymbol{p}_i(\boldsymbol{q}) \right] = \frac{\partial \boldsymbol{\varphi}}{\partial \boldsymbol{q}} \frac{\partial \boldsymbol{p}_i}{\partial \boldsymbol{q}} + \begin{bmatrix} \boldsymbol{p}_i^T \boldsymbol{B}_1 \\ \vdots \\ \boldsymbol{p}_i^T \boldsymbol{B}_k \end{bmatrix} = 0 \quad \text{for } i = 1, 2$$

where $\boldsymbol{B}_i(\boldsymbol{q})$ is the Hessian of the i-th component of $\varphi(\boldsymbol{q})$ and given by

$$\boldsymbol{B}_i(\boldsymbol{q}) = \frac{\partial}{\partial \boldsymbol{q}} \left[\frac{\partial \varphi_i}{\partial \boldsymbol{q}} \right]^T.$$

Generally, $\partial \varphi_i / \partial q_j \partial q_l = \partial \varphi_i / \partial q_l \partial q_j$ and the Hessian is a symmetric matrix. By substituting these equations into Eq. (3.11), the relation can be rewritten as

$$\boldsymbol{A}[\boldsymbol{p}_1, \boldsymbol{p}_2] = \begin{bmatrix} \boldsymbol{p}_1^T \boldsymbol{B}_1 \\ \vdots \\ \boldsymbol{p}_1^T \boldsymbol{B}_k \end{bmatrix} \boldsymbol{p}_2 - \begin{bmatrix} \boldsymbol{p}_2^T \boldsymbol{B}_1 \\ \vdots \\ \boldsymbol{p}_2^T \boldsymbol{B}_k \end{bmatrix} \boldsymbol{p}_1 = 0. \tag{3.13}$$

The Eq. (3.13) implies $[\boldsymbol{p}_1, \boldsymbol{p}_2] \in TM$.

Since the Jacobian and Hassian do not exist in the case of the non-holonomic constraint, in general, it will be possible that the vector $[\boldsymbol{p}_1, \boldsymbol{p}_2]$ goes beyond the $n - k$ dimension hyperplane TM.

Figure 3.3 shows the difference in motion between the holonomic and non-holonomic constraints. There exists a manifold satisfying

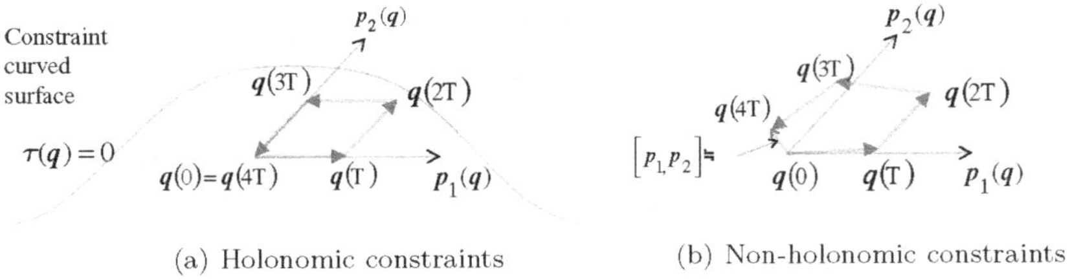

Fig. 3.3. Kinematics of holonomic and non-holonomic constraints.

the constraints in the case of the holonomic constraints, but no such manifold exists in the the case of the non-holonomic constraints. Therefore, there is a possibility of movement in the space of the n dimensional generalized coordinates even if there are k number of constraints represented by Eq. (3.5).

3.3. The Basic Differential Geometry

3.3.1. *Coordinate Transformation and Diffeomorphism*

The coordinate transformation $\varphi : U \subseteq \mathbb{R}^n \rightarrow V \subseteq \mathbb{R}^m$ means that it converts a state $\boldsymbol{q} \in \mathbb{R}^n$ to a state $\varphi(\boldsymbol{q}) \in \mathbb{R}^m$. This is also called a **map**. The map φ is said to be a **smooth map** if all partial derivatives of φ, of any order, exist and are continuous. Assume that the new mapped state is $\boldsymbol{z} \in \mathbb{R}^m$, then

$$\boldsymbol{z} = \varphi(\boldsymbol{q}). \tag{3.14}$$

For a linear system, a transformation matrix $\boldsymbol{T} \in \mathbb{R}^{m \times n}$ is generally used for the coordinate transformation, which is represented by

$$\boldsymbol{z} = \boldsymbol{T}\boldsymbol{q}. \tag{3.15}$$

If $n = m$ and the linear system with respect to $\boldsymbol{z}$ is equivalent to that with respect to $\boldsymbol{q}$, then there should exist the inverse transformation:

$$\boldsymbol{q} = \boldsymbol{T}^{-1}\boldsymbol{z}. \tag{3.16}$$

Therefore, the matrix $\boldsymbol{T}$ must be non-singular. Similarly, the coordinate transformation $\boldsymbol{z} = \varphi(\boldsymbol{q})$ in the non-linear system requires the existence of the inverse function $\varphi^{-1}(\boldsymbol{z})$ for the original non-linear system to be equivalent to the mapped system, and this is represented formally by

$$\boldsymbol{q} = \varphi^{-1}(\boldsymbol{z}). \tag{3.17}$$

Then, the map is said to be bijective. Hereafter, it is assumed that $n = m$ when considering the bijective map.

Definition 3.1. (Diffeomorphism)

The map $\varphi : U \to V$ is called a **diffeomorphism** if it is bijective ($U = V$) and if both φ and φ^{-1} are smooth.

If the set $\Omega(= U = V)$ is the entire space $\mathbb{R}^n$, φ is called **global diffeomorphism**. If the set Ω is not the entire space $\mathbb{R}^n$, φ is called **local diffeomorphism**.

Theorem 3.1. (Inverse function theorem)

Assume that $\varphi(\boldsymbol{q}) : \Omega \to \Omega$ is a smooth map. If and only if the Jacobian matrix $\partial\varphi(\boldsymbol{q})/\partial\boldsymbol{q}$ is non-singular in the set $\Omega, \varphi(\boldsymbol{q})$ is a local diffeomorphism in the set Ω.

This theorem is used to determine whether the given map is a coordinate transformation having smooth inverse transformation. Consider the non-linear control system with state vector $\boldsymbol{q} \in \Omega \subseteq \mathbb{R}^n$ given by

$$\begin{aligned} \dot{\boldsymbol{q}} &= \boldsymbol{f}(\boldsymbol{q}) + \boldsymbol{g}(\boldsymbol{q})\boldsymbol{u} \\ y &= h(\boldsymbol{q}), \end{aligned} \tag{3.18}$$

where $\boldsymbol{u} \in \mathbb{R}^m$ is the control input, and $y \in \mathbb{R}^1$ is the output. $\boldsymbol{f}$ and $\boldsymbol{g}$ are vector functions, and h is a scalar function. Assume the vector function $\varphi(\boldsymbol{q}) \in \Omega$ is smooth and there exists $\partial\varphi(\boldsymbol{q})/\partial\boldsymbol{q} \in \Omega$. Let the mapped vector be represented by

$$\boldsymbol{z} = \varphi(\boldsymbol{q}).$$

Using the relationship of $\boldsymbol{q} = \varphi^{-1}(\boldsymbol{z})$, the derivative of $\boldsymbol{z}$ is represented by

$$\begin{aligned} \dot{\mathbf{z}} = \frac{\partial\varphi}{\partial\boldsymbol{q}}\dot{q} = \frac{\partial\varphi}{\partial\boldsymbol{q}}(\boldsymbol{f}(\boldsymbol{q}) + \boldsymbol{g}(\boldsymbol{q})\boldsymbol{u}) &= \boldsymbol{f}(\boldsymbol{z})^* + \boldsymbol{g}(\boldsymbol{z})^*\boldsymbol{u} \\ y &= h^*(\boldsymbol{z}), \end{aligned} \tag{3.19}$$

where

$$\boldsymbol{f}(\boldsymbol{z})^* = \frac{\partial \boldsymbol{\varphi}}{\partial \boldsymbol{q}} \boldsymbol{f}(\boldsymbol{\varphi}^{-1}(\boldsymbol{z})), \quad \boldsymbol{g}(\boldsymbol{z})^* = \frac{\partial \boldsymbol{\varphi}}{\partial \boldsymbol{q}} \boldsymbol{g}(\boldsymbol{\varphi}^{-1}(\boldsymbol{z})),$$

$$h^*(\boldsymbol{z}) = h^*(\boldsymbol{\varphi}^{-1}(\boldsymbol{z})).$$

If $\partial \boldsymbol{\varphi}(\boldsymbol{q})/\partial \boldsymbol{q}$ is non-singular in the set Ω, the non-linear system of Eq. (3.19) is equivalent to the non-linear system of Eq. (3.18) in the set Ω.

Example 3.1. For $\boldsymbol{q} = [q_1, q_2]^T$, consider the following non-linear vector function:

$$\boldsymbol{\varphi}(\boldsymbol{q}) = \begin{bmatrix} q_1 + q_2 \\ \sin q_2 \end{bmatrix}. \tag{1}$$

The Jacobian matrix of this function is given by

$$\frac{\partial \boldsymbol{\varphi}(\boldsymbol{q})}{\partial \boldsymbol{q}} = \begin{bmatrix} 1 & 1 \\ 0 & \cos q_2 \end{bmatrix}. \tag{2}$$

Since this matrix is non-singular in a set $\Omega = \{(q_1, q_2) : |q_2| < \pi\}$, $\boldsymbol{\varphi}$ is a local diffeomorphism in the set Ω. However, it is not a local diffeomorphism outside this region.

3.3.2. *Vector Fields and Lie Derivative*

A **vector field** $\boldsymbol{p}$ on $\mathbb{R}^n$ is a vector-valued function that assigns a vector $\boldsymbol{q} \in \mathbb{R}^n$ to the tangent vector $\dot{\boldsymbol{q}}$ as follows:

$$\dot{\boldsymbol{q}} = \boldsymbol{p}(\boldsymbol{q}). \tag{3.20}$$

A vector field is smooth if all elements of $\boldsymbol{p}(\boldsymbol{q})$ are smooth. The **smooth vector fields** are discussed in the following. The derivative of a smooth scalar function $\phi(\boldsymbol{q})$ along the flow of the vector field $\boldsymbol{p}$ is represented by

$$\dot{\phi}(\boldsymbol{q}) = \frac{\partial \phi}{\partial \boldsymbol{q}} \dot{\boldsymbol{q}} = \frac{\partial \phi}{\partial \boldsymbol{q}} \boldsymbol{p}(\boldsymbol{q}), \tag{3.21}$$

where $\partial \phi/\partial \boldsymbol{q} = [\partial \phi/\partial q_1, \ldots, \partial \phi/\partial q_n]$ represents a row vector and is called the **gradient** of ϕ. It is referred to as the **Lie derivative** of ϕ along $\boldsymbol{p}$, and is denoted $L_p \phi$.

Definition 3.2. (Lie derivative)

The Lie derivative of the smooth scalar function $\phi(\boldsymbol{q})$ along vector field $p(\boldsymbol{q})$ is defined by

$$L_{\boldsymbol{p}}\phi(\boldsymbol{q}) = \frac{\partial \phi}{\partial \boldsymbol{q}}\boldsymbol{p}(\boldsymbol{q}). \tag{3.22}$$

When repeating the Lie derivative, use the symbols in the following recurrence formula:

$$L_{\boldsymbol{p}}^{0}\phi(\boldsymbol{q}) = \phi(\boldsymbol{q}), L_{\boldsymbol{p}}^{i+1}\phi(\boldsymbol{q}) = L_{\boldsymbol{p}}L_{\boldsymbol{p}}^{i}\phi(\boldsymbol{q}). \tag{3.23}$$

The Lie derivative is used to calculate the derivative of certain types of functions. Consider an output system expressed by

$$\begin{aligned} \dot{\boldsymbol{q}} &= \boldsymbol{p}(\boldsymbol{q}) \\ y &= \phi(\boldsymbol{q}). \end{aligned}$$

The first and second derivatives of the output are represented by

$$\begin{aligned} \dot{y} &= \frac{\partial \phi}{\partial \boldsymbol{q}}\dot{\boldsymbol{q}} = L_{\boldsymbol{p}}\phi \\ \ddot{y} &= \frac{\partial (L_{\boldsymbol{p}}\phi)}{\partial \boldsymbol{q}}\dot{\boldsymbol{q}} = L_{\boldsymbol{p}}L_{\boldsymbol{p}}\phi = L_{\boldsymbol{p}}^{2}\phi. \end{aligned}$$

As shown above, the derivatives are expressed simply by the Lie derivatives. The derivative of higher order is also written $y^{(j)} = L_{\boldsymbol{p}}^{j}\phi$, similar to the treatment above. The Lie derivatives have the following characteristics:

Property 3.1. (Lie derivatives)

Given vector fields $\boldsymbol{p}(\boldsymbol{q}), \boldsymbol{p}_1(\boldsymbol{q}), \boldsymbol{p}_2(\boldsymbol{q})$ on $\mathbb{R}^n$, smooth scalar functions $\phi_1(\boldsymbol{q}), \phi_2(\boldsymbol{q})$, and arbitral real numbers α_1, α_2. Then, the Lie derivatives have the following properties:

(1) $$L_{\boldsymbol{p}}(\phi_1(\boldsymbol{q}) + \phi_2(\boldsymbol{q})) = L_{\boldsymbol{p}}\phi_1(\boldsymbol{q}) + L_{\boldsymbol{p}}\phi_2(\boldsymbol{q}) \tag{3.24}$$

(2) $$L_{\alpha_1\boldsymbol{p}_1+\alpha_2\boldsymbol{p}_2}(\phi_1(\boldsymbol{q})) = \alpha_1 L_{\boldsymbol{p}_1}\phi_1(\boldsymbol{q}) + \alpha_2 L_{\boldsymbol{p}_2}\phi_1(\boldsymbol{q}) \tag{3.25}$$

(3) $$L_{\boldsymbol{p}}(\phi_1(\boldsymbol{q})\phi_2(\boldsymbol{q})) = \phi_1(\boldsymbol{q})L_{\boldsymbol{p}}\phi_2(\boldsymbol{q}) + \phi_2(\boldsymbol{q})L_{\boldsymbol{p}}\phi_1(\boldsymbol{q}). \tag{3.26}$$

These properties can be easily demonstrated by the definition of (3.22).

3.3.3. *Lie Bracket and Frobenius Theorem*

The **Lie bracket** $[\boldsymbol{p}_1, \boldsymbol{p}_2]$ defined in Eq. (3.10) is a basis of differential geometry. We review our above treatment.

Definition 3.3. (Lie brackets)

The Lie bracket of two vector fields $\boldsymbol{p}_1(\boldsymbol{q}), \boldsymbol{p}_2(\boldsymbol{q})$ is a vector field that is defined by

$$[\boldsymbol{p}_1, \boldsymbol{p}_2](\boldsymbol{q}) = \frac{\partial \boldsymbol{p}_2}{\partial \boldsymbol{q}}(\boldsymbol{q})\boldsymbol{p}_1(\boldsymbol{q}) - \frac{\partial \boldsymbol{p}_1}{\partial \boldsymbol{q}}(\boldsymbol{q})\boldsymbol{p}_2(\boldsymbol{q}). \tag{3.27}$$

The Lie bracket is notated as

$$[\boldsymbol{p}_1, \boldsymbol{p}_2] = ad_{\boldsymbol{p}_1}\boldsymbol{p}_2. \tag{3.28}$$

Repeating Lie brackets can be represented by the recurrence formula

$$ad^0_{\boldsymbol{p}_1}\boldsymbol{p}_2 = \boldsymbol{p}_2, ad^j_{\boldsymbol{p}_1}\boldsymbol{p}_2 = [\boldsymbol{p}_1, ad^{j-1}_{\boldsymbol{p}_1}\boldsymbol{p}_2]. \tag{3.29}$$

Lie brackets have the following properties:

Property 3.2. (Lie brackets)

Given vector fields $\boldsymbol{p}_1(\boldsymbol{q}), \boldsymbol{p}_2(\boldsymbol{q}), \boldsymbol{p}_3(\boldsymbol{q}), \mathbf{r}_1(\boldsymbol{q}), \mathbf{r}_2(\boldsymbol{q}) \in \mathbb{R}^n$ and constant scalars α_1, α_2. Then, Lie brackets have the following properties:

(1) Bilinearity:

$$[\alpha_1\boldsymbol{p}_1 + \alpha_2\boldsymbol{p}_2, \mathbf{r}_1] = \alpha_1[\boldsymbol{p}_1, \mathbf{r}_1] + \alpha_2[\boldsymbol{p}_2, \mathbf{r}_1] \tag{3.30}$$

$$[\boldsymbol{p}_1, \alpha_1\mathbf{r}_1 + \alpha_2\mathbf{r}_2] = \alpha_1[\boldsymbol{p}_1, \mathbf{r}_1] + \alpha_2[\boldsymbol{p}_1, \mathbf{r}_2] \tag{3.31}$$

(2) Skew symmetry:

$$[\boldsymbol{p}_1, \boldsymbol{p}_2] = -[\boldsymbol{p}_2, \boldsymbol{p}_1] \tag{3.32}$$

(3) Jacobian identity:

$$[\boldsymbol{p}_1, [\boldsymbol{p}_2, \boldsymbol{p}_3]] + [\boldsymbol{p}_2, [\boldsymbol{p}_3, \boldsymbol{p}_1]] + [\boldsymbol{p}_3, [\boldsymbol{p}_1, \boldsymbol{p}_2]] = 0. \tag{3.33}$$

These properties can be verified from the definition of Eq. (3.27).

Definition 3.4. (Distribution)

A linear subspace which is spanned by a set of smooth vector fields $\boldsymbol{p}_i (i = 1, \ldots, n-k)$ at point $\boldsymbol{q} \in \mathbb{R}^n$ is called distribution, and is expressed by

$$\Delta(\boldsymbol{q}) = span(\boldsymbol{p}_1(\boldsymbol{q}), \ldots, \boldsymbol{p}_k(\boldsymbol{q})). \tag{3.34}$$

The distribution is said to be **regular** if the dimension of $\Delta(\boldsymbol{q})$ does not vary with each point $\boldsymbol{q} \in \mathbb{R}^n$.

Definition 3.5. (Involutive)

A distribution $\Delta(\boldsymbol{q})$ is said to be **involutive** if the Lie bracket of any two vector fields is contained in the distribution, i.e.,

$$\Delta(\boldsymbol{q}) \text{ is involutive } \Leftrightarrow [\boldsymbol{p}_1, \boldsymbol{p}_2] \in \Delta(\boldsymbol{q}) \text{ for any } \boldsymbol{p}_1, \boldsymbol{p}_2 \in \Delta(\boldsymbol{q}).$$

From the properties of Lie brackets (1) and (2), whether a distribution $\Delta(\boldsymbol{q})$ is involutive can be concluded by checking that the Lie brackets of all the combinations of the basis elements are contained in the distribution. The distributions have the following properties:

- The distribution made from constant vector fields is always involutive, because if $\boldsymbol{p}_1, \boldsymbol{p}_2 \in \Delta(\boldsymbol{q})$ are constant vector fields, then $[\boldsymbol{p}_1, \boldsymbol{p}_2] = 0 \in \Delta(\boldsymbol{q})$.
- The distribution made from only a vector field $\boldsymbol{p}$ is involutive, because $[\boldsymbol{p}, \boldsymbol{p}] = 0 \in \Delta(\boldsymbol{q})$.

Definition 3.6. (Complete integrability)

If an $n - k$ dimension distribution $\Delta(\boldsymbol{q})$ is regular and there exist scalar functions $\phi_i(\boldsymbol{q})(i = 1, \ldots, k)$ such that the linear space spanned by $\partial\phi_i/\partial\boldsymbol{q}(i = 1, \ldots, k)$ is always orthogonal to $\Delta(\boldsymbol{q})$, the distribution $\Delta(\boldsymbol{q})$ is said to be **completely integrable**.

This definition indicates that the distribution $\Delta(\boldsymbol{q})$ is completely integrable if it is a tangent hyperplane TM of the manifold $M(= \{\boldsymbol{q} : \phi_i(\boldsymbol{q}) = 0\ (i = 1, \ldots, k)\})$. In addition, a set M is called an **integral manifold** for the distribution. Using the concept above, the **Frobenius theorem** is derived as follows.

Theorem 3.1. (Frobenius theorem)

Consider a distribution $\Delta(\boldsymbol{q})$ that is spanned by a set of vector fields $\boldsymbol{p}_i(i = 1, \ldots, n - k)$ at a point $\boldsymbol{q} \in \mathbb{R}^n$. A necessary and sufficient condition for the distribution $\Delta(\boldsymbol{q})$ to be non-singular and completely integrable is that $\Delta(\boldsymbol{q})$ is involutive.

The Frobenius theorem shows that, if $\Delta(\boldsymbol{q})$ is completely integrable, then there exist $\phi_i = 0$ $(i = 1, \dots, k)$ such that $\partial\phi_i/\partial\boldsymbol{q}$ $(i = 1, \dots, k)$ are orthogonal to $\Delta(\boldsymbol{q})$, and the constraint $\phi_i = 0$ $(i = 1, \dots, k)$ is a holonomic constraint. On the other hand, taking the contrapositive of this theorem, the non-holonomic constraint can be confirmed by demonstrating that $\Delta(\boldsymbol{q})$ is not involutive.

Example 3.2. Consider the motion of the independently driven two-wheel vehicle shown in Fig. 3.2(a). In this example, the generalized coordinate is given by $\boldsymbol{q} = [x, y, \theta]^T$. For any $\boldsymbol{q}$, there are two independent velocity vectors that can represent the movement of the vehicle. Since any spin motion at the contact point is possible, a velocity is represented by

$$\dot{\boldsymbol{q}} = \begin{bmatrix} 0 \\ 0 \\ 1 \end{bmatrix}. \tag{1}$$

Since translational motion with angle θ is also possible, its velocity is represented by

$$\dot{\boldsymbol{q}} = \begin{bmatrix} \cos\theta \\ \sin\theta \\ 0 \end{bmatrix}. \tag{2}$$

Thus, assume the linearly independent vector fields that represent the basis vectors of the constraint as

$$\boldsymbol{p}_1 = \begin{bmatrix} 0 \\ 0 \\ 1 \end{bmatrix}, \quad \boldsymbol{p}_2 = \begin{bmatrix} \cos\theta \\ \sin\theta \\ 0 \end{bmatrix}. \tag{3}$$

The Pfaffian constraint can be represented as

$$\boldsymbol{A}(\boldsymbol{q}) = [\sin\theta, -\cos\theta, 0]. \tag{4}$$

Obviously, $\boldsymbol{A}(\boldsymbol{q})\boldsymbol{p}_1 = \boldsymbol{A}(\boldsymbol{q})\boldsymbol{p}_2 = 0$. The partial derivative of $\boldsymbol{p}_i$ with respect to $\boldsymbol{q}$ is

$$\frac{\partial \boldsymbol{p}_1}{\partial \boldsymbol{q}} = 0, \quad \frac{\partial \boldsymbol{p}_2}{\partial \boldsymbol{q}} = \begin{bmatrix} 0 & 0 & -\sin\theta \\ 0 & 0 & \cos\theta \\ 0 & 0 & 0 \end{bmatrix}.$$

Therefore, the Lie bracket $[\boldsymbol{p}_1, \boldsymbol{p}_2]$ is given by

$$[\boldsymbol{p}_1, \boldsymbol{p}_2] = \begin{bmatrix} -\sin\theta \\ \cos\theta \\ 0 \end{bmatrix}. \tag{5}$$

This equation indicates that the Lie bracket $[\boldsymbol{p}_1, \boldsymbol{p}_2]$ is orthogonal to the direction with angle θ at the contact point and does not exist on $\Delta(\boldsymbol{q})$ spanned by $\boldsymbol{p}_1$ and $\boldsymbol{p}_2$. Consequently, $\Delta(\boldsymbol{q})$ is not involutive, and the two-wheel vehicle moves under the non-holonomic constraint. Therefore, it is possible to move to any $\boldsymbol{q} \in \mathbb{R}^3$ because of the effect of vector $[\boldsymbol{p}_1, \boldsymbol{p}_2]$.

Example 3.3. The hemisphere fingertip with radius ρ, as shown in Fig. 2.7, makes a pure rolling contact motion on a plane. Let us show that this constraint is non-holonomic. This system contains the Pfaffian constraint shown in Eq. (3.6). As seen in Exercise 2.7, since ω_x, ω_y are regarded as variables, the basis vectors of the constraint are obtained as

$$\boldsymbol{p}_1 = \begin{bmatrix} 0 \\ \sec q_1 \\ -\rho\sin q_5 \\ -\rho\cos q_5 \\ -\tan q_1 \end{bmatrix}, \quad \boldsymbol{p}_2 = \begin{bmatrix} -1 \\ 0 \\ -\rho\cos q_5 \\ \rho\sin q_5 \\ 0 \end{bmatrix}. \tag{1}$$

The distribution spanned by these two vector fields is given by $\Delta(\boldsymbol{q}) = span(\boldsymbol{p}_1(\boldsymbol{q}), \boldsymbol{p}_2(\boldsymbol{q}))$. On the other hand, the Lie bracket $[\boldsymbol{p}_1, \boldsymbol{p}_2]$ is calculated as

$$\boldsymbol{p}_3 = [\boldsymbol{p}_1, \boldsymbol{p}_2] = \frac{\partial \boldsymbol{p}_2}{\partial \boldsymbol{q}}\boldsymbol{p}_1 - \frac{\partial \boldsymbol{p}_1}{\partial \boldsymbol{q}}\boldsymbol{p}_2 = \begin{bmatrix} 0 \\ \sec q_1 \tan q_1 \\ -\rho\tan q_1 \sin q_5 \\ -\rho\tan q_1 \cos q_5 \\ -\sec^2 q_1 \end{bmatrix}. \tag{2}$$

Obviously, since $\boldsymbol{p}_3$ does not exist in the linear space spanned by $\boldsymbol{p}_1$ and $\boldsymbol{p}_2$, the distribution is not involutive. As a result, this is a non-holonomic system.

3.4. Controllability of the System

3.4.1. *Concept of Controllability*

When Eq. (3.5) is a differential equation representing holonomic constraints, the kinematic motion is limited to the manifold M satisfying $\phi_i = 0$ $(i = 1, \ldots, k)$, and thus cannot move from this manifold. On the other hand, when Eq. (3.5) is a differential equation representing non-holonomic constraints, there is no such manifold. This means that if the constraint is non-holonomic, there is a possibility of the existence of a trajectory that starts from any initial value in the n dimensional generalized coordinate space and reaches any end point in time.

Assume that the linearly independent vector fields $\boldsymbol{p}_i(\boldsymbol{q})(i = 1, 2, \ldots, n-k)$ are found for the constraint given by Eq. (3.5). Define the following matrix $\boldsymbol{P}(\boldsymbol{q})$ using these vector fields:

$$\boldsymbol{P}(\boldsymbol{q}) = [\boldsymbol{p}_1(\boldsymbol{q}), \ldots, \boldsymbol{p}_{n-k}(\boldsymbol{q})] \in \mathbb{R}^{n \times (n-k)}. \tag{3.35}$$

The following relationship is clearly satisfied:

$$\boldsymbol{A}(\boldsymbol{q})\boldsymbol{P}(\boldsymbol{q}) = 0. \tag{3.36}$$

Then, using arbitral vector $\boldsymbol{u} \in \mathbb{R}^{n-k}$, the velocity vector $\dot{\boldsymbol{q}}$ satisfying the constraint can be represented by

$$\dot{q} = \boldsymbol{P}(\boldsymbol{q})\boldsymbol{u}. \tag{3.37}$$

Equation (3.37) is considered to be the non-linear control system where $\boldsymbol{q}$ is the state variable and $\boldsymbol{u}$ is the input.

Now, consider a non-linear system that is linear with respect to the input $\boldsymbol{u}$ and non-linear with respect to the state variable $\boldsymbol{x}$, as expressed by

$$\dot{\boldsymbol{x}} = \boldsymbol{f}(\boldsymbol{x}) + \boldsymbol{P}(\boldsymbol{x})\boldsymbol{u}. \tag{3.38}$$

This non-linear system is called an **affine system**. When the non-linear system

$$\dot{\boldsymbol{x}} = \boldsymbol{h}(\boldsymbol{x}, \boldsymbol{u}) \tag{3.39}$$

satisfies

$$\boldsymbol{h}(\boldsymbol{x}, -\boldsymbol{u}) = -\boldsymbol{h}(\boldsymbol{x}, \boldsymbol{u}), \tag{3.40}$$

the non-linear system is called a **symmetric system**. Eq. (3.38) and Eq. (3.37) match if $\boldsymbol{x} = \boldsymbol{q}, \boldsymbol{f}(\boldsymbol{x}) = 0$, which satisfies the relationship of Eq. (3.40). Therefore, the non-linear system of Eq. (3.37) is a symmetric affine system. Next, we will consider the basic concepts involved in the affine system.

Definition 3.7. (Controllability)

The system of Eq. (3.39) is said to be controllable if there exists an admissible input $\boldsymbol{u}(t)$ such that the state vector $\boldsymbol{x}(t)$ with any initial condition $\boldsymbol{x}(t_0) = \boldsymbol{x}_0$ reaches to any another state vector $\boldsymbol{x}(t_1) = \boldsymbol{x}_1 (t_1 > t_0)$ in a finite time interval.

Definition 3.8. (Concept of neighborhood)

A subspace $U \subset \mathbb{R}^n$ containing $\boldsymbol{x}_0$ as an interior point is called a neighborhood of $\boldsymbol{x}_0$.

It should be noted that the neighborhood of $\boldsymbol{x}_0$ is sometimes big and sometimes small. It should also be noted that $\boldsymbol{x}_0$ is not on the boundary of U. If $\phi(\boldsymbol{x})$ exists in a neighborhood U of $\boldsymbol{x}_0$ for any $\boldsymbol{x} = \boldsymbol{x}_0$, then it is said that '$\phi(\boldsymbol{x})$ locally exists.'

Definition 3.9. (Reachable set)

A set of all $\boldsymbol{x}$ for which there exists an admissible control u such that there is a trajectory of Eq. (3.39) from $\boldsymbol{x}(t_0) = \boldsymbol{x}_0$ to $\boldsymbol{x}(t) = \boldsymbol{x}$ with $0 \leq t \leq T$ is called a reachable set from $\boldsymbol{x}_0$ in time T.

Definition 3.10. (Local reachability)

If a reachable set from $\boldsymbol{x}_0$ in time T for the system of Eq. (3.39) is a non-empty open set, the system is said to be **locally reachable** from $\boldsymbol{x}_0$ in time T.

The local reachability of a symmetric affine system is consistent with the controllability because it always has $\boldsymbol{x}(t)$ within the open set. The reachable open set in the neighborhood of $\boldsymbol{x}$ is called the **accessibility distribution**, which is defined as follows.

Definition 3.11. (Accessibility distribution)

The linear space $\bar{\Delta}(\boldsymbol{q})$ which is spanned by the vector fields $\boldsymbol{p}_i(\boldsymbol{q}) = \boldsymbol{0}$ $(i = 1, 2, \ldots, n-k)$ and the vector fields made by the Lie brackets given by

$$[\mathbf{r}_j, [\mathbf{r}_{j-1}, [\cdots, [\mathbf{r}_2, \mathbf{r}_1]\cdots]]] \quad j = 2, 3, \ldots$$

is called the accessibility distribution, where $\boldsymbol{r}_j \in \{\boldsymbol{p}_1(\boldsymbol{q}), \ldots, \boldsymbol{p}_k(\boldsymbol{q})\}$.

By using this definition, there is a controllability theorem for the symmetric affine systems.

Theorem 3.2. (Controllability for symmetric affine systems) *Consider a symmetric affine system with the accessibility distribution* $\bar{\Delta}(\boldsymbol{q})$. *If* $dim\ \bar{\Delta}(\boldsymbol{q}) = n$ *for all* $\boldsymbol{q} \in \mathbb{R}^n$, *the system is controllable.*

Example 3.4. Examine the controllability of the system in Example 3.2 by determining the accessibility distribution of the system. The vector fields and the Lie brackets are represented by

$$\boldsymbol{p}_1 = \begin{bmatrix} 0 \\ 0 \\ 1 \end{bmatrix}, \quad \boldsymbol{p}_2 = \begin{bmatrix} \cos\theta \\ \sin\theta \\ 0 \end{bmatrix}, \quad \boldsymbol{p}_3 = [\boldsymbol{p}_1, \boldsymbol{p}_2] = \begin{bmatrix} -\sin\theta \\ \cos\theta \\ 0 \end{bmatrix}. \quad (1)$$

Therefore, the accessibility distribution is obtained by

$$\bar{\Delta}(\boldsymbol{q}) = span\{\boldsymbol{p}_1, \boldsymbol{p}_2, \boldsymbol{p}_3\} = span\left\{\begin{bmatrix} 0 \\ 0 \\ 1 \end{bmatrix}, \begin{bmatrix} \cos\theta \\ \sin\theta \\ 0 \end{bmatrix}, \begin{bmatrix} -\sin\theta \\ \cos\theta \\ 0 \end{bmatrix}\right\}. \quad (2)$$

Clearly, $dim\ \bar{\Delta}(\boldsymbol{q}) = 3$ for any $\boldsymbol{q} = (x, y, \theta)^T$. Therefore, the system is controllable.

Example 3.5. Figure 3.4 shows a two-wheel bicycle equipped with a handle that can be used to change the direction of the front wheel,

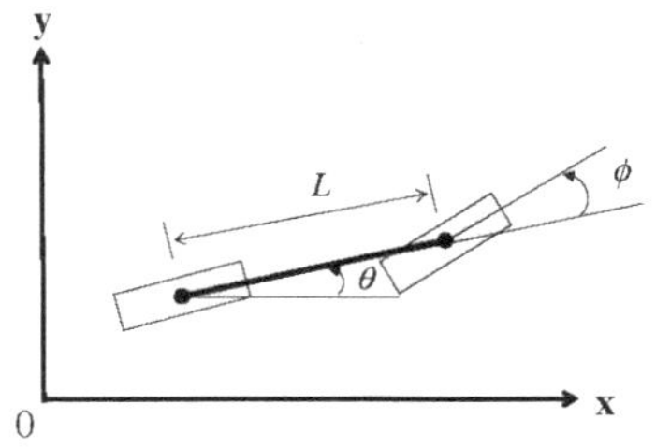

Fig. 3.4. Kinematics of the bicycle.

where the position of the rear wheel is $[x, y]^T$, the angle of the vehicle body for the horizontal is θ, the steering angle of the front wheel is ϕ, and the distance between the front and rear wheel is L. Let us show that the two-wheel bicycle is controllable, provided that the wheel does not slip.

The generalized coordinates of the bicycle can be given by $\boldsymbol{q} = [x, y, \theta, \phi]^T$ and the position of the front wheel is $[x + L\sin\theta, y + L\cos\theta]^T$. For the rear wheel, since the direction orthogonal to the traveling direction is zero velocity, there is the constraint of

$$\dot{x}\sin\theta - \dot{y}\cos\theta = 0. \tag{1}$$

Since the velocity of the front wheel along the traveling direction is $[\dot{x} - L\dot{\theta}\sin\theta, \dot{y} + L\dot{\theta}\cos\theta]^T$, we have

$$\begin{aligned}&(\dot{x} - L\dot{\theta}\sin\theta)\sin(\theta+\phi) - (\dot{y} + L\dot{\theta}\cos\theta)\cos(\theta+\phi)\\ &\quad = \dot{x}\sin(\theta+\phi) - \dot{y}\cos(\theta+\phi) - L\dot{\theta}\cos\phi = 0.\end{aligned} \tag{2}$$

By denoting the traveling velocity of the rear wheel by u_1 and using the relation of $\dot{x} = u_1\cos\theta$, the following is obtained:

$$L\dot{\theta} = \frac{\dot{x}\sin(\theta+\phi) - \dot{y}\cos(\theta+\phi)}{\cos\phi} = \frac{\dot{x}\tan\phi}{\cos\theta} = u_1\tan\phi. \tag{3}$$

By selecting the traveling velocity u_1 of the rear wheel and the steering angular velocity u_2 as the input of the control system, the derivative of the generalized coordinates can be represented by

$$\begin{bmatrix}\dot{x}\\ \dot{y}\\ \dot{\theta}\\ \dot{\phi}\end{bmatrix} = \begin{bmatrix}\cos\theta\\ \sin\theta\\ \dfrac{1}{L}\tan\phi\\ 0\end{bmatrix} u_1 + \begin{bmatrix}0\\ 0\\ 0\\ 1\end{bmatrix} u_2. \tag{4}$$

From the above, the vector fields are obtained by

$$\boldsymbol{p}_1 = \begin{bmatrix}\cos\theta\\ \sin\theta\\ \dfrac{\tan\phi}{L}\\ 0\end{bmatrix}, \quad \boldsymbol{p}_2 = \begin{bmatrix}0\\ 0\\ 0\\ 1\end{bmatrix}.$$

Next, find the Lie brackets of $\boldsymbol{p}_1$ and $\boldsymbol{p}_2$. Since the partial derivatives of the vector fields with respect to $\boldsymbol{q}$ are

$$\frac{\partial \boldsymbol{p}_1}{\partial \boldsymbol{q}} = \begin{bmatrix} 0 & 0 & -\sin\theta & 0 \\ 0 & 0 & \cos\theta & 0 \\ 0 & 0 & 0 & \dfrac{1}{L\cos^2\phi} \\ 0 & 0 & 0 & 0 \end{bmatrix}, \quad \frac{\partial \boldsymbol{p}_2}{\partial \boldsymbol{q}} = \begin{bmatrix} 0 & 0 & 0 & 0 \\ 0 & 0 & 0 & 0 \\ 0 & 0 & 0 & 0 \\ 0 & 0 & 0 & 0 \end{bmatrix}, \tag{5}$$

the Lie brackets of $\boldsymbol{p}_1$ and $\boldsymbol{p}_2$ can be obtained by

$$\boldsymbol{p}_3 = [\boldsymbol{p}_2, \boldsymbol{p}_1] = \frac{\partial \boldsymbol{p}_1}{\partial \boldsymbol{q}}\boldsymbol{p}_2 - \frac{\partial \boldsymbol{p}_2}{\partial \boldsymbol{q}}\boldsymbol{p}_1 = \begin{bmatrix} 0 \\ 0 \\ \dfrac{1}{L\cos^2\phi} \\ 0 \end{bmatrix}. \tag{6}$$

This system is non-holonomic because $\boldsymbol{p}_3$ is independent of $\boldsymbol{p}_1$ and $\boldsymbol{p}_2$. Since the partial derivative of $\boldsymbol{p}_3$ with respect to $\boldsymbol{q}$ is

$$\frac{\partial \boldsymbol{p}_3}{\partial \boldsymbol{q}} = \begin{bmatrix} 0 & 0 & 0 & 0 \\ 0 & 0 & 0 & 0 \\ 0 & 0 & 0 & -\dfrac{2\tan\phi}{L\cos^2\phi} \\ 0 & 0 & 0 & 0 \end{bmatrix}, \tag{7}$$

the Lie brackets of $\boldsymbol{p}_3$ and $\boldsymbol{p}_1$ can be obtained by

$$\boldsymbol{p}_4 = [\boldsymbol{p}_3, \boldsymbol{p}_1] = \begin{bmatrix} -\dfrac{\sin\theta}{L\cos^2\phi} \\ \dfrac{\cos\theta}{L\cos^2\phi} \\ 0 \\ 0 \end{bmatrix}. \tag{8}$$

Thus, the accessibility distribution of this system is obtained by

$\bar{\Delta}(\boldsymbol{q}) = span\{\boldsymbol{p}_1, \boldsymbol{p}_2, \boldsymbol{p}_3, \boldsymbol{p}_4\}$

$$= span \left\{ \begin{bmatrix} \cos\theta \\ -\sin\theta \\ \frac{1}{L}\tan\phi \\ 0 \end{bmatrix}, \begin{bmatrix} 0 \\ 0 \\ 0 \\ 1 \end{bmatrix}, \begin{bmatrix} 0 \\ 0 \\ \frac{1}{L\cos^2\phi} \\ 0 \end{bmatrix}, \begin{bmatrix} -\frac{\sin\theta}{L\cos^2\phi} \\ \frac{\cos\theta}{L\cos^2\phi} \\ 0 \\ 0 \end{bmatrix} \right\}. \tag{9}$$

Obviously, $\boldsymbol{p}_1, \boldsymbol{p}_2, \boldsymbol{p}_3, \boldsymbol{p}_4$ are independent in the range of $-\pi/2 < \phi < \pi/2$. Therefore, $\dim \bar{\Delta}(\boldsymbol{q}) = 4$. Since $\boldsymbol{q} = [x, y, \theta, \phi]^T \in \mathbb{R}^4$, this system is controllable in the range of $-\pi/2 < \phi < \pi/2$.

Example 3.6. Consider the pure rolling contact motion on a plane by the hemisphere fingertips with radius ρ, as shown in Example 3.3. Let us show that the system is controllable. The generalized coordinates of the system can be given by $\boldsymbol{q} = [u_f, v_f, u_o, v_o, \phi]^T = \{q_i\}$. Consider that ω_x, ω_y are variables. Then the derivatives of the generalized coordinates are represented by

$$\dot{\boldsymbol{q}} = \boldsymbol{P}(\boldsymbol{q})\,\boldsymbol{u}, \tag{1}$$

where

$$\boldsymbol{P}(\boldsymbol{q}) = \begin{bmatrix} 0 & -1 \\ \sec q_1 & 0 \\ -\rho\sin q_5 & -\rho\cos q_5 \\ -\rho\cos q_5 & \rho\sin q_5 \\ -\tan q_1 & 0 \end{bmatrix}, \quad \boldsymbol{u} = \begin{bmatrix} \omega_x \\ \omega_y \end{bmatrix}.$$

This system is a symmetric affine system. From Eq. (1) in Example 3.3,

$$\boldsymbol{p}_1 = \begin{bmatrix} 0 \\ \sec q_1 \\ -\rho\sin q_5 \\ -\rho\cos q_5 \\ -\tan q_1 \end{bmatrix}, \quad \boldsymbol{p}_2 = \begin{bmatrix} -1 \\ 0 \\ -\rho\cos q_5 \\ \rho\sin q_5 \\ 0 \end{bmatrix} \tag{2}$$

and the Lie brackets of $\boldsymbol{p}_1$ and $\boldsymbol{p}_2$ are

$$\boldsymbol{p}_3 = [\boldsymbol{p}_1, \boldsymbol{p}_2] = \begin{bmatrix} 0 \\ \sec q_1 \tan q_1 \\ -\tan q_1 \rho \sin q_5 \\ -\tan q_1 \rho\cos q_5 \\ -\sec^2 q_1 \end{bmatrix}, \tag{3}$$

while the Lie brackets of $\boldsymbol{p}_3$ and $\boldsymbol{p}_1$ are

$$\boldsymbol{p}_4 = [\boldsymbol{p}_1, \boldsymbol{p}_3] = \begin{bmatrix} 0 \\ 0 \\ -\rho\cos q_5 \\ \rho\sin q_5 \\ 0 \end{bmatrix}. \tag{4}$$

In addition, the Lie brackets of $\boldsymbol{p}_2$ and $\boldsymbol{p}_3$ are

$$\boldsymbol{p}_5 = [\boldsymbol{p}_2, \boldsymbol{p}_3] = \begin{bmatrix} 0 \\ \dfrac{\cos^2 q_1 - 2}{\cos^3 q_1} \\ \dfrac{2\rho\sin q_5}{\cos^2 q_1} \\ \dfrac{2\rho\cos q_5}{\cos^2 q_1} \\ \dfrac{2\sin q_1}{\cos^3 q_1} \end{bmatrix}. \tag{5}$$

From these vector fields, the accessibility distribution of this system is expressed by

$$\begin{aligned} \bar{\Delta}(\boldsymbol{q}) &= span\{\boldsymbol{p}_1, \boldsymbol{p}_2, \boldsymbol{p}_3, \boldsymbol{p}_4, \boldsymbol{p}_5\} \\ &= span\left\{ \begin{bmatrix} 0 \\ \sec q_1 \\ -\rho\sin q_5 \\ -\rho\cos q_5 \\ -\tan q_1 \end{bmatrix}, \begin{bmatrix} -1 \\ 0 \\ -\rho\cos q_5 \\ \rho\sin q_5 \\ 0 \end{bmatrix}, \begin{bmatrix} 0 \\ \sec q_1 \tan q_1 \\ -\tan q_1 \rho\sin q_5 \\ -\tan q_1 \rho\cos q_5 \\ -\sec^2 q_1 \end{bmatrix}, \right. \\ &\qquad \left. \begin{bmatrix} 0 \\ 0 \\ -\rho\cos q_5 \\ \rho\sin q_5 \\ 0 \end{bmatrix}, \begin{bmatrix} 0 \\ \dfrac{\cos^2 q_1 - 2}{\cos^3 q_1} \\ \dfrac{2\rho\sin q_5}{\cos^2 q_1} \\ \dfrac{2\rho\cos q_5}{\cos^2 q_1} \\ \dfrac{2\sin q_1}{\cos^3 q_1} \end{bmatrix} \right\} \end{aligned} \tag{6}$$

The determinants of the matrix $[\boldsymbol{p}_1, \boldsymbol{p}_2, \boldsymbol{p}_3, \boldsymbol{p}_4, \boldsymbol{p}_5]$ are obtained by

$$\det([\boldsymbol{p}_1, \boldsymbol{p}_2, \boldsymbol{p}_3, \boldsymbol{p}_4, \boldsymbol{p}_5]) = \frac{\rho^2}{\cos q_1}. \tag{7}$$

In the range of $-\pi/2 < q_1 < \pi/2$, $\det([\boldsymbol{p}_1, \boldsymbol{p}_2, \boldsymbol{p}_3, \boldsymbol{p}_4, \boldsymbol{p}_5]) \neq 0$. That is, $\boldsymbol{p}_1, \boldsymbol{p}_2, \boldsymbol{p}_3, \boldsymbol{p}_4, \boldsymbol{p}_5$ are linearly independent and $dim\bar{\Delta}(\boldsymbol{q}) = 5$.

Therefore, this system is controllable in the range of $-\pi/2 < q_1 < \pi/2$.

3.4.2. *Controllability of the Linear System*

Consider the linear control system

$$\dot{\boldsymbol{x}} = \boldsymbol{A}\boldsymbol{x} + \boldsymbol{b}\boldsymbol{u}, \tag{3.41}$$

where $\boldsymbol{x} \in \mathbb{R}^n$ is the state vector, $\boldsymbol{u} \in \mathbb{R}^1$ is the input, $\boldsymbol{A} \in \mathbb{R}^{n \times n}$ is a constant matrix, and $\boldsymbol{b} \in \mathbb{R}^n$ is a constant vector. The controllability matrix of the linear system is defined by

$$\boldsymbol{M}_c = [\boldsymbol{b}, \boldsymbol{A}\boldsymbol{b}, \ldots, \boldsymbol{A}^k\boldsymbol{b}]. \tag{3.42}$$

It is well known that if $\boldsymbol{M}_c$ has full rank, then the linear system is controllable. Now, let us find the Lie brackets for this system in order.

$$ad_{\boldsymbol{A}\boldsymbol{x}}\boldsymbol{b} = [\boldsymbol{A}\boldsymbol{x}, \boldsymbol{b}] = \left(\frac{\partial}{\partial \boldsymbol{x}}\boldsymbol{b}\right)\boldsymbol{A}\boldsymbol{x} - \left(\frac{\partial}{\partial \boldsymbol{x}}\boldsymbol{A}\boldsymbol{x}\right)\boldsymbol{b} = -\boldsymbol{A}\boldsymbol{b}$$

$$ad^2_{\boldsymbol{A}\boldsymbol{x}}\boldsymbol{b} = [\boldsymbol{A}\boldsymbol{x}, -\boldsymbol{A}\boldsymbol{b}] = \left(\frac{\partial}{\partial \boldsymbol{x}} - \boldsymbol{A}\boldsymbol{b}\right)\boldsymbol{A}\boldsymbol{x} - \left(\frac{\partial}{\partial \boldsymbol{x}}\boldsymbol{A}\boldsymbol{x}\right)(-\boldsymbol{A}\boldsymbol{b}) = \boldsymbol{A}^2\boldsymbol{b}$$

$$\vdots$$

$$ad^k_{\boldsymbol{A}\boldsymbol{x}}\boldsymbol{b} = (-1)^k\boldsymbol{A}^k\boldsymbol{b}$$

As shown above, the Lie brackets correspond to the column vectors of the controllability matrix $\boldsymbol{M}_c$ of the linear system. The accessibility distribution corresponding to the space spanned by column vectors of the controllability matrix is given by

$$\bar{\Delta}(\boldsymbol{x}) = span\{\boldsymbol{b}, \boldsymbol{A}\boldsymbol{b}, \ldots, \boldsymbol{A}^k\boldsymbol{b}\}. \tag{3.43}$$

Exercises

3.1. Prove (3.24)∼(3.26).

3.2. Prove (3.30)∼(3.33).

3.3. Examine the controllability of a four-wheel vehicle with no handles in Fig. 3.2(b).

Chapter 4

Robot Dynamics

This chapter reviews basic methodologies for creating **dynamic models** of robot hands. A dynamic model is a representation of the relationship between the joint torques and the dynamic motion of the robotic hand. The dynamic model is used widely in the simulation of robot motion, analysis of robot hand structures, and design of control algorithms. In this chapter, we will explore the dynamic model of a rigid body using the Newton–Euler method. Next, we will determine the dynamic model of a multi-joint link mechanism using the Lagrange method. When a robot hand with multiple fingers grasps an object, the robotic hand motion is restricted in order to maintain contact with the object. The dynamics model with constraint is derived.

4.1. Motion of a Rigid Body

According to the **Newton-Euler equation** (1642–1727) and **Euler** (1707–1783) equations, the motion of a rigid body where gravity acts in three-dimensional (3D) space can be written as

$$\boldsymbol{f} = m_O\ddot{\boldsymbol{p}} - m_O\tilde{\boldsymbol{g}}, \tag{4.1}$$

$$\boldsymbol{n} = \frac{d}{dt}(\hat{\boldsymbol{I}}_O\boldsymbol{\omega}_O) = \hat{\boldsymbol{I}}_O\dot{\boldsymbol{\omega}}_O + \boldsymbol{\omega}_O \times \hat{\boldsymbol{I}}_O\boldsymbol{\omega}_O, \tag{4.2}$$

where $\boldsymbol{p} \in \mathbb{R}^3$ is a position vector at the center of mass, $\boldsymbol{f} \in \mathbb{R}^3$ is an applied force to the center of mass of the object, $\boldsymbol{n} \in \mathbb{R}^3$ is an applied moment to the center of mass of the moment, $\boldsymbol{\omega}_O \in \mathbb{R}^3$ is the angle velocity of the object, $\tilde{\boldsymbol{g}} \in \mathbb{R}^3$ is the gravitational acceleration, and $\hat{\boldsymbol{I}}_O \in \mathbb{R}^{3\times3}$ is an inertia tensor around the center of mass of the object. These are referred with the reference coordinate system. It is not convenient to deal with the inertia tensor, $\hat{\boldsymbol{I}}_O$, since its value alters with changes in the orientation of the object.

Consider the object coordinate system Σ_C, which is attached to the center of mass. The inertia tensor of the rigid body referred to the object coordinate system ${}^C\hat{\boldsymbol{I}}_O$ is represented by

$$ {}^C\hat{\boldsymbol{I}}_O = \boldsymbol{R}_C^T \hat{\boldsymbol{I}}_O \boldsymbol{R}_C, \tag{4.3} $$

where $\boldsymbol{R}_C \in \mathbb{R}^{3\times3}$ is a rotation matrix that represents the transformation matrix of vector coordinates in the object coordinate system Σ_C into the coordinates of the same vector in the reference coordinate system. The dynamic equation of the object using ${}^C\hat{\boldsymbol{I}}_O$ will be easily handled, because ${}^C\hat{\boldsymbol{I}}_O$ is the constant. A relation of $\hat{\boldsymbol{I}}_O = \boldsymbol{R}_C {}^C\hat{\boldsymbol{I}}_O \boldsymbol{R}_C^T$ is derived from Eq. (4.3). By substituting this into Eq. (4.2), then putting it together with Eq. (4.1), it is expressed as

$$ \begin{bmatrix} \boldsymbol{f} \\ \boldsymbol{n} \end{bmatrix} = \begin{bmatrix} m_O \boldsymbol{I}_{3\times3} & 0 \\ 0 & \boldsymbol{R}_C {}^C\hat{\boldsymbol{I}}_O \boldsymbol{R}_C^T \end{bmatrix} \begin{bmatrix} \ddot{\boldsymbol{p}} \\ \dot{\boldsymbol{\omega}}_O \end{bmatrix} + \begin{bmatrix} 0 \\ \boldsymbol{\omega}_O \times \boldsymbol{R}_C {}^C\hat{\boldsymbol{I}}_O \boldsymbol{R}_C^T \boldsymbol{\omega}_O \end{bmatrix} + \begin{bmatrix} -m_O \tilde{\boldsymbol{g}} \\ 0 \end{bmatrix}. \tag{4.4} $$

This is the equation of motion using the inertia tensor at the center of mass of the object. See the right-hand side of the equation. The first term represents the influence of the inertia force term, the second represents the quadratic velocity term, consisting of the centrifugal force and the Coriolis force, and the third term represents the gravitational term. Next, we set a new object coordinate system Σ_O that is transferred to a position different from the center of mass of the object, as shown in Fig. 4.1. Using the origin of the new object coordinate system as a position vector $\boldsymbol{p}_O \in \mathbb{R}^3$ and an inertia tensor of the object around the origin of new object coordinate system ${}^O\boldsymbol{I}_O \in \mathbb{R}^{3\times3}$ that is referred to Σ_O, let us derive the equation of object motion using ${}^O\boldsymbol{I}_O$. The position of origin of the center of mass is expressed as

$$ \boldsymbol{p} = \boldsymbol{p}_O + \boldsymbol{R}_O {}^O\boldsymbol{s}_{OC}, \tag{4.5} $$

where ${}^O\boldsymbol{s}_{OC}$ is a position vector from the origin of Σ_O to the origin of Σ_C with respect to Σ_O. Since ${}^O\boldsymbol{s}_{OC}$ is a constant vector, the first- and

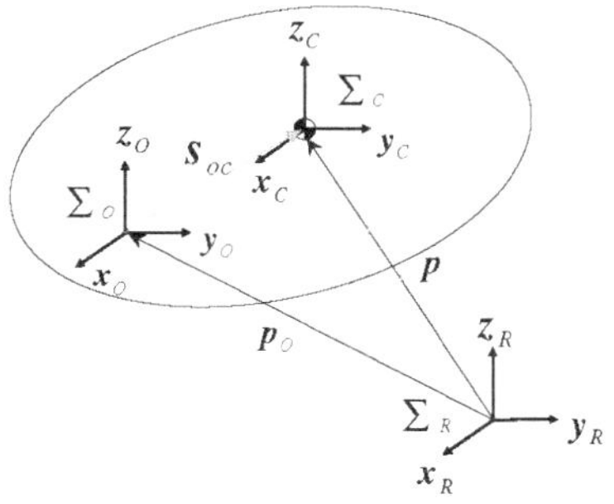

Fig. 4.1. Coordinate system to be set for the target object.

second-order time differentials of Eq. (4.5) are expressed by

$$\dot{\boldsymbol{p}} = \dot{\boldsymbol{p}}_O + \boldsymbol{\omega}_O \times \boldsymbol{R}_O {}^O\boldsymbol{s}_{OC}, \tag{4.6}$$

$$\ddot{\boldsymbol{p}} = \ddot{\boldsymbol{p}}_O + \dot{\boldsymbol{\omega}}_O \times \boldsymbol{R}_O {}^O\boldsymbol{s}_{OC} + \boldsymbol{\omega}_O \times (\boldsymbol{\omega}_O \times \boldsymbol{R}_O {}^O\boldsymbol{s}_{OC}). \tag{4.7}$$

The relationships between force and moment acting on the center of mass and the equivalent force $\boldsymbol{f}_O$ and moment $\boldsymbol{n}_O$ acting on the origin of the coordinate system Σ_O are given by

$$\boldsymbol{f}_O = \boldsymbol{f}, \tag{4.8}$$

$$\boldsymbol{n}_O = \boldsymbol{n} + \boldsymbol{s}_{OC} \times \boldsymbol{f}. \tag{4.9}$$

Moreover, according to the parallel axis theorem regarding the inertia tensor, there is a relationship as follows

$${}^O\boldsymbol{I}_O = {}^C\hat{\boldsymbol{I}}_O + m_O({}^O\boldsymbol{s}_{OC}^T {}^O\boldsymbol{s}_{OC}\boldsymbol{I}_{3\times 3} - {}^O\boldsymbol{s}_{OC} {}^O\boldsymbol{s}_{OC}^T). \tag{4.10}$$

By substituting Eq. (4.6)–Eq. (4.10) into Eq. (4.4), it is expressed as

$$\boldsymbol{M}_O\dot{\boldsymbol{v}}_O + \boldsymbol{C}_O\boldsymbol{v}_O + \boldsymbol{g}_O = \boldsymbol{F}_O, \tag{4.11}$$

where

$$\boldsymbol{v}_O = \begin{bmatrix} \dot{\boldsymbol{p}}_O \\ \boldsymbol{\omega}_O \end{bmatrix}, \quad \boldsymbol{M}_O = \begin{bmatrix} m_O\boldsymbol{I}_{3\times 3} & m_O[\boldsymbol{R}_O {}^O\boldsymbol{s}_{OC}\times]^T \\ m_O[\boldsymbol{R}_O {}^O\boldsymbol{s}_{OC}\times] & \boldsymbol{R}_O {}^O\boldsymbol{I}_O\boldsymbol{R}_O^T \end{bmatrix},$$

$$\boldsymbol{C}_O = \begin{bmatrix} 0 & [(\boldsymbol{\omega}_O \times m_O\boldsymbol{R}_O {}^O\boldsymbol{s}_{OC})\times] \\ 0 & [(\boldsymbol{R}_O {}^O\boldsymbol{I}_O\boldsymbol{R}_O^T\boldsymbol{\omega}_O)\times] \end{bmatrix},$$

$$\boldsymbol{g}_O = \begin{bmatrix} -m_O\tilde{\boldsymbol{g}} \\ -m_O[\boldsymbol{R}_O {}^O\boldsymbol{s}_{OC}\times]\tilde{\boldsymbol{g}} \end{bmatrix}, \quad \boldsymbol{F}_O = \begin{bmatrix} \boldsymbol{f}_O \\ \boldsymbol{n}_O \end{bmatrix}.$$

For a more detailed derivation of the above equation, see Exercise 4.1. Eq. (4.11) is the equation of object motion established by the force and moment acting on the origin of the coordinate system Σ_O. When the orientation of the object is represented by the *zyz*-Euler angle $\boldsymbol{\eta}_O = [\phi, \theta, \varphi]^T$, the position and orientation is expressed as $\boldsymbol{r}_O = [\boldsymbol{p}_O^T, \boldsymbol{\eta}_O^T]^T$, and its time derivative is expressed by $\dot{\boldsymbol{r}}_O = [\dot{\boldsymbol{p}}_O^T, \dot{\boldsymbol{\eta}}_O^T]^T$. On the other hand, the object velocity can be expressed by $\boldsymbol{v}_O = [\dot{\boldsymbol{p}}_O^T, \boldsymbol{\omega}_O^T]^T$ when the object angular velocity $\boldsymbol{\omega}_O$ is used. Both have a relation of

$$\boldsymbol{v}_O = \boldsymbol{T}_O \dot{\boldsymbol{r}}_O, \tag{4.12}$$

where

$$\boldsymbol{T}_O = block\ diag\left[\boldsymbol{I}_{3\times3}, \begin{bmatrix} 0 & -S\phi & C\phi S\theta \\ 0 & C\phi & S\phi S\theta \\ 1 & 0 & C\theta \end{bmatrix}\right] \in \mathbb{R}^{6\times6}. \tag{4.13}$$

Note that $S*$ and $C*$ are abbreviations of sin($*$) and cos($*$), respectively. Using this relation, Eq. (4.11) can be represented by

$$\tilde{\boldsymbol{M}}_O(\boldsymbol{r}_O)\ddot{\boldsymbol{r}}_O + \tilde{\boldsymbol{C}}_O(\boldsymbol{r}_O, \dot{\boldsymbol{r}}_O)\dot{\boldsymbol{r}}_O + \tilde{\boldsymbol{g}}_O(\boldsymbol{r}_O) = \tilde{\boldsymbol{F}}_O, \tag{4.14}$$

where

$$\tilde{\boldsymbol{M}}_O = \boldsymbol{T}_O^T \boldsymbol{M}_O \boldsymbol{T}_O, \quad \tilde{\boldsymbol{C}}_O = \boldsymbol{T}_O^T(\tilde{\boldsymbol{C}}_O \boldsymbol{T}_O + \boldsymbol{M}_O \dot{\mathrm{T}}_O),$$
$$\tilde{\boldsymbol{g}}_O = \boldsymbol{T}_O^T \boldsymbol{g}_O, \quad \tilde{\boldsymbol{F}}_O = \boldsymbol{T}_O^T \boldsymbol{F}_O.$$

4.2. Lagrange's Equation

The geometric position of a mass point moving freely in 3D space is often represented by the orthogonal coordinate system. However, the geometric position can also be expressed by other coordinate systems, such as the cylindrical coordinate system or the polar coordinate system, by using three independent variables. Furthermore, when the movement is bound to a two-dimensional (2D) surface, the geometric position is represented with two independent variables.

A number of these independent variables are called the **degree of freedom (DOF)**. The DOF n of a mass system consisting of

N mass points is $n = 3N - h$, where h is a number of constraint conditions. The geometric position of the mass system with n DOF can be represented by n independent variable $q_i(t)(i = 1, \ldots, n)$. The $q_i(t)$ is called the **generalized coordinate**. The time derivative of the generalized coordinate $\dot{q}_i(t)$ is called the **generalized velocity**. Now, let $\boldsymbol{x}_j$ be a 3D position vector of mass point P in the inertial system. It can then be expressed by

$$\boldsymbol{x}_j = \boldsymbol{x}_j(q_1, q_2, \ldots, q_n, t). \tag{4.15}$$

Newton's equation of motion of the mass point is represented by

$$\boldsymbol{F}_j = m_j \ddot{\boldsymbol{x}}_j, \tag{4.16}$$

where m_j is the mass of the mass point P_j, and $\boldsymbol{F}_j$ is the force applied to the mass point P_j. From Eq. (4.15), the time differentiation of x_j can be expressed by

$$\dot{\boldsymbol{x}}_j = \sum_{i=1}^{n} \frac{\partial \boldsymbol{x}_j}{\partial q_i} \dot{q}_i + \frac{\partial \boldsymbol{x}_j}{\partial t}, \tag{4.17}$$

and the following relation is obtained by the partial derivative of this expression with $\dot{q}_i$:

$$\frac{\partial \dot{\boldsymbol{x}}_j}{\partial \dot{q}_i} = \frac{\partial \boldsymbol{x}_j}{\partial q_i}. \tag{4.18}$$

Moreover, by changing the subscript in the right side of Eq. (4.17) from i to k, and the partial derivative of both sides of the equation with q_i, it becomes

$$\begin{aligned} \frac{\partial \dot{\boldsymbol{x}}_j}{\partial q_i} &= \sum_{k=1}^{n} \frac{\partial}{\partial q_k} \left[\frac{\partial \boldsymbol{x}_j}{\partial q_i} \right] \dot{q}_k + \frac{\partial}{\partial t} \left[\frac{\partial \boldsymbol{x}_j}{\partial q_i} \right] \\ &= \frac{d}{dt} \left[\frac{\partial \boldsymbol{x}_j}{\partial q_i} \right] \end{aligned} \tag{4.19}$$

The following relation is then obtained by using Eq. (4.19).

$$\begin{aligned} \frac{d}{dt} \left[\frac{\partial \dot{\boldsymbol{x}}_j^T \dot{\boldsymbol{x}}_j}{\partial \dot{q}_i} \right] &= \frac{d}{dt} \left[2 \dot{\boldsymbol{x}}_j^T \frac{\partial \dot{\boldsymbol{x}}_j}{\partial \dot{q}_i} \right] \\ &= 2 \ddot{\boldsymbol{x}}_j^T \frac{\partial \boldsymbol{x}_j}{\partial q_i} + 2 \dot{\boldsymbol{x}}_j^T \frac{\partial \dot{\boldsymbol{x}}_j}{\partial q_i} \end{aligned} \tag{4.20}$$

Furthermore, to obtain the inner product of both sides of Eq. (4.16) as well as $\partial \boldsymbol{x}_j / \partial q_i$ leads to:

$$\sum_{j=1}^{n} \boldsymbol{F}_j^T \frac{\partial \boldsymbol{x}_j}{\partial q_i} = \sum_{j=1}^{n} m_j \ddot{\boldsymbol{x}}_j^T \frac{\partial \boldsymbol{x}_j}{\partial q_i} \tag{4.21}$$

By substituting Eq. (4.21) into Eq. (4.20), Eq. (4.22) is then derived as follows

$$\frac{d}{dt}\left[\frac{\partial K}{\partial \dot{q}_i}\right] - \frac{\partial K}{\partial q_i} = Q_i, \tag{4.22}$$

where K is the **kinematic energy** of the mass system represented by

$$K = \frac{1}{2}\sum_{j=1}^{n} m_j \dot{\boldsymbol{x}}_j^T \dot{\boldsymbol{x}}_j, \tag{4.23}$$

and Q_i is the **generalized force** corresponding to q_i represented by

$$Q_i = \sum_{j=1}^{n} \boldsymbol{F}_j^T \frac{\partial \boldsymbol{x}_j}{\partial q_i} \quad (i = 1, 2, \ldots, n). \tag{4.24}$$

The force $\boldsymbol{F}_j$ can be separated into the conservative force $\boldsymbol{F}_j''$ and the other force $\boldsymbol{F}_j'$. The conservative force is represented as

$$\boldsymbol{F}_j'' = -\frac{\partial U}{\partial \boldsymbol{x}_j}, \tag{4.25}$$

where U is a potential energy of the mass system and does not depend on the velocity; therefore, $\partial U / \partial \dot{q}_j = 0$. Now, let the **Lagrangian function** L be defined as

$$L = K - U. \tag{4.26}$$

This derives

$$\frac{d}{dt}\left[\frac{\partial L}{\partial \dot{q}_i}\right] - \frac{\partial L}{\partial q_i} = Q_i' \quad (i = 1, 2, \ldots, n), \tag{4.27}$$

where

$$Q_i' = \sum_{j=1}^{n} \boldsymbol{F}_j'^T \frac{\partial \boldsymbol{x}_j}{\partial q_i} \quad (i = 1, 2, \ldots, n). \tag{4.28}$$

Equation (4.27) is the **Lagrange's equation** (Lagrange, 1736–1813).

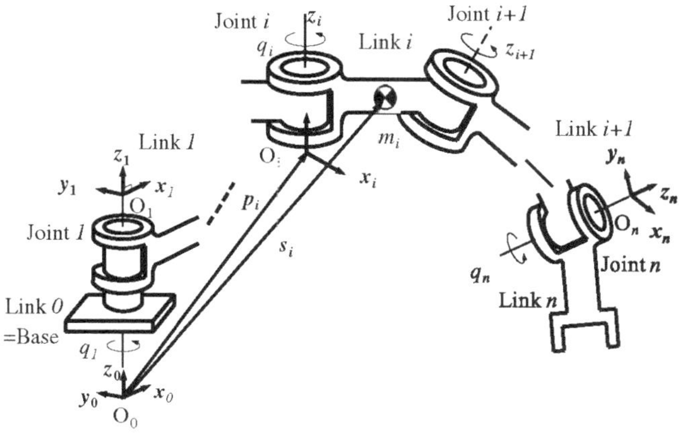

Fig. 4.2. n DOF for robotic arm.

4.3. Dynamics Model of a Multi-Link Mechanism

The dynamics model of a serial link mechanism connected with multi-links can be derived in a systematic way using Lagrange's equation. This section describes the essential derivation of motion of a robotic arm with n DOF, as shown in Fig. 4.2. In the figure, the links are numbered from Link 0, which is the base link, to Link n, which is the end-link; Joint i is located between Link $i-1$ and Link i. The i-th coordinate system, $\Sigma_i = \{\boldsymbol{x}_i, \boldsymbol{y}_i, \boldsymbol{z}_i\}$, is set at Link i in order for the $\boldsymbol{z}_i$ axis to fit with the joint axis. The $q_{\rm i}$ represents the i-th joint displacement. Let us denote the kinematic energy of the entire arm as K, and the potential energy of the whole arm as P. The Lagrange function is represented as $L = K - P$. Let the joint variables $q_i (i = 1, \ldots, n)$ be selected as the generalized coordinates. Then the generalized forces associated with the generalized coordinates are the joint torque $\tau_i (i = 1, \ldots, n)$. The Lagrange's equations are then represented by

$$\tau_i = \frac{d}{dt}\left(\frac{\partial L}{\partial \dot{q}_i}\right) - \frac{\partial L}{\partial q_i}, \quad i = 1, \ldots, n. \tag{4.29}$$

Since the potential energy does not depend on $\dot{q}_i$, this is rewritten by

$$\tau_i = \frac{d}{dt}\left(\frac{\partial K}{\partial \dot{q}_i}\right) - \frac{\partial K}{\partial q_i} + \frac{\partial P}{\partial q_i}, \quad i = 1, \ldots, n. \tag{4.30}$$

The kinetic energy of Link i is represented by

$$K_i = \frac{1}{2} m_i \hat{\boldsymbol{v}}_i^T \hat{\boldsymbol{v}}_i + \frac{1}{2} \boldsymbol{\omega}_i^T \hat{\boldsymbol{I}}_i \boldsymbol{\omega}_i, \tag{4.31}$$

where m_i is the mass of Link i, $\hat{\boldsymbol{I}}_i \in \mathbb{R}^{3\times3}$ is the inertia tensor of the link center of mass with respect to the reference coordinate system, $\hat{\boldsymbol{v}}_i \in \mathbb{R}^3$ is the linear velocity of the center of mass of Link i, and $\boldsymbol{\omega}_i \in \mathbb{R}^3$ is the angular velocity of Link i. The first term on the right-hand side in Eq. (4.31) is the kinematic energy due to the linear velocity, and the second term is the kinematic energy due to the angular velocity of the link. The total kinematic energy of the arm is expressed by

$$K = \sum_{i=1}^{n} K_i. \tag{4.32}$$

The translational velocity $\hat{\boldsymbol{v}}_i$ and the angular velocity $\boldsymbol{\omega}_i$ can be expressed using the generalized coordinates. When a link is considered as an end-effector of the arm, the **Jacobian** matrix can be applied to find the relation between the joint velocity and the link velocity. That is,

$$\begin{aligned} \hat{\boldsymbol{v}}_i &= \boldsymbol{J}_{p1}^{(i)} \dot{q}_1 + \cdots + \boldsymbol{J}_{pi}^{(i)} \dot{q}_i = \boldsymbol{J}_p^{(i)} \dot{\boldsymbol{q}} \\ \boldsymbol{\omega}_i &= \boldsymbol{J}_{o1}^{(i)} \dot{q}_1 + \cdots + \boldsymbol{J}_{oi}^{(i)} \dot{q}_i = \boldsymbol{J}_o^{(i)} \dot{\boldsymbol{q}} \end{aligned} \tag{4.33}$$

where $\boldsymbol{J}_p^{(i)}$ and $\boldsymbol{J}_o^{(i)} \in \mathbb{R}^{3\times3}$ are the Jacobian matrices of the position and orientation related to Link i, and $\boldsymbol{J}_{pj}^{(i)}$ and $\boldsymbol{J}_{oj}^{(i)} \in \mathbb{R}^3$ are the j-th columns of $\boldsymbol{J}_p^{(i)}$ and $\boldsymbol{J}_o^{(i)}$. Because the motion of Link i depends only on the motions from Joint 1 to Joint i, the column is the zero vector when $j > i$. Therefore, $\boldsymbol{J}_p^{(i)}$ and $\boldsymbol{J}_o^{(i)}$ are represented by

$$\boldsymbol{J}_p^{(i)} = \left[\boldsymbol{J}_{p1}^{(i)} \cdots \boldsymbol{J}_{pi}^{(i)} 0 \cdots 0 \right] \tag{4.34-1}$$

$$\boldsymbol{J}_o^{(i)} = \left[\boldsymbol{J}_{o1}^{(i)} \cdots \boldsymbol{J}_{oi}^{(i)} 0 \cdots 0 \right] \tag{4.34-2}$$

The columns of the matrices are expressed by

$$\boldsymbol{J}_{pj}^{(i)} = \begin{cases} \boldsymbol{z}_j & \text{for translational joint} \\ \boldsymbol{z}_j \times (\boldsymbol{s}_i - \boldsymbol{p}_j) & \text{for rotational joint} \end{cases} \tag{4.35)-1}$$

$$\boldsymbol{J}_{oj}^{(i)} = \begin{cases} 0 & \text{if translational joint} \\ \boldsymbol{z}_j & \text{if rotational joint} \end{cases} \tag{4.35)-2}$$

where $\boldsymbol{z}_i$ is the unit vector along the axis of Joint i, $\boldsymbol{s}_i$ is a position vector from the origin of the reference coordinate system to the center of mass of Link i, and $\boldsymbol{p}_i$ is the position vector from the origin of the reference coordinate system to the origin of the i-th coordinate system. These position vectors are referred to the reference coordinate system. Then the total kinematic energy is written by

$$K = \frac{1}{2}\dot{\boldsymbol{q}}^T \boldsymbol{M} \dot{\boldsymbol{q}}, \tag{4.36}$$

where $\boldsymbol{M} \in \mathbb{R}^{n \times n}$ is the robot inertia matrix given by

$$\boldsymbol{M} = \sum_{i=1}^{n} (m_i \boldsymbol{J}_p^{(i)T} \boldsymbol{J}_p^{(i)} + \boldsymbol{J}_o^{(i)T} \hat{\boldsymbol{I}}_i \boldsymbol{J}_o^{(i)}). \tag{4.37}$$

This robot inertia matrix is symmetric and positive, definite, and varies with the position and orientation of the end-effector. Let M_{ij} be as the (i, j) element of $\boldsymbol{M}$; then, the total kinetic energy is represented by

$$K = \frac{1}{2} \sum_{i=1}^{n} \sum_{j=1}^{n} M_{ij} \dot{q}_i \dot{q}_j. \tag{4.38}$$

The total potential energy of the robot is represented by

$$P = \sum_{i=1}^{n} m_i \tilde{\boldsymbol{g}}^T \boldsymbol{s}_i, \tag{4.39}$$

where $\tilde{\boldsymbol{g}} \in \mathbb{R}^3$ is the gravity acceleration vector with respect to the reference coordinate system, and $\boldsymbol{s}_i \in \mathbb{R}^3$ is the position vector of the center of mass of Link i with respect to the reference coordinate system. The $\boldsymbol{s}_i$ depends on the joint angle, and the total potential energy is a function only of $q_i (i = 1, \ldots, n)$, and not of joint velocities.

Let these relations be substituted into Lagrange's equation of motion. Since M_{ij} is a function of the joint displacement, its time derivative is

$$\frac{dM_{ij}}{dt} = \sum_{k=1}^{n} \frac{\partial M_{ij}}{\partial q_k} \dot{q}_k. \tag{4.40}$$

From this relation, the first term of the right-hand side of Eq. (4.30) can be rewritten as

$$\begin{aligned} \frac{d}{dt}\left(\frac{\partial K}{\partial \dot{q}_i}\right) &= \frac{d}{dt}\left(\sum_{j=1}^{n} M_{ij}\dot{q}_j\right) = \sum_{j=1}^{n} M_{ij}\ddot{q}_j + \sum_{j=1}^{n} \frac{dM_{ij}}{dt}\dot{q}_j \\ &= \sum_{j=1}^{n} M_{ij}\ddot{q}_j + \sum_{j=1}^{n}\sum_{k=1}^{n} \frac{\partial M_{ij}}{\partial q_k}\dot{q}_k\dot{q}_j \end{aligned} \tag{4.41}$$

If the subscript i and j in the right-hand side of Eq. (4.38) are replaced with j and k, respectively, the second term of the right-hand side of Eq. (4.30) becomes,

$$\frac{\partial K}{\partial q_i} = \frac{1}{2}\sum_{j=1}^{n}\sum_{k=1}^{n} \frac{\partial M_{jk}}{\partial q_i}\dot{q}_j\dot{q}_k. \tag{4.42}$$

By substituting Eq. (4.39) into the right-hand side of the third term of Eq. (4.30) and using the relation $\partial \boldsymbol{s}_j/\partial q_i = \boldsymbol{J}_{pi}^{(i)}$, the third term can be represented by

$$\frac{\partial P}{\partial q_i} = \sum_{j=1}^{n} m_j \tilde{\boldsymbol{g}}^T \frac{\partial \boldsymbol{s}_j}{\partial q_i} = \sum_{j=1}^{n} m_j \tilde{\boldsymbol{g}}^T \boldsymbol{J}_{pi}^{(j)}. \tag{4.43}$$

By substituting Eq. (4.43) into Eq. (4.30), it is rewritten as

$$\tau_i = \sum_{j=1}^{n} M_{ij}\ddot{q}_j + \sum_{j=1}^{n}\sum_{k=1}^{n} h_{ijk}\dot{q}_j\dot{q}_k + \sum_{j=1}^{n} m_j \tilde{\boldsymbol{g}}^T \boldsymbol{J}_{pi}^{(j)} \quad (i = 1, \ldots, n), \tag{4.44}$$

where

$$h_{ijk} = \frac{\partial M_{ij}}{\partial q_k} - \frac{1}{2}\frac{\partial M_{jk}}{\partial q_i}. \tag{4.45}$$

The first term on the right-hand side of Eq. (4.44) represents the inertia force, and $M_{ii}\ddot{q}_i$ represents the inertial force term due to the acceleration of Joint i; meanwhile, $M_{ij}\ddot{q}_j(i \neq j)$ represents the interference term applied to Joint i by the acceleration of Joint j. The second term represents a quadratic velocity term, in which $h_{ijj}\dot{q}_j^2$ is the **centrifugal force** induced on Joint i by the velocity of Joint j, and $h_{ijk}\dot{q}_j\dot{q}_k$ is the **Coriolis force** induced on Joint i by the velocities of Joint j and Joint k. In addition, $h_{iii} = 0$, and the third term represents the gravitational term. The equations of motion can be rewritten in a compact matrix form to represent the joint space dynamic model as follows:

$$\boldsymbol{M}(\mathrm{q})\ddot{\boldsymbol{q}} + \boldsymbol{C}(\boldsymbol{q}, \dot{\boldsymbol{q}})\dot{\boldsymbol{q}} + \boldsymbol{g}(\boldsymbol{q}) = \boldsymbol{\tau}, \tag{4.46}$$

where $\boldsymbol{C}(\boldsymbol{q}, \dot{\boldsymbol{q}}) \in \mathbb{R}^{n\times n}$ is a suitable matrix such that its element, C_{ij}, satisfy the following relation

$$\sum_{j=1}^{n} C_{ij}\dot{q}_j = \sum_{j=1}^{n}\sum_{k=1}^{n} h_{ijk}\dot{q}_j\dot{q}_k. \tag{4.47}$$

Therefore, the representation of C_{ij} is not unique. Also, let the i-th element of $\boldsymbol{g}(\boldsymbol{q}) \in \mathbb{R}^n$ be denoted as g_i. It can then be represented by

$$g_i = \sum_{j=1}^{n} m_j \tilde{\boldsymbol{g}}^T \boldsymbol{J}_{Pi}^{(j)}. \tag{4.48}$$

When a wrench $\boldsymbol{F}_H$ is applied on the end-effector of a robotic arm, the joint torque balanced with this is $\boldsymbol{J}^T(\boldsymbol{q})\boldsymbol{F}_H$, and the equation of motion is represented by

$$\boldsymbol{M}\ddot{\boldsymbol{q}} + \boldsymbol{C}(\boldsymbol{q}, \dot{\boldsymbol{q}})\dot{\boldsymbol{q}} + \boldsymbol{g}(\boldsymbol{q}) = \boldsymbol{\tau} - \boldsymbol{J}^T(\boldsymbol{q})\boldsymbol{F}_H. \tag{4.49}$$

Moreover, when the **viscous friction** that is proportional to the joint velocity and the **Coulomb friction** that depends only on the sign of the joint velocity are applied, these act as forces to reduce the joint driving force. Therefore, the equation of motion is expressed by

$$\boldsymbol{M}\ddot{\boldsymbol{q}} + \boldsymbol{C}(\boldsymbol{q}, \dot{\boldsymbol{q}})\dot{\boldsymbol{q}} + \boldsymbol{g}(\boldsymbol{q}) = \boldsymbol{\tau} - \boldsymbol{\tau}_{friction} - \boldsymbol{J}^T(\boldsymbol{q})\boldsymbol{F}_H, \tag{4.50}$$

where $\boldsymbol{\tau}_{friction} \in \mathbb{R}^n$ is the friction torque and its i-th element can be given by

$$\tau_{friction_i} = d_i\dot{q}_i + fr_i \,\mathrm{sgn}\,(\dot{q}_i), \tag{4.51}$$

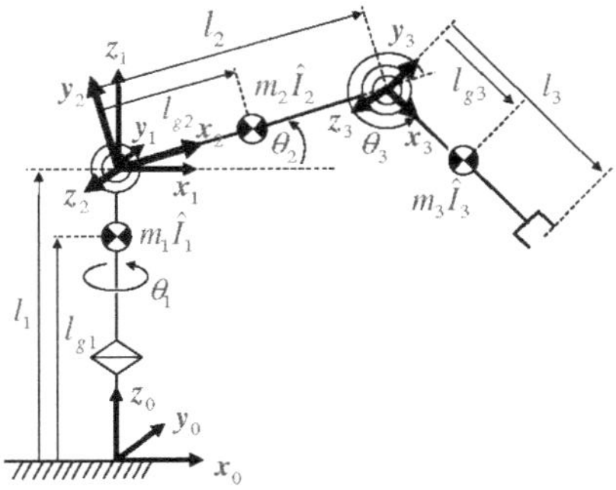

Fig. 4.3. Three-jointed robot finger.

where d_i is the viscous friction coefficient of the Joint i, fr_i is the Coulomb friction force of the Joint i, and sgn$(*)$ is the sign function.

Example 4.1. Find the motion equation of a robot with a three-jointed finger, as shown in Fig. 4.3. Note that coordinate systems are set in each link where the mass of Link i is m_i, and the center of mass of the link is located on the line connecting adjacent origins of the link coordinate systems. Given that the inertia tensor at the center of mass of Link i is $\hat{\boldsymbol{I}}_i$, the length of Link i is l_i, and the length from the origin of the i-th link coordinate to the center of mass of Link i is l_{gi}. Let us suppose that the inertia tensor with respect to the i-th link coordinate system ${}^i\hat{\boldsymbol{I}}_i$ is a diagonal matrix.

Position vectors of the center of mass of Links 1, 2, and 3, that is, $\boldsymbol{s}_1$, $\boldsymbol{s}_2$, $\boldsymbol{s}_3$ are given by

$$\boldsymbol{s}_1 = \begin{bmatrix} 0 \\ 0 \\ l_{g1} \end{bmatrix}, \quad \boldsymbol{s}_2 = \begin{bmatrix} l_{g2}C_1C_2 \\ l_{g2}S_1C_2 \\ l_1 + l_{g2}S_2 \end{bmatrix}, \quad \boldsymbol{s}_3 = \begin{bmatrix} l_2C_1C_2 + l_{g3}C_1C_{23} \\ l_2S_1C_2 + l_{g3}S_1C_{23} \\ l_1 + l_2S_2 + l_{g3}S_{23} \end{bmatrix}, \tag{1}$$

where abbreviations are used as follows: $C_{23} = \cos(\theta_2 + \theta_3)$ and $S_{23} = \sin(\theta_2 + \theta_3)$. In addition, the velocity vectors $\hat{\boldsymbol{v}}_1$, $\hat{\boldsymbol{v}}_2$, $\hat{\boldsymbol{v}}_3$ of Links 1, 2, and 3 are represented by

$$\hat{\boldsymbol{v}}_1 = \boldsymbol{J}_P^{(1)}\dot{\boldsymbol{q}}, \quad \hat{\boldsymbol{v}}_2 = \boldsymbol{J}_P^{(2)}\dot{\boldsymbol{q}}, \quad \hat{\boldsymbol{v}}_3 = \boldsymbol{J}_P^{(3)}\dot{\boldsymbol{q}}, \tag{2}$$

where $\dot{\boldsymbol{q}} = [\dot{\theta}_1\ \dot{\theta}_2\ \dot{\theta}_3]^T$. The Jacobian matrices $\boldsymbol{J}_p^{(i)} (i = 1, 2, 3)$ are

$$\boldsymbol{J}_p^{(1)} = \begin{bmatrix} 0 & 0 & 0 \\ 0 & 0 & 0 \\ 0 & 0 & 0 \end{bmatrix}, \quad \boldsymbol{J}_p^{(2)} = \begin{bmatrix} -l_{g2}S_1C_2 & -l_{g2}C_1S_2 & 0 \\ l_{g2}C_1C_2 & -l_{g2}S_1S_2 & 0 \\ 0 & l_{g2}C_2 & 0 \end{bmatrix},$$

$$\boldsymbol{J}_p^{(3)} = \begin{bmatrix} -l_2S_1C_2 - l_{g3}S_1C_{23} & -l_2C_1S_2 - l_{g3}C_1S_{23} & -l_{g3}C_1S_{23} \\ l_2C_1C_2 + l_{g3}C_1C_{23} & -l_2S_1S_2 - l_{g3}S_1S_{23} & -l_{g3}S_1S_{23} \\ 0 & l_2C_2 + l_{g3}C_{23} & l_{g3}C_{23} \end{bmatrix}. \tag{3}$$

The angular velocity vectors $\boldsymbol{\omega}_1$, $\boldsymbol{\omega}_2$, $\boldsymbol{\omega}_3$ of Links 1, 2, and 3 are represented by

$$\boldsymbol{\omega}_1 = \boldsymbol{J}_o^{(1)}\dot{\boldsymbol{q}}, \boldsymbol{\omega}_2 = \boldsymbol{J}_o^{(2)}\dot{\boldsymbol{q}}, \boldsymbol{\omega}_3 = \boldsymbol{J}_o^{(3)}\dot{\boldsymbol{q}}, \tag{4}$$

where

$$\boldsymbol{J}_o^{(1)} = \begin{bmatrix} 0 & 0 & 0 \\ 0 & 0 & 0 \\ 1 & 0 & 0 \end{bmatrix}, \boldsymbol{J}_o^{(2)} = \begin{bmatrix} 0 & C_1 & 0 \\ 0 & S_1 & 0 \\ 1 & 0 & 0 \end{bmatrix}, \boldsymbol{J}_o^{(3)} = \begin{bmatrix} 0 & C_1 & C_1 \\ 0 & S_1 & S_1 \\ 1 & 0 & 0 \end{bmatrix}. \tag{5}$$

By substituting the Jacobian matrix $\boldsymbol{J}_P^{(i)}$ and $\boldsymbol{J}_o^{(i)}$ $(i = 1, 2, 3)$ into Eq. (4.37), the inertia matrix $\boldsymbol{M} = (M_{ii})$ is represented by

$$\boldsymbol{M} = \begin{bmatrix} M_{11} & 0 & 0 \\ 0 & M_{22} & M_{23} \\ 0 & M_{32} & M_{33} \end{bmatrix}. \tag{6}$$

Let the (j, k) element of $\hat{\boldsymbol{I}}_i$ be denoted as $\hat{I}_{ijk}$, and the dynamic parameter be defined as

$$\boldsymbol{\sigma} = \begin{bmatrix} \hat{I}_{1\mathrm{zz}} + \hat{I}_{2yy} + \hat{I}_{3yy} + m_3l_2^2 \\ m_2l_{g2} + m_3l_2 \\ \hat{I}_{2\mathrm{xx}} - \hat{I}_{2yy} - \hat{I}_{3yy} - m_3l_2^2 \\ \hat{I}_{2zz} + m_2l_{g2}^2 + m_3l_2^2 \\ m_3l_{g3} \\ \hat{I}_{3xx} - \hat{I}_{3yy} \\ \hat{I}_{3zz} + m_3l_{g3}^2 \end{bmatrix}. \tag{7}$$

Then, elements of the inertia matrix are expressed by

$$\begin{aligned} M_{11} &= \sigma_1 + S_2^2\sigma_3 + 2C_{23}C_2l_2\sigma_5 + S_{23}^2\sigma_6 \\ M_{22} &= \sigma_4 + 2C_3l_2\sigma_5 + \sigma_7 \\ M_{23} &= M_{32} = l_2C_3\sigma_5 + \sigma_7 \\ M_{33} &= \sigma_7, \end{aligned} \tag{8}$$

where σ_i is the i-th element of $\boldsymbol{\sigma}$. From Eq. (4.47), the quadratic velocity term $\boldsymbol{C}\dot{\boldsymbol{q}} = [C_1\ C_2\ C_3]^T$ is expressed by

$$\boldsymbol{C}\dot{\boldsymbol{q}} = \begin{bmatrix} 2S_2C_2\sigma_3\dot{\theta}_1\dot{\theta}_2 + 2l_2((-2C_2S_{23} + S_3)\dot{\theta}_1\dot{\theta}_2 + C_2S_{23}\dot{\theta}_1\dot{\theta}_3)\sigma_5 \\ \quad + 2(2C_2C_3S_{23} - C_2S_2 - C_3S_3)(\dot{\theta}_2 + \dot{\theta}_3)\dot{\theta}_1\sigma_6 \\ \quad - S_2C_2\sigma_3\dot{\theta}_1^2 + ((-S_3 + 2C_2S_{23})\dot{\theta}_1^2 - 2S_3\dot{\theta}_2\dot{\theta}_3 - S_3\dot{\theta}_3^2)l_2\sigma_5 \\ \quad + (-2C_2C_3S_{23} + C_2S_2 + C_3S_3)\dot{\theta}_1^2\sigma_6 \\ (C_2S_{23}\dot{\theta}_1^2 + S_3\dot{\theta}_2^2)l_2\sigma_5 + (-2C_2C_3S_{23} + C_2S_2 + C_3S_3)\dot{\theta}_1^2\sigma_6 \end{bmatrix}. \tag{9}$$

From Eq. (4.48), the gravity term $\boldsymbol{g} = [g_1\ g_2\ g_3]^T$ is expressed by

$$\boldsymbol{g} = \begin{bmatrix} 0 \\ -C_2\sigma_2 - C_{23}\sigma_5 \\ -C_{23}\sigma_5 \end{bmatrix} \tilde{\boldsymbol{g}}, \tag{10}$$

where $\tilde{\boldsymbol{g}} = [0\ 0\ -g]^T$ is the acceleration of gravity. Since an external force is not applied, the equation of motion can be obtained by substituting $\boldsymbol{M}$, $\boldsymbol{C}$, and $\boldsymbol{g}$ into Eq. (4.49).

4.4. The Basic Properties of the Dynamic Model

Let a matrix be defined as bounded if all the elements of the matrix are bounded. Then, the equation of motion for robots has the following essential propositions.

Proposition 4.1.

The inertia matrix of the robot $\boldsymbol{M}(\boldsymbol{q})$ is a symmetric and positive matrix.

Proof. The inertia matrix of the robot is represented by Eq. (4.37). Clearly, it is a non-negative symmetric matrix. The $\boldsymbol{M}(\boldsymbol{q})$ is positive as long as m_i and $\hat{\boldsymbol{I}}_i (i = 1, \ldots, n)$ are not zero. Therefore, the inertia matrix, $\boldsymbol{M}(\boldsymbol{q})$, is generally a symmetric and positive matrix.

Proposition 4.2.

The inertia matrix of the robot $\boldsymbol{M}(\boldsymbol{q})$ is bounded when the robot structure satisfies one of the following conditions:

(1) *All joints of the robot are rotational joints.*
(2) *Joint axis $\boldsymbol{z}_k$ of the arbitrary translational joint k is parallel to the joint axes of all the rotational joints j $(j < k)$on the base side.*

Proof. According to Eq. (4.37), the inertia matrix is bounded if the $\boldsymbol{J}_{pj}^{(i)}$ and $\boldsymbol{J}_{oj}^{(i)}$ are bounded for all $1 \leq i \leq n$. According to Eq. (4.35)-1, $\boldsymbol{J}_{pj}^{(i)}$ and $\boldsymbol{J}_{oj}^{(i)}$ are bounded if $\boldsymbol{z}_j$ and $\boldsymbol{z}_j \times (\boldsymbol{s}_i - \boldsymbol{p}_j)$ is bounded. Under condition (1), $\boldsymbol{z}_j$ and $(\boldsymbol{s}_k - \boldsymbol{p}_j)$ are bounded because these are described by triangle functions with joint variables. Under condition (2), $\boldsymbol{z}_j$ is bounded and $\boldsymbol{z}_j \times (\boldsymbol{s}_k - \boldsymbol{p}_j) = 0$ for case of $j < k$. Therefore, the inertia matrix $\boldsymbol{M}(\boldsymbol{q})$ is bounded under condition (1) or (2).

In addition, according to Ghorbel *et al.* (1998), the upper and lower limits of the matrix norm of bounded inertia matrix have been shown.

Proposition 4.3.

By defining $\boldsymbol{C}(\boldsymbol{q}, \dot{\boldsymbol{q}})$ appropriately, $\dot{\boldsymbol{M}}(\boldsymbol{q}) - 2\boldsymbol{C}(\boldsymbol{q}, \dot{\boldsymbol{q}})$ is a skewed symmetric matrix.

Proof. There is arbitrariness in the representation where the element C_{ij} of $\boldsymbol{C}(\boldsymbol{q}, \dot{\boldsymbol{q}})$ satisfies Eq. (4.47). Now, it can be expressed by

$$C_{ij} = \sum_{k=1}^{n} c_{ijk} \dot{q}_k, \tag{4.52}$$

where

$$c_{ijk} = \frac{1}{2} \left(\frac{\partial M_{ij}}{\partial q_k} + \frac{\partial M_{ik}}{\partial q_j} - \frac{\partial M_{jk}}{\partial q_i} \right). \tag{4.53}$$

In this case, it is confirmed that C_{ij} satisfies the relations of Eq. (4.47) (see Exercise 4.2). c_{ijk} is called a **Christoffel symbol of the**

first type. It is $c_{ijk} = c_{ikj}$ due to the symmetry of $\boldsymbol{M}(\boldsymbol{q})$. When C_{ij} is given by Eq. (4.52), the (i, j) element of $\dot{\boldsymbol{M}}(\boldsymbol{q}) - 2\boldsymbol{C}(\boldsymbol{q}, \dot{\boldsymbol{q}})$ is

$$\dot{M}_{ij} - 2C_{ij} = \sum_{k=1}^{n} \frac{\partial M_{ij}}{\partial q_k}\dot{q}_k - \sum_{k=1}^{n} \left(\frac{\partial M_{ij}}{\partial q_k} + \frac{\partial M_{ik}}{\partial q_j} - \frac{\partial M_{jk}}{\partial q_i} \right) \dot{q}_k$$

$$= \sum_{k=1}^{n} \left(-\frac{\partial M_{ik}}{\partial q_j} + \frac{\partial M_{jk}}{\partial q_i} \right) \dot{q}_k$$

It is clear that $\dot{M}_{ii} - 2C_{ii} = 0$, and $\dot{M}_{ij} - 2C_{ij} = -\dot{M}_{ji} + 2C_{ji}$. Therefore, $\dot{\boldsymbol{M}}(\boldsymbol{q}) - 2\boldsymbol{C}(\boldsymbol{q}, \dot{\boldsymbol{q}})$ is a skewed symmetric matrix, which satisfies $\boldsymbol{x}^T(\dot{\boldsymbol{M}}(\boldsymbol{q}) - 2\boldsymbol{C}(\boldsymbol{q}, \dot{\boldsymbol{q}}))\boldsymbol{x} = 0$ for an arbitrary vector $\boldsymbol{x} \in \mathbb{R}^n$.

This proposition leads to the following proposition, 4.4. Proposition 4.3 is often referred to as the **passivity property** of robots. In addition, when $\dot{\boldsymbol{q}}$ is used instead of $\boldsymbol{x}$, the following relation can be proven regardless of the definition of $\boldsymbol{C}(\boldsymbol{q}, \dot{\boldsymbol{q}})$ for any $\dot{\boldsymbol{q}}$:

$$\dot{\boldsymbol{q}}^T(\dot{\boldsymbol{M}}(\boldsymbol{q}) - 2\boldsymbol{C}(\boldsymbol{q}, \dot{\boldsymbol{q}}))\dot{\boldsymbol{q}} = 0 \tag{4.54}$$

This is verified by the following relationship. That is, the time derivative of the kinetic energy using Eq. (4.36) can be expressed as

$$\begin{aligned} \frac{d}{dt}K &= \frac{1}{2}\dot{\boldsymbol{q}}^T\dot{\boldsymbol{M}}\dot{\boldsymbol{q}} + \dot{\boldsymbol{q}}^T\boldsymbol{M}\ddot{\mathrm{q}} \\ &= \frac{1}{2}\dot{\boldsymbol{q}}^T(\dot{\boldsymbol{M}} - 2\boldsymbol{C})\dot{\boldsymbol{q}} + \dot{\boldsymbol{q}}^T(\boldsymbol{\tau} - \boldsymbol{g} - \boldsymbol{J}^T(\boldsymbol{q})\boldsymbol{F}_H). \end{aligned} \tag{4.55}$$

On the other hand, from the energy conservation law, change of kinetic energy balances work generated by all forces applied to the joints of the robot, and therefore the relationship of $\delta K = \delta\boldsymbol{q}^T(\boldsymbol{\tau} - \boldsymbol{g} - \boldsymbol{J}^T(\boldsymbol{q})\boldsymbol{F}_H)$ is established and leads to:

$$\frac{d}{dt}K = \dot{\boldsymbol{q}}^T(\boldsymbol{\tau} - \boldsymbol{g} - \boldsymbol{J}^T(\boldsymbol{q})\boldsymbol{F}_H) \tag{4.56}$$

Eq. (4.54) is derived from Eq. (4.55) and Eq. (4.56).

Proposition 4.4.

Robot dynamics, according to Eq. (4.46), have passivity regarding the input–output relation such that the joint torque is input and the joint velocity is output. That is, the following is established:

$$\int_0^t \dot{\boldsymbol{q}}(s)^T \boldsymbol{\tau}(s)ds \geq -\gamma_0, \tag{4.57}$$

where γ_0 is an appropriate positive constant.

Proof. Define the sum of the kinetic energy K and the potential energy P as $E_r = K + P(\geq 0)$, and then calculate the time derivative along the trajectory. Refer to Eq. (4.55) and Proposition 4.3, and we get

$$\dot{E}_r(\boldsymbol{q}, \dot{\boldsymbol{q}}) = \dot{\boldsymbol{q}}^T \boldsymbol{M} \ddot{\boldsymbol{q}} + \frac{1}{2}\dot{\boldsymbol{q}}^T \dot{\mathbf{M}}\dot{\mathbf{q}} + \frac{\partial P}{\partial \boldsymbol{q}}\dot{\boldsymbol{q}} = \dot{\boldsymbol{q}}^T \boldsymbol{\tau}.$$

From this relationship,

$$\begin{aligned}\int_0^t \dot{\boldsymbol{q}}(s)^T \boldsymbol{\tau}(s)ds &= \int_0^t \dot{E}_r ds \\ &= E_r(t) - E_r(0) \geq -E_r(0) = -\gamma_0\end{aligned}$$

The passivity represented by Eq. (4.57) has thus been proven.

Proposition 4.5.

The dynamics model is linear with respect to dynamic parameters that are chosen appropriately. That is, the dynamics model is expressed by

$$\boldsymbol{M}(\boldsymbol{q})\ddot{\mathbf{q}}_r + \boldsymbol{C}(\boldsymbol{q}, \dot{\boldsymbol{q}})\dot{\boldsymbol{q}}_r + \boldsymbol{g}(\boldsymbol{q}) = \mathbf{Y}(\boldsymbol{q}, \dot{\boldsymbol{q}}, \dot{\boldsymbol{q}}_r, \ddot{\mathbf{q}}_r)\boldsymbol{\sigma}, \tag{4.58}$$

where $\dot{\boldsymbol{q}}_r \in \mathbb{R}^n$ is any velocity vector, $\boldsymbol{\sigma} \in \mathbb{R}^p$ is a dynamic parameter vector, and $\boldsymbol{Y}(\boldsymbol{q}, \dot{\boldsymbol{q}}, \dot{\boldsymbol{q}}_r, \ddot{\boldsymbol{q}}_r) \in \mathbb{R}^{n \times p}$ is a regressor on σ.

See Kawasaki and Nishimura (1986) and Kawasaki *et al.* (1996) for the proof of this proposition. Given that the dynamic parameters of Link i are

$$\boldsymbol{\sigma}_i = \begin{bmatrix} m_i & ms_{ix} & ms_{iy} & ms_{iz} & I_{ixx} & I_{ixy} & I_{ixz} & I_{iyy} & I_{iyz} & I_{izz} \end{bmatrix} \in \mathbb{R}^{10}, \tag{4.59}$$

it can be shown that the force and moment applied to Link i are linear with respect to $\boldsymbol{\sigma}_i$. The joint torque τ_i is proportional to the force and moment, and therefore, the joint torque $\boldsymbol{\tau}$ becomes linear with respect to parameters $\boldsymbol{\sigma} = [\sigma_1^T \sigma_2^T \cdots \sigma_n^T]^T$.

The dynamic parameters of Eq. (4.59) can be classified into three categories — i.e., a parameter that does not appear in the dynamic model, a parameter that appears in combined form, and a parameter that appears individually. Parameters that consist of the minimum number of parameters to represent a dynamic model are called **base parameters**. Base parameters are required for parameter identification and for control, based on the parameter estimates. Let it be noted that it is convenient to use the library Robot Symbolic Analysis System (ROSAM) (Kawasaki and Shimizu, 2000) for base parameter analyses.

Example 4.2. Consider the Selective Compliance Assembly Robot Arm (SCARA) robot shown in Fig. 4.4. Find the equation of motion using the Lagrange method and verify Propositions 4.3 and 4.5. In the interest of simplification, the center of mass of each link is located at a base side joint of the link, and the inertia tensor of the link should be a diagonal matrix in which the coordinate system is set to the base of the link.

The position of the mass center of each link is expressed by

$$\boldsymbol{s}_1 = \begin{bmatrix} 0 \\ 0 \\ l_0 \end{bmatrix}, \quad \boldsymbol{s}_2 = \begin{bmatrix} l_1 C\theta_1 \\ l_1 S\theta_1 \\ l_0 \end{bmatrix}, \quad \boldsymbol{s}_3 = \begin{bmatrix} l_1 C\theta_1 + l_2 C\theta_{12} \\ l_1 S\theta_1 + l_2 S\theta_{12} \\ l_0 + d_3 \end{bmatrix}. \tag{1}$$

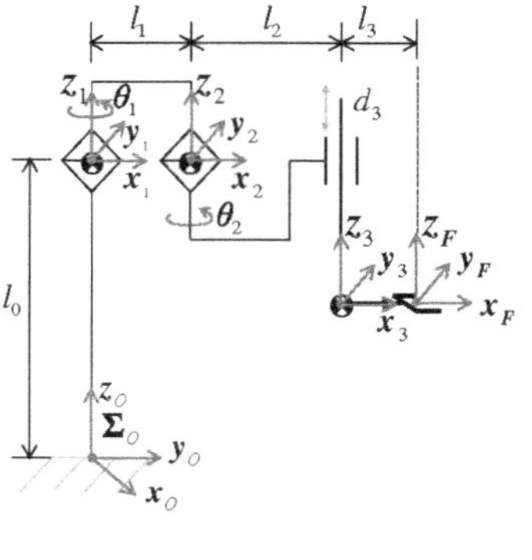

Fig. 4.4. SCARA robot.

Given $\boldsymbol{q} = [\theta_1\ \theta_2\ d_3]^T$, the velocity of the mass center is represented by

$$\hat{\boldsymbol{v}}_1 = \begin{bmatrix} 0 & 0 & 0 \\ 0 & 0 & 0 \\ 0 & 0 & 0 \end{bmatrix} \dot{\boldsymbol{q}}, \quad \hat{\boldsymbol{v}}_2 = \begin{bmatrix} -l_1 S\theta_1 & 0 & 0 \\ l_1 C\theta_1 & 0 & 0 \\ 0 & 0 & 0 \end{bmatrix} \dot{\boldsymbol{q}},$$

$$\hat{\boldsymbol{v}}_3 = \begin{bmatrix} -l_1 S\theta_1 - l_2 S\theta_{12} & -l_2 S\theta_{12} & 0 \\ l_1 C\theta_1 + l_2 C\theta_{12} & l_2 C\theta_{12} & 0 \\ 0 & 0 & 1 \end{bmatrix} \dot{\boldsymbol{q}}. \tag{2}$$

Therefore, kinematic Jacobian relationship to link positions can be represented by

$$\boldsymbol{J}_P^{(1)} = 0, \quad \boldsymbol{J}_P^{(2)} = \begin{bmatrix} -l_1 S\theta_1 & 0 & 0 \\ l_1 C\theta_1 & 0 & 0 \\ 0 & 0 & 0 \end{bmatrix},$$

$$\boldsymbol{J}_P^{(3)} = \begin{bmatrix} -l_1 S\theta_1 - l_2 S\theta_{12} & -l_2 S\theta_{12} & 0 \\ l_1 C\theta_1 + l_2 C\theta_{12} & l_2 C\theta_{12} & 0 \\ 0 & 0 & 1 \end{bmatrix}. \tag{3}$$

From the geometric relationship, the angular velocities of each link are

$$\boldsymbol{\omega}_1 = \begin{bmatrix} 0 & 0 & 0 \\ 0 & 0 & 0 \\ 1 & 0 & 0 \end{bmatrix} \dot{\boldsymbol{q}}, \quad \boldsymbol{\omega}_2 = \begin{bmatrix} 0 & 0 & 0 \\ 0 & 0 & 0 \\ 1 & 1 & 0 \end{bmatrix} \dot{\boldsymbol{q}}, \quad \boldsymbol{\omega}_3 = \begin{bmatrix} 0 & 0 & 0 \\ 0 & 0 & 0 \\ 1 & 1 & 0 \end{bmatrix} \dot{\boldsymbol{q}}, \tag{4}$$

and the kinematic Jacobian relationships to link orientation are represented by

$$\boldsymbol{J}_o^{(1)} = \begin{bmatrix} 0 & 0 & 0 \\ 0 & 0 & 0 \\ 1 & 0 & 0 \end{bmatrix}, \quad \boldsymbol{J}_o^{(2)} = \begin{bmatrix} 0 & 0 & 0 \\ 0 & 0 & 0 \\ 1 & 1 & 0 \end{bmatrix}, \quad \boldsymbol{J}_o^{(3)} = \begin{bmatrix} 0 & 0 & 0 \\ 0 & 0 & 0 \\ 1 & 1 & 0 \end{bmatrix}. \tag{5}$$

Let the (j, k) element of the inertia tensor ${}^i\boldsymbol{I}_i$ be defined as I_{ijk}, and the kinetic parameter vector as

$$\boldsymbol{\sigma} = \begin{bmatrix} I_{1zz} + m_2 l_1^2 \\ I_{2zz} + I_{3zz} \\ m_3 \end{bmatrix}, \tag{6}$$

where ${}^{i}\boldsymbol{I}_i$ is expressed in the i-th link coordinate system. From Eq. (4.37), the inertia matrix $\boldsymbol{M} = (M_{ij})$ is represented by

$$\boldsymbol{M} = \begin{bmatrix} \sigma_1 + \sigma_2 + (l_1^2 + l_2^2 + 2l_1 l_2 C\theta_2)\sigma_3 & \sigma_2 + (l_2^2 + 2l_1 l_2 C\theta_2)\sigma_3 & 0 \\ \sigma_2 + (l_2^2 + 2l_1 l_2 C\theta_2)\sigma_3 & \sigma_2 + l_2^2\sigma_3 & 0 \\ 0 & 0 & \sigma_3 \end{bmatrix}. \tag{7}$$

From Eq. (4.53), the Christoffel symbol of the first type is expressed by

$$c_{112} = c_{121} = c_{122} = c_{211} = -\sigma_3 l_1 S\theta_2, \tag{8}$$

and the other c_{ijk} are zero. From these relations and Eq. (4.52), $\boldsymbol{C}(q, \dot{q})$ can be represented by

$$\boldsymbol{C} = \begin{bmatrix} -l_1 l_2 \sigma_3 S\theta_2 \dot{\theta}_2 & -l_1 l_2 \sigma_3 S\theta_2(\dot{\theta}_1 + \dot{\theta}_2) & 0 \\ l_1 l_2 \sigma_3 S\theta_2 \dot{\theta}_1 & 0 & 0 \\ 0 & 0 & 0 \end{bmatrix}. \tag{9}$$

Given that the gravitational acceleration vector is $\tilde{\boldsymbol{g}} = [0\ 0\ -g]^T$, from the relationship in Eq. (4.48), the gravitational term is represented as

$$\boldsymbol{g} = \begin{bmatrix} 0 \\ 0 \\ -\sigma_3 g \end{bmatrix}. \tag{10}$$

When an external force is not applied, the equation of motion is represented by

$$\boldsymbol{M}(\boldsymbol{q})\ddot{q} + \boldsymbol{C}(\boldsymbol{q}, \dot{\boldsymbol{q}})\dot{\boldsymbol{q}} + \boldsymbol{g}(\boldsymbol{q}) = \boldsymbol{\tau}, \tag{11}$$

where $\boldsymbol{M}(\boldsymbol{q})$, $\boldsymbol{C}(\boldsymbol{q}, \dot{\boldsymbol{q}})$ and $\boldsymbol{g}(\boldsymbol{q})$ are given by Eq. (7), (9) and (10), respectively.

Next, let us verify Proposition 4.3. The time differentiation of the elements of the inertia matrix is

$$\dot{M}_{11} = \dot{M}_{12} = \dot{M}_{21} = -2l_1 l_2 S\theta_2 \sigma_3 \dot{\theta}_2, \tag{12}$$

and the other $\dot{M}_{ij}$ are zero. Therefore,

$$\dot{\boldsymbol{M}}(q) - 2\boldsymbol{C}(\boldsymbol{q}, \dot{\boldsymbol{q}}) = \begin{bmatrix} 0 & l_1 l_2 S\theta_2(\dot{\theta}_1 + \dot{\theta}_2) & 0 \\ -l_1 l_2 S\theta_2(\dot{\theta}_1 + \dot{\theta}_2) & 0 & 0 \\ 0 & 0 & 0 \end{bmatrix}. \tag{13}$$

Obviously, $\dot{\boldsymbol{M}}(q) - 2\boldsymbol{C}(\boldsymbol{q}, \dot{\boldsymbol{q}})$ is a skew-symmetric matrix.

Finally, let us verify Proposition 4.5. The regressor of Eq. (6) is

$$\boldsymbol{Y} = \begin{bmatrix} \ddot{\theta}_1 & \ddot{\theta}_1 + \ddot{\theta}_2 & (l_1^2 + l_2^2 + 2l_1 l_2 C_2)\ddot{\theta}_1 + (l_1 l_2 C\theta_2 + l_2^2)\ddot{\theta}_2 \\ & & -2l_1 l_2 \dot{\theta}_1 \dot{\theta}_2 S\theta_2 - l_1 l_2 \dot{\theta}_2^2 S\theta_2 \\ 0 & \ddot{\theta}_1 + \ddot{\theta}_2 & (l_2^2 + 2l_1 l_2 C\theta_2)\ddot{\theta}_1 + l_2^2 \ddot{\theta}_2 + l_1 l_2 \dot{\theta}_1^2 S\theta_2 \\ 0 & 0 & -g + \ddot{d}_3 \end{bmatrix} \tag{14}$$

Consequently, this shows that the equation of motion is linear with respect to the dynamic parameter vector $\boldsymbol{\sigma}$. All of the column vectors of the regressor are independent, because no column vector can be expressed by the linear combination of the other column vectors. Hence, the dynamic parameters given by Eq. (6) are the base parameters.

4.5. Dynamics Model in Cartesian Space

Let us convert the equation of motion in the joint space into the equation of motion in the **Cartesian space**. Let the relationship between the position and orientation of the end-effector be given by $\boldsymbol{r} \in \mathbb{R}^m$ and the joint variables $\boldsymbol{q} \in \mathbb{R}^n$ by $\boldsymbol{r} = \boldsymbol{f}(\boldsymbol{q})$. Then the time derivative is represented by

$$\dot{\boldsymbol{r}} = \boldsymbol{J}(\boldsymbol{q})\dot{\boldsymbol{q}}, \quad \ddot{\boldsymbol{r}} = \boldsymbol{J}(\boldsymbol{q})\ddot{q} + \dot{\boldsymbol{J}}(\boldsymbol{q})\dot{\boldsymbol{q}}, \tag{4.60}$$

where $\boldsymbol{J}(\boldsymbol{q}) = \partial \boldsymbol{f}/\partial \boldsymbol{q}^T$ is a **Jacobian** matrix. When the Jacobian matrix $\boldsymbol{J}(\boldsymbol{q})$ is non-singular, the following equations are derived

$$\dot{\boldsymbol{q}} = \boldsymbol{J}(\boldsymbol{q})^{-1}\dot{\boldsymbol{r}}, \tag{4.61}$$

$$\ddot{\boldsymbol{q}} = \boldsymbol{J}(\boldsymbol{q})^{-1}(\ddot{\boldsymbol{r}} - \dot{\boldsymbol{J}}(\boldsymbol{q})\boldsymbol{J}^{-1}(\boldsymbol{q})\dot{\boldsymbol{r}}). \tag{4.62}$$

By substituting these equations into Eq. (4.46), it can be represented by

$$\boldsymbol{M}\boldsymbol{J}^{-1}(\ddot{\boldsymbol{r}} - \dot{\boldsymbol{J}}\boldsymbol{J}^{-1}\dot{\boldsymbol{r}}) + \boldsymbol{C}\boldsymbol{J}^{-1}\dot{\boldsymbol{r}} + \boldsymbol{g}(\boldsymbol{q}) = \boldsymbol{\tau}.$$

By multiplying $\boldsymbol{J}^{-T}$ from the left to both sides of this equation, it can be expressed by

$$\boldsymbol{M}_r(\boldsymbol{q})\ddot{\boldsymbol{r}} + \boldsymbol{C}_r(\boldsymbol{q}, \dot{\boldsymbol{q}})\dot{\boldsymbol{r}} + \boldsymbol{g}_r(\boldsymbol{q}) = \boldsymbol{J}(\boldsymbol{q})^{-T}\boldsymbol{\tau}, \tag{4.63}$$

where $\boldsymbol{J}(\boldsymbol{q})^{-T}$ represents the transpose of the inverse matrix of $\boldsymbol{J}(\boldsymbol{q})$, and

$$\begin{aligned} \boldsymbol{M}_r &= \boldsymbol{J}^{-T}\mathbf{M}\mathbf{J}^{-1}, \\ \boldsymbol{C}_r &= \boldsymbol{J}^{-T}(\boldsymbol{C} - \boldsymbol{M}_r\dot{\boldsymbol{J}})\boldsymbol{J}^{-1}, \\ \boldsymbol{g}_r &= \boldsymbol{J}^{-T}\boldsymbol{g}. \end{aligned} \tag{4.64}$$

The dynamics in Cartesian space maintain the basic propositions of the dynamics in the joint space, as shown in Sec. 4.3. However, it is important to note that several variables often become infinite in the vicinity of the singular point of the robot.

When the robot is redundant, that is $n > m$, the inverse of the Jacobian does not exist. In such cases, the joint velocity to achieve the end-effector's velocity is given by

$$\dot{\boldsymbol{q}} = \boldsymbol{J}^{+}\dot{\boldsymbol{r}} + (\boldsymbol{I}_{n\times n} - \boldsymbol{J}^{+}\boldsymbol{J})\boldsymbol{w}, \tag{4.65}$$

where $\boldsymbol{J}^{+}$ is the generalized inverse matrix of $\boldsymbol{J}$ given by

$$\boldsymbol{J}^{+} = \boldsymbol{J}^{T}(\boldsymbol{J}\boldsymbol{J}^{T})^{-1},$$

and $\boldsymbol{w} \in \mathbb{R}^n$ is an arbitrary vector. This means that there are countless solutions, because the joint number is redundant to determine the end-effector's velocity. The vector $\boldsymbol{w}$ is available to use for optimization of performances such as the joint torques or the joint velocity. When $\boldsymbol{w} = 0$, the equation of motion in Cartesian space should be determined by substituting $\boldsymbol{J}^{+}$ instead of $\boldsymbol{J}^{-1}$ in Eq. (4.61)–Eq. (4.64).

Example 4.3. Find the equation of motion in Cartesian space that also satisfies Proposition 4.3.

By substituting $\boldsymbol{M}_r = \boldsymbol{J}^{-T}\boldsymbol{M}\boldsymbol{J}^{-1}$ and $\boldsymbol{C}_r = \boldsymbol{J}^{-T}(\boldsymbol{C}-\boldsymbol{M}_r\dot{\boldsymbol{J}})\boldsymbol{J}^{-1}$ into $\dot{\boldsymbol{M}}_r - 2\boldsymbol{C}_r$, we get

$$\begin{aligned}\dot{\boldsymbol{M}}_r - 2\boldsymbol{C}_r &= \frac{d}{dt}(\boldsymbol{J}^{-T}\boldsymbol{M}\boldsymbol{J}^{-1}) - 2\boldsymbol{J}^{-T}(\boldsymbol{C} - \boldsymbol{M}_r\dot{\boldsymbol{J}})\boldsymbol{J}^{-1}\\ &= \dot{\boldsymbol{J}}^{-T}\boldsymbol{M}\boldsymbol{J}^{-1} + \boldsymbol{J}^{-T}\dot{\boldsymbol{M}}\boldsymbol{J}^{-1} + \boldsymbol{J}^{-T}\boldsymbol{M}\dot{\boldsymbol{J}}^{-1} - 2\boldsymbol{J}^{-T}\boldsymbol{C}\boldsymbol{J}^{-1}\\ &\quad + 2\boldsymbol{J}^{-T}\boldsymbol{M}\dot{\boldsymbol{J}}\boldsymbol{J}^{-1}. \end{aligned} \tag{1}$$

The time differentiation of $\boldsymbol{J}^{-1}\boldsymbol{J} = \boldsymbol{I}$ is $\dot{\boldsymbol{J}}^{-1}\boldsymbol{J} + \boldsymbol{J}^{-1}\dot{\boldsymbol{J}} = 0$. This derives $\dot{\boldsymbol{J}}^{-1} = -\boldsymbol{J}^{-1}\dot{\mathbf{J}}\mathbf{J}^{-1}$. By substituting this into Eq. (1), it eventually becomes,

$$\dot{\boldsymbol{M}}_r - 2\boldsymbol{C}_r = \boldsymbol{J}^{-T}(\dot{\boldsymbol{M}} - 2\boldsymbol{C})\boldsymbol{J}^{-1} + \dot{\boldsymbol{J}}^{-T}\boldsymbol{M}\boldsymbol{J}^{-1} - \boldsymbol{J}^{-T}\boldsymbol{M}\dot{\boldsymbol{J}}^{-1}. \tag{2}$$

A quadratic form of this matrix is represented as

$$\begin{aligned}&\boldsymbol{x}^T(\dot{\boldsymbol{M}}_r - 2\boldsymbol{C}_r + \dot{\boldsymbol{J}}^{-T}\boldsymbol{M}\boldsymbol{J}^{-1} - \boldsymbol{J}^{-T}\boldsymbol{M}\dot{\boldsymbol{J}}^{-1})\boldsymbol{x}\\ &\quad = (\boldsymbol{J}^{-1}\boldsymbol{x})^T(\dot{\boldsymbol{M}} - 2\boldsymbol{C})\boldsymbol{J}^{-1}\boldsymbol{x} = 0.\end{aligned} \tag{3}$$

From these relations, it is shown that $\dot{\boldsymbol{M}}_r - 2\boldsymbol{C}_r$ is a skew-symmetric matrix.

4.6. Dynamics of the Robot with Holonomic Constraints

4.6.1. *Principle of Virtual Work*

Let it be supposed that forces are applied to a rigid system consisting of n links, and so these balance. The balance gives constraints to the displacement of the rigid body. The force that brings the bind is called the **force of constraint**. When the force $\boldsymbol{F}_i$ applied to link i balances the force of constraint $\mathbf{S}_i$, we get

$$\boldsymbol{F}_i + \boldsymbol{S}_i = 0. \tag{4.66}$$

Let us now consider an infinitely small displacement, $\delta\boldsymbol{r}_i$, of link i, which does not break the constraint conditions of the system. This is an arbitrary displacement that is considered virtually in satisfying constraint conditions and is called the **virtual displacement**. When a virtual displacement is generated, the work, δW, that is performed by all forces including the force of the constraints is called the **virtual work**, which is expressed by

$$\delta W = \sum_{i=1}^{n} (\boldsymbol{F}_i + \boldsymbol{S}_i)^T \delta \boldsymbol{r}_i.$$

From Eq. (4.66), it is obviously

$$\delta W = \sum_{i=1}^{n} (\boldsymbol{F}_i + \boldsymbol{S}_i)^T \delta \boldsymbol{r}_i = 0. \tag{4.67}$$

On the other hand, if Eq. (4.67) is satisfied with an arbitrary $\delta \boldsymbol{r}_i$, the system is in a state of balance. This is called the **principle of virtual work**. Here, consider a case in which the work by the forces of constraint is zero for any $\delta \boldsymbol{r}_i$; that is, the energy of the system is neither released nor injected by the force of constraint. Such constraint is said to be **smooth**. When the constraint is smooth, the principle of virtual work is expressed as

$$\delta W = \sum_{i=1}^{n} \boldsymbol{F}_i^T \delta \boldsymbol{r}_i = 0, \tag{4.68}$$

and the forces of constraint do not appear explicitly.

4.6.2. *D'Alembert's Principles*

An equation of motion of the link mechanism is shown in Eq. (4.29) and/or Eq. (4.46). These equations can be re-written as follows:

$$\tau_i - \frac{d}{dt}\left(\frac{\partial L}{\partial \dot{q}_i}\right) + \frac{\partial L}{\partial q_i} = 0, \quad i = 1, \ldots, n \tag{4.69-1}$$

$$\boldsymbol{\tau} - \boldsymbol{M}(\boldsymbol{q})\ddot{\boldsymbol{q}} - \boldsymbol{C}(\boldsymbol{q}, \dot{\boldsymbol{q}})\dot{\boldsymbol{q}} - \boldsymbol{g}(\boldsymbol{q}) = 0. \tag{4.69-2}$$

These indicate that all forces, including the inertia force, Coriolis force, centrifugal force, the sum forces of gravitational terms, and the joint torque, are in balance, and the problem of dynamics can be considered one of statics. This is called **D'Alembert's principle**. When the principle of virtual work is applied to the issues of balance based on D'Alembert's principle, we obtain

$$\sum_{i=1}^{n} \left(\tau_i - \frac{d}{dt}\left(\frac{\partial L}{\partial \dot{q}_i}\right) + \frac{\partial L}{\partial q_i}\right) \delta q_i = 0. \tag{4.70}$$

This can then be expressed by a vector as follows:

$$\left(\boldsymbol{\tau} - \frac{d}{dt}\left(\frac{\partial L}{\partial \dot{\boldsymbol{q}}}\right) + \frac{\partial L}{\partial \boldsymbol{q}}\right)^T \delta \boldsymbol{q} = 0, \tag{4.71}$$

where $\delta \boldsymbol{q} = [\delta q_1, \ldots, \delta q_n]^T$.

4.6.3. *Undetermined Multipliers and the Lagrange's Equation*

Let's consider the dynamics model in a case where the robotic hand is constrained by the environment. Given that the fingertips are subjected to constraints, the k holonomic constraints can be expressed by

$$\varphi_j(q_1, \ldots, q_n) = 0, \quad j = 1, 2, \ldots, k. \tag{4.72}$$

A condition such as that described above is called the **constraint condition**. Assume that these constraints are smooth and differentiable by $q_i (i = 1, \ldots, n)$. Now, let's consider the virtual displacement δq_i of the displacement q_i. Because this virtual displacement should not break the constraints of Eq. (4.72), variation $\delta \varphi_j$ of φ_j must be zero when q_i is varied to $q_i + \delta q_i$. Therefore, according to the formula of total differential, it becomes

$$\delta \varphi_j = \sum_{i=1}^{n} \frac{\partial \varphi_j}{\partial q_i} \delta q_i = 0. \tag{4.73}$$

Since the outcome is zero even if both sides of Eq. (4.73) are multiplied by the arbitrary multiplier λ_j, by adding this to Eq. (4.70), we get

$$\sum_{i=1}^{n} \left(\tau_i - \frac{d}{dt}\left(\frac{\partial L}{\partial \dot{q}_i}\right) + \frac{\partial L}{\partial q_i} + \sum_{j=1}^{k} \lambda_j \frac{\partial \varphi_j}{\partial q_i} \right) \delta q_i = 0. \tag{4.74}$$

Since δq_i must satisfy Eq. (4.73), the number of elements that can be independently chosen from n elements is $n - k$,and the rest of the k elements can be expressed using the first $n - k$ elements by Eq. (4.73). Hence, we renumber joint variables properly such that k

elements that are not independent are selected as $\delta q_1, \ldots, \delta q_k$, and an undetermined λ_j is selected, which satisfies the following equation:

$$\tau_i - \frac{d}{dt}\left(\frac{\partial L}{\partial \dot{q}_i}\right) + \frac{\partial L}{\partial q_i} + \sum_{j=1}^{l} \lambda_j \frac{\partial \varphi_j}{\partial q_i} = 0, \quad i = 1, 2, \ldots, k. \tag{4.75}$$

In this case, Eq. (4.74) becomes

$$\sum_{i=k+1}^{n} \left(\tau_i - \frac{d}{dt}\left(\frac{\partial L}{\partial \dot{q}_i}\right) + \frac{\partial L}{\partial q_i} + \sum_{j=1}^{l} \lambda_j \frac{\partial \varphi_j}{\partial q_i}\right) \delta q_i = 0. \tag{4.76}$$

It should be noted that the sum starts from $k+1$. Also, since $\delta q_i (i = k+1, l+2, \ldots, n)$ can be chosen independently, the parentheses must always be zero in order to satisfy Eq. (4.76).

$$\tau_i - \frac{d}{dt}\left(\frac{\partial L}{\partial \dot{q}_i}\right) + \frac{\partial L}{\partial q_i} + \sum_{j=1}^{l} \lambda_j \frac{\partial \varphi_j}{\partial q_i} = 0, \quad i = k+1, l+2, \ldots, n. \tag{4.77}$$

The combination of Eq. (4.75) and Eq. (4.77) is given by

$$\frac{d}{dt}\left(\frac{\partial L}{\partial \dot{q}_i}\right) - \frac{\partial L}{\partial q_i} = \tau_i + \sum_{j=1}^{l} \lambda_j \frac{\partial \varphi_j}{\partial q_i} = 0, \quad i = 1, 2, \ldots, n. \tag{4.78}$$

By using notations of $\boldsymbol{\lambda} = [\lambda_1, \ldots, \lambda_k]^T$ and $\boldsymbol{\varphi} = [\varphi_1, \ldots, \varphi_k]^T$, the above equations are expressed by a vector form:

$$\frac{d}{dt}\left(\frac{\partial L}{\partial \dot{\boldsymbol{q}}}\right) - \frac{\partial L}{\partial \boldsymbol{q}} = \boldsymbol{\tau} + \boldsymbol{A}(\boldsymbol{q})^T \boldsymbol{\lambda}, \tag{4.79}$$

where

$$\boldsymbol{A}(\boldsymbol{q}) = \frac{\partial \boldsymbol{\varphi}}{\partial \boldsymbol{q}} \in \mathbb{R}^{k \times n}, \tag{4.80}$$

and $\boldsymbol{\lambda}$ is called the **Lagrange's undetermined multiplier**. The second term on the right-hand side of Eq. (4.79),

$$\boldsymbol{\tau}_C = \boldsymbol{A}^T \boldsymbol{\lambda}, \tag{4.81}$$

is the force of constraint that generates the constraint of Eq (4.72). Eq. (4.79) is the equation of motion under the holonomic constraints given by Eq. (4.72). The time derivative of Eq. (4.72) is expressed by

$$\boldsymbol{A}(\boldsymbol{q})\dot{\boldsymbol{q}} = 0. \tag{4.82}$$

It can thus be seen that the equation of motion is limited in the following $\mathbb{R}^{2n}$ -dimensional constraint manifold S.

$$S = \{(\boldsymbol{q}, \dot{\boldsymbol{q}}) : \varphi(\boldsymbol{q}) = 0, \boldsymbol{A}(\boldsymbol{q})\dot{\boldsymbol{q}} = 0\}. \tag{4.83}$$

Furthermore, by multiplying $\boldsymbol{\lambda}^T$ to Eq. (4.82) from the left, it becomes

$$\boldsymbol{\tau}_C^T \dot{\boldsymbol{q}} = 0. \tag{4.84}$$

This equation indicates that the force of constraint and the velocity in the joint space is orthogonal.

The dynamic equation of the n-link mechanism is derived in the same way as described in Sec. 4.3, and is represented by

$$\boldsymbol{M}(\boldsymbol{q})\ddot{q} + \boldsymbol{C}(\boldsymbol{q}, \dot{q})\dot{q} + \boldsymbol{g}(\boldsymbol{q}) = \boldsymbol{\tau} + \boldsymbol{A}(\boldsymbol{q})^T \boldsymbol{\lambda}. \tag{4.85}$$

Let's find the force of constraint under such conditions. By differentiating Eq. (4.82),

$$\boldsymbol{A}(\boldsymbol{q})\ddot{q} + \dot{\boldsymbol{A}}(\boldsymbol{q})\dot{q} = 0.$$

By deriving $\ddot{\boldsymbol{q}}$ from Eq. (4.85) and substituting it into the equation above, it becomes

$$\boldsymbol{A}\boldsymbol{M}^{-1}\boldsymbol{A}^T\boldsymbol{\lambda} = \boldsymbol{A}\boldsymbol{M}^{-1}(\boldsymbol{C}\dot{\boldsymbol{q}} + \boldsymbol{g} - \boldsymbol{\tau}) - \dot{\boldsymbol{A}}(\boldsymbol{q})\dot{\boldsymbol{q}}.$$

When the constraint is independent, $\boldsymbol{A}\boldsymbol{M}^{-1}\boldsymbol{A}^T$ is non-singular. Therefore,

$$\boldsymbol{\lambda} = (\boldsymbol{A}\boldsymbol{M}^{-1}\boldsymbol{A}^T)^{-1}(\boldsymbol{A}\boldsymbol{M}^{-1}(\boldsymbol{C}\dot{\boldsymbol{q}} + \boldsymbol{g} - \boldsymbol{\tau}) - \dot{\boldsymbol{A}}(\boldsymbol{q})\dot{\boldsymbol{q}}). \tag{4.86}$$

This equation shows that force of constraint can be calculated when the state vectors $\boldsymbol{q}$ and $\dot{\boldsymbol{q}}$ are given, and the equation of motion of the system is provided by Eq. (4.85) using $\boldsymbol{\lambda}$ given by Eq. (4.86).

Finally, by assuming that the variables $\boldsymbol{q}$ and $\boldsymbol{\lambda}$ are $n+k$ elements of generalized coordinates, and the Lagrangian function for these generalized coordinates is defined by

$$L' = L + \boldsymbol{\lambda}^T \boldsymbol{\varphi}, \tag{4.87}$$

then, Eq. (4.79) and Eq. (4.72) are rewritten as

$$\frac{d}{dt}\left(\frac{\partial L'}{\partial \dot{\boldsymbol{q}}}\right) - \frac{\partial L'}{\partial \boldsymbol{q}} = \boldsymbol{\tau}, \tag{4.88}$$

$$\frac{d}{dt}\left(\frac{\partial L'}{\partial \dot{\lambda}}\right) - \frac{\partial L'}{\partial \boldsymbol{\lambda}} = 0, \tag{4.89}$$

where L' is an extended Lagrangian function rather than a simple difference between the kinetic energy and the potential energy. By introducing the extended Lagrangian function, Lagrange's equations, in which the undetermined multipliers are all considered as generalized coordinates, can be derived.

Example 4.4. Let's find the equation of motion when the position of end-effector of the SCARA robot in Example 4.2 is constrained at $y = l$, as shown in Fig. 4.5.

The constraint condition of the end-effector is expressed by

$$\varphi(q) = l_1 S\theta_1 + (l_2 + l_3) S\theta_{12} - l = 0. \tag{1}$$

Its time derivative is represented by

$$\boldsymbol{A}\dot{\boldsymbol{q}} = 0, \tag{2}$$

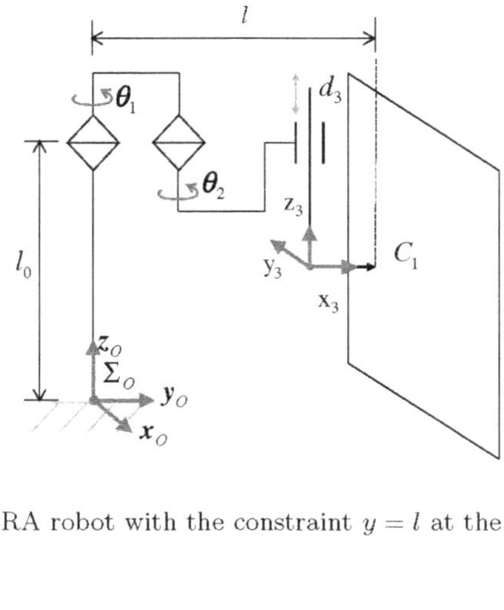

Fig. 4.5. SCARA robot with the constraint $y = l$ at the hand position.

where

$$A = \begin{bmatrix} l_1 C\theta_1 + (l_2 + l_3) C\theta_{12} & (l_2 + l_3) C\theta_{12} & 0 \end{bmatrix}. \tag{3}$$

Therefore, the equation of motion is

$$M(q)\ddot{q} + C(q, \dot{q})\dot{q} + g(q) = \tau + A(q)^T \lambda,$$

where $M(q)$, $C(q, \dot{q})$, and $g(q)$ are given by Eq. (7), (9), and (10) of Example 4.2.

4.6.4. *Low-Dimension Model*

When the k constraints are given, the DOF of the hand is $n - k$. So, the joint vector is appropriately divided into two parts: an independent variable and a dependent variable;

$$q = \begin{bmatrix} q_1 \\ q_2 \end{bmatrix}, \tag{4.90}$$

where $q_1 \in \mathbb{R}^{n-k}$ is a independent variable and $q_2 \in \mathbb{R}^k$ is a dependent variable. The time derivative of the constraint condition derives

$$\frac{d\varphi(q)}{dt} = \frac{\partial \varphi(q)}{\partial q_1}\dot{q}_1 + \frac{\partial \varphi(q)}{\partial q_2}\dot{q}_2 = 0.$$

From these relations, if $\partial\varphi(q)/\partial q_2 \in \mathbb{R}^{k \times k}$ is non-singular, we obtain

$$\dot{q} = L(q)\dot{q}_1, \tag{4.91}$$

where

$$L(q) = \begin{bmatrix} I_{(n-k)\times(n-k)} \\ -\left(\dfrac{\partial\varphi(q)}{\partial q_2}\right)^{-1} \dfrac{\partial\varphi(q)}{\partial q_1} \end{bmatrix} \in \mathbb{R}^{n \times (n-k)}. \tag{4.92}$$

Equation (4.91) shows that the dependent variable is determined when the independent variable is determined. So, it is assumed that there exists a relationship such that the dependent variable is a function of the independent variable as given by

$$q_2 = \Omega(q_1). \tag{4.93}$$

On the other hand, from Eq. (4.84) and (4.91), we obtain $\boldsymbol{\tau}_c^T \boldsymbol{L}(\boldsymbol{q})$ $\dot{\boldsymbol{q}}_1 = 0$. Since this is true for any $\dot{\boldsymbol{q}}_1$, the following is derived:

$$\boldsymbol{\tau}_c^T \boldsymbol{L}(\boldsymbol{q}) = 0.$$

Also, the time derivative of Eq. (4.91) is expressed by

$$\ddot{q} = \boldsymbol{L}(\boldsymbol{q})\ddot{\boldsymbol{q}}_1 + \dot{\mathbf{L}}(\boldsymbol{q})\dot{\boldsymbol{q}}_1. \tag{4.94}$$

By substituting these into Eq. (4.85), and multiplying $\boldsymbol{L}(\boldsymbol{q})^T$ from the left-hand side of the equation, it becomes

$$\boldsymbol{M}_1(\boldsymbol{q}_1)\ddot{\boldsymbol{q}}_1 + \boldsymbol{C}_1(\boldsymbol{q}_1, \dot{\boldsymbol{q}}_1)\dot{\boldsymbol{q}}_1 + \boldsymbol{g}_1(\boldsymbol{q}_1) = \boldsymbol{\tau}_1, \tag{4.95}$$

where $\boldsymbol{M}_1 = \boldsymbol{L}^T \boldsymbol{M}\boldsymbol{L}$, $\boldsymbol{C}_1 = \boldsymbol{L}^T(\boldsymbol{M}\dot{\boldsymbol{L}} + \boldsymbol{C}\boldsymbol{L})$, $\boldsymbol{g}_1 = \boldsymbol{L}^T \boldsymbol{g}$, and $\boldsymbol{\tau}_1 = \boldsymbol{L}^T \boldsymbol{\tau}$.

The vectors and matrices above can be rewritten only with variable $\boldsymbol{q}_1$ using Eq. (4.91). Eq. (4.95) is the equation of motion by the independent variables in which force of constraints does not appear explicitly, and it is a low-dimension dynamic model of the robot with the constraints of the end-effector. When the joint torque is given, the motion of $\boldsymbol{q}_1$ is calculated from the motion of equation given by Eq. (4.95) and the motion of $\boldsymbol{q}_2$ by Eq. (4.93).

Example 4.5. Find a low-dimensional dynamic model of the SCARA robot with the constraint as shown in Example 4.4.

The constraint of the end-effector is expressed by

$$\phi(q) = l_1 S\theta_1 + (l_2 + l_3)S\theta_{12} - l = 0. \tag{1}$$

Given $q_1 = (\theta_1, d_3)^{\mathrm{T}}$ as the independent variable, and $q_2 = \theta_2$ as the dependent variable, then

$$\frac{\partial\varphi(q)}{\partial q_1} = \begin{bmatrix} l_1 C\theta_1 + (l_2 + l_3)C\theta_{12} & 0 \end{bmatrix},$$

$$\frac{\partial\varphi(q)}{\partial q_2} = (l_2 + l_3)C\theta_{12}.$$

Therefore, from Eq. (4.92),

$$\boldsymbol{L}(q) = \begin{bmatrix} 1 & 0 \\ 0 & 1 \\ h & 0 \end{bmatrix}, \tag{2}$$

where

$$h = -\frac{l_1 C\theta_1}{(l_2 + l_3)C\theta_{12}} - 1. \tag{3}$$

The equation of motion without constraint can be seen in Example 4.2. Giving the inertia matrix of the robot without constraint as $\boldsymbol{M} = (M_{ij})$, the i-th element of the joint torque $\boldsymbol{\tau}$ as τ_i, and finding $\boldsymbol{M}_1 = \boldsymbol{L}^T \boldsymbol{M}\boldsymbol{L}$, $\boldsymbol{C}_1 = \boldsymbol{L}^T(\boldsymbol{M}\dot{\boldsymbol{L}} + \boldsymbol{C}\boldsymbol{L})$, and $\boldsymbol{g}_1 = \boldsymbol{L}^T \boldsymbol{g}$, the following are derived:

$$\boldsymbol{M}_1 = \begin{bmatrix} M_{11} + h^2 M_{33} & M_{12} \\ M_{21} & M_{22} \end{bmatrix},$$

$$\boldsymbol{C}_1 = \begin{bmatrix} M_{11} + C_{11} + M_{33}\dot{h} & M_{12} + C_{12} \\ M_{21} + C_{21} & M_{22} + C_{22} \end{bmatrix},$$

$$\boldsymbol{g}_1 = \begin{bmatrix} -m_3 \tilde{g} h \\ 0 \end{bmatrix}, \quad \boldsymbol{\tau}_1 = \begin{bmatrix} \tau_1 + h\tau_3 \\ \tau_2 \end{bmatrix}, \tag{4}$$

where

$$\dot{h} = \frac{l_1 \dot{\theta}_1 S\theta_1}{(l_2 + l_3)C\theta_{12}} - \frac{l_1(\dot{\theta}_1 + \dot{\theta}_2)C\theta_1 S\theta_{12}}{(l_2 + l_3)C\theta_{12}^2}. \tag{5}$$

Using these relations, the low-dimensional dynamic model is expressed by

$$\boldsymbol{M}_1(\boldsymbol{q}_1)\ddot{\boldsymbol{q}}_1 + \boldsymbol{C}_1(\boldsymbol{q}_1, \dot{\boldsymbol{q}}_1)\dot{\boldsymbol{q}}_1 + \boldsymbol{g}_1(\boldsymbol{q}_1) = \boldsymbol{\tau}_1. \tag{6}$$

4.7. Robot Dynamics with Non-Holonomic Constraints

Let us consider robot dynamics with non-holonomic constraints, where the Pfaffian constraint is non-integrable, as seen in Sec. 3.2 of Chapter 3. It is assumed that the velocity constraint is given by

$$\boldsymbol{A}(\boldsymbol{q})\dot{\boldsymbol{q}} = 0, \tag{4.96}$$

where $\boldsymbol{A} \in \mathbb{R}^{k \times n}$. At this time, the constraint force $\boldsymbol{\tau}_C$ is added to the equation of motion. When the constraint is smooth, the virtual work is zero along the virtual displacement $\delta\boldsymbol{q}$, and therefore $\boldsymbol{\tau}_C^T \delta\boldsymbol{q} = 0$. On the other hand, according to the definition of virtual displacement,

Eq. (4.96) must be satisfied in the case of $d\boldsymbol{q} \to \delta\boldsymbol{q}$, $dt \to 0$, this leads to

$$\boldsymbol{A}(\boldsymbol{q})\delta\boldsymbol{q} = 0. \tag{4.97}$$

Consequently, the constraint force is rewritten as

$$\boldsymbol{\tau}_C = \boldsymbol{A}(\boldsymbol{q})^T\boldsymbol{\lambda}, \tag{4.98}$$

where $\boldsymbol{\lambda} \in \mathbb{R}^k$ is the undetermined multiplier. By adding $\boldsymbol{\tau}_C^T\delta\boldsymbol{q} = 0$ to Eq. (4.71), the same equation as in Eq. (4.79) or Eq. (4.85) is derived. This indicates that the equation of motion with non-holonomic constraints can be expressed in the same way as the equation of motion with holonomic constraints, although the Lagrange's equation using the expanded Lagrangian function expressed by Eq. (4.88) and Eq. (4.89) cannot be derived.

4.8. Robot Hand Dynamics

In order to achieve dexterous hand manipulation to perform coordinated control, it is essential that a multi-fingered robotic hand be able to grasp and manipulate the target object. In this section, we consider the dynamics model of a robotic hand consisting of k fingers, as shown in Fig. 4.6. In the figure, each symbol or equation represents the following: Σ_R is the reference coordinate system, Σ_O is the object coordinate system that is set to the object, Σ_i is the i-th fingertip coordinate system to be set in the fingertip link of the i-th finger, $\boldsymbol{p}_O \in \mathbb{R}^3$ is the position vector of the origin of the object coordinate system, $\boldsymbol{p}_i \in \mathbb{R}^3$ is the position vector of the origin of the i-th fingertips coordinate system, $\boldsymbol{p}_{OC_i} \in \mathbb{R}^3$ is the position vector from the origin of the object coordinate system to the contact point of the i-th finger, and $\boldsymbol{p}_{FC_i} \in \mathbb{R}^3$ is the position vector from the origin of the i-th fingertip coordinate system to the contact point of the i-th finger. Moreover, $\boldsymbol{\eta}_O \in \mathbb{R}^3$ is the orientation of the object coordinate system, $\boldsymbol{\eta}_i \in \mathbb{R}^3$ is the orientation of the i-th fingertip coordinate system, $\boldsymbol{r}_O = [\boldsymbol{p}_O^T, \boldsymbol{\eta}_O^T]^T$ is the position and orientation of the object, $\boldsymbol{v}_O = [\dot{\boldsymbol{p}}_O^T, \boldsymbol{\omega}_O^T]^T$ is the object velocity, $\boldsymbol{r}_i = [\boldsymbol{p}_i^T, \boldsymbol{\eta}_i^T]^T$ is the position and orientation of the i-th fingertip, and $\boldsymbol{v}_i = [\dot{\boldsymbol{p}}_i^T, \boldsymbol{\omega}_i^T]^T$

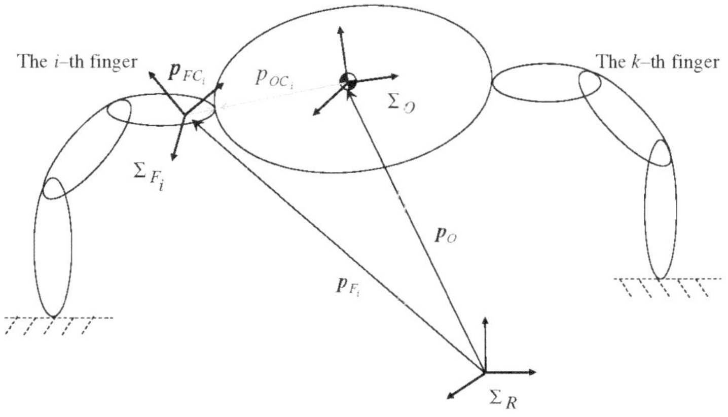

Fig. 4.6. Manipulation by multiple fingers.

is the velocity of the i-th fingertip. In addition, all these vectors are expressed with respect to the reference coordinate system.

The equation of motion of the entire system consisting of the hand and object can be obtained by combining each of those of the hand and the object under the constraint of grasp. In general, the equation of motion for the object is expressed as shown in Eq. (4.11).

$$\boldsymbol{M}_O(\boldsymbol{r}_O)\dot{\boldsymbol{v}}_O + \boldsymbol{C}_O(\boldsymbol{r}_O, \dot{\boldsymbol{r}}_O)\boldsymbol{v}_O + \boldsymbol{g}_O(\boldsymbol{r}_O) = \boldsymbol{F}_O \tag{4.99}$$

It is noted here that $\boldsymbol{M}_O \in \mathbb{R}^{6\times6}$ is a symmetric and positive inertia matrix, $\boldsymbol{C}_O\boldsymbol{v}_O \in \mathbb{R}^6$ is a quadratic velocity term including the Coriolis force and centrifugal force, $\boldsymbol{g}_O \in \mathbb{R}^6$ is a gravitational term, and $\boldsymbol{F}_O \in \mathbb{R}^6$ is an external force applied to the object. The fundamental propositions are explained in Sec. 4.4. They can be listed as follows:

Proposition 4.6.

(1) *$\boldsymbol{M}_O$ is a symmetric and positive matrix.*
(2) *By the appropriate definition of $\boldsymbol{C}_O, \dot{\boldsymbol{M}}_O - 2\boldsymbol{C}_O$ is a skewed symmetric matrix.*
(3) *The object dynamics model is linear with respect to dynamic parameters that are chosen appropriately. That is,*

$$\begin{aligned}\boldsymbol{M}_O(\boldsymbol{r}_O)\dot{\boldsymbol{v}}_{Or} + \boldsymbol{C}_O(\boldsymbol{r}_O, \dot{\boldsymbol{r}}_O)\boldsymbol{v}_{Or} + \boldsymbol{g}_O(\boldsymbol{r}_O)\\ = \boldsymbol{Y}_O(\boldsymbol{r}_O, \dot{\boldsymbol{r}}_O, \boldsymbol{v}_{Or}, \dot{\boldsymbol{v}}_{Or})\boldsymbol{\sigma}_O,\end{aligned} \tag{4.100}$$

where $\boldsymbol{v}_{Or} \in \mathbb{R}^6$ is the arbitrary defined object velocity, $\boldsymbol{\sigma}_O \in \mathbb{R}^p$ is the dynamic parameter of the object, and $\boldsymbol{Y}_O(\boldsymbol{r}_O, \dot{\boldsymbol{r}}_O, \boldsymbol{v}_{Or}, \dot{\boldsymbol{v}}_{Or}) \in \mathbb{R}^{6\times p}$ is the regressor related to $\boldsymbol{\sigma}_O$.

According to Lagrange's equation, the equation of motion of the i-th finger without constraint is written as

$$\boldsymbol{M}_{F_i}\ddot{\boldsymbol{q}}_{F_i} + \boldsymbol{C}_{F_i}(\boldsymbol{q}_{F_i}, \dot{\boldsymbol{q}}_{F_i})\dot{\boldsymbol{q}}_{F_i} + \boldsymbol{g}_{F_i}(\boldsymbol{q}_{F_i}) = \boldsymbol{\tau}_{F_i} \quad (i = 1, \ldots, k), \tag{4.101}$$

where $\boldsymbol{q}_{F_i}$ is the joint displacement of i-th finger and $\boldsymbol{\tau}_{F_i}$ is the joint torque of i-th finger. In combining all fingers together, we obtain

$$\boldsymbol{M}_F\ddot{\boldsymbol{q}}_F + \boldsymbol{C}_F\dot{\boldsymbol{q}}_F + \boldsymbol{g}_F = \boldsymbol{\tau}_F, \tag{4.102}$$

where $\boldsymbol{q}_F = [\boldsymbol{q}_{F_1}^T, \ldots, \boldsymbol{q}_{F_k}^T]^T, \boldsymbol{M}_F = block\ diag[\boldsymbol{M}_{F_1}, \ldots, \boldsymbol{M}_{F_k}]$, $\boldsymbol{C}_F = block\ diag[\boldsymbol{C}_{F_1}, \ldots, \boldsymbol{C}_{F_k}]$, $\boldsymbol{g}_F = [\boldsymbol{g}_{F_i}^T, \ldots, \boldsymbol{g}_{F_k}^T]^T$, and $\boldsymbol{\tau}_F = [\boldsymbol{\tau}_{F_i}^T, \cdots, \boldsymbol{\tau}_{F_k}^T]^T$. The dynamic model also has fundamental propositions that are explained in Sec. 4.4.

Proposition 4.7.

(1) *$\boldsymbol{M}_F$ is a symmetric and positive matrix.*
(2) *By the appropriate definition of $\boldsymbol{C}_{F_i}$ for all $i, \dot{\boldsymbol{M}}_F - 2\boldsymbol{C}_F$ is a skewed symmetric matrix.*
(3) *The hand dynamic model is linear with respect to dynamic parameter vector $\boldsymbol{\sigma}_F$ when chosen appropriately. That is,*

$$\boldsymbol{M}_F\ddot{\boldsymbol{q}}_{Fr} + \boldsymbol{C}_F\dot{\boldsymbol{q}}_{Fr} + \boldsymbol{g}_F = \boldsymbol{Y}_F(\boldsymbol{q}_F, \dot{\boldsymbol{q}}_F, \dot{\boldsymbol{q}}_{Fr}, \ddot{\boldsymbol{q}}_{Fr})\boldsymbol{\sigma}_F, \tag{4.103}$$

where $\dot{\boldsymbol{q}}_{Fr} \in \mathbb{R}^6$ is the arbitrary defined finger velocity and $\boldsymbol{Y}_F(\boldsymbol{q}_F, \dot{\boldsymbol{q}}_F, \dot{\boldsymbol{q}}_{Fr}, \ddot{\boldsymbol{q}}_{Fr})$ is the regressor related to $\boldsymbol{\sigma}_F$.

Next, let us consider the constraint of grasp. The constraint of grasp is provided by Eq. (2.45) in Sec. 2.5.1 of Chapter 2;

$$\boldsymbol{G}_C^T\boldsymbol{v}_O = \boldsymbol{J}_C\dot{\boldsymbol{q}}_F. \tag{4.104}$$

By defining $\dot{\boldsymbol{q}} = [\boldsymbol{v}_O^T, \dot{\boldsymbol{q}}_F^T]^T$, the above equation can be represented by

$$\boldsymbol{A}(\boldsymbol{q})\dot{\boldsymbol{q}} = 0, \tag{4.105}$$

where

$$\boldsymbol{A} = [\boldsymbol{G}_C^T, -\boldsymbol{J}_C]. \tag{4.106}$$

Equation (4.105) shows the constraint of grasp. Hereafter, the constraint conditions of grasp satisfy following conditions:

Assumptions of Grasp Conditions

A1. The grasp is a force closure and manipulable grasp. That is,

(i) $rank\boldsymbol{G}_C = 6$.

(ii) $\boldsymbol{G}_C\boldsymbol{f}_{FC} = \boldsymbol{F}_O$ has a solution of $\boldsymbol{f}_{FC} > 0$.

(iii) $Im(\boldsymbol{G}_C^T) \subseteq Im(\boldsymbol{J}_C)$.

A2. The contact force is in the friction cone at all contact points. This indicates that there is no slippage at the contact point.

A3. $\boldsymbol{J}_C$ is a non-singular matrix. That is, the DOF of the finger is not redundant, and there is sufficient DOF to manipulate the object.

A4. The constraint is smooth. In other words, the work is not generated by the constraint.

Under these assumptions, let us find the dynamics model of the entire system using the Lagrangian method. The Lagrangian function of the entire system is expressed by

$$L = \frac{1}{2}\dot{\boldsymbol{q}}_F^T\boldsymbol{M}_F\dot{\boldsymbol{q}}_F - P_F(\boldsymbol{q}_F) + \frac{1}{2}\boldsymbol{v}_C^T\boldsymbol{M}_O\boldsymbol{v}_O - P_O(\boldsymbol{r}_O), \qquad (4.107)$$

where $P_F(\boldsymbol{q}_F)$ and $P_O(\boldsymbol{r}_O)$ represent the potential of the finger and the object, respectively. This is given by the relationship of $\boldsymbol{v}_O = d\boldsymbol{r}_O/dt$, and the generalized coordinates of $\boldsymbol{q} = [\boldsymbol{r}_O^T, \boldsymbol{q}_F^T]^T$. In the same manner as in Eq. (4.85), the Lagrange's equation with the constraint given by Eq. (4.105) is

$$\frac{d}{dt}\left(\frac{\partial L}{\partial \dot{\boldsymbol{q}}}\right) + \frac{\partial L}{\partial \boldsymbol{q}} + \boldsymbol{A}(\boldsymbol{q})^T\boldsymbol{\lambda} - \begin{bmatrix} 0 \\ \boldsymbol{\tau}_F \end{bmatrix} = 0, \qquad (4.108)$$

where $\boldsymbol{\lambda}$ is an undetermined multiplier. This equation can be expanded as follows:

$$\begin{bmatrix} \dfrac{d}{dt}\left(\dfrac{\partial L}{\partial \dot{\boldsymbol{r}}_O}\right) + \dfrac{\partial L}{\partial \boldsymbol{r}_O} + \boldsymbol{G}_C\boldsymbol{\lambda} \\ \dfrac{d}{dt}\left(\dfrac{\partial L}{\partial \dot{\boldsymbol{q}}_F}\right) + \dfrac{\partial L}{\partial \boldsymbol{q}_F} - \boldsymbol{J}_C^T\boldsymbol{\lambda} - \boldsymbol{\tau}_F \end{bmatrix} = 0.$$

By substituting the Lagrangian function into this formula, these are represented by

$$\boldsymbol{M}_O(\boldsymbol{r}_O)\dot{\boldsymbol{v}}_O + \boldsymbol{C}_O(\boldsymbol{r}_O, \dot{\boldsymbol{r}}_O)\boldsymbol{v}_O + \boldsymbol{g}_O(\boldsymbol{r}_O) = \boldsymbol{G}_C\boldsymbol{\lambda}, \tag{4.109}$$

$$\boldsymbol{M}_F\ddot{\boldsymbol{q}}_F + \boldsymbol{C}_F(\boldsymbol{q}_F, \dot{\boldsymbol{q}}_F)\dot{\boldsymbol{q}}_F + \boldsymbol{g}_F(\boldsymbol{q}_F) = \boldsymbol{\tau}_F - \boldsymbol{J}_C^T\boldsymbol{\lambda}. \tag{4.110}$$

Equation (4.109) is an equation of motion of the target object. Meanwhile, Eq. (4.110) is the equation of motion of the finger. At this time, the external force applied to the target object is given as

$$\boldsymbol{F}_O = \boldsymbol{G}_C\boldsymbol{\lambda}. \tag{4.111}$$

According to the assumption, $\boldsymbol{J}_C$ is non-singular, and therefore the undetermined multiplier is represented as

$$\boldsymbol{\lambda} = \boldsymbol{J}_C^{-T}(\boldsymbol{\tau}_F - \boldsymbol{M}_F\ddot{\boldsymbol{q}}_F - \boldsymbol{C}_F(\boldsymbol{q}_F, \dot{\boldsymbol{q}}_F)\dot{\boldsymbol{q}}_F - \boldsymbol{g}_F(\boldsymbol{q}_F)). \tag{4.112}$$

By substituting this relation into Eq. (4.109), we get

$$\begin{aligned}&\boldsymbol{M}_O(\boldsymbol{r}_O)\dot{\boldsymbol{v}}_O + \boldsymbol{C}_O(\boldsymbol{r}_O, \dot{\boldsymbol{r}}_O)\boldsymbol{v}_O + \boldsymbol{g}_O(\boldsymbol{r}_O)\\ &\quad + \boldsymbol{G}_C\boldsymbol{J}_C^{-T}(\boldsymbol{M}_F\ddot{\boldsymbol{q}}_F + \boldsymbol{C}_F(\boldsymbol{q}_F, \dot{\boldsymbol{q}}_F)\dot{\boldsymbol{q}}_F + \boldsymbol{g}_F(\boldsymbol{q}_F)) = \boldsymbol{G}_C\boldsymbol{J}_C^{-T}.\end{aligned} \tag{4.113}$$

Equation (4.113) is a dynamics model in which the constraints are not represented explicitly, but there are the constraints of Eq. (4.105). Then, by substituting $\dot{\boldsymbol{q}}_F = \boldsymbol{J}_C^{-1}\boldsymbol{G}_C^T\boldsymbol{v}_O$, $\ddot{\boldsymbol{q}}_F = \boldsymbol{J}_C^{-1}\boldsymbol{G}_C^T\dot{\boldsymbol{v}}_O + (d(\boldsymbol{J}_C^{-1}\boldsymbol{G}_C^T)/dt)\boldsymbol{v}_O$ into above equation, it becomes

$$\tilde{\boldsymbol{M}}(\boldsymbol{q})\dot{\boldsymbol{v}}_O + \tilde{\boldsymbol{C}}(\boldsymbol{q}, \dot{\boldsymbol{q}})\boldsymbol{v}_O + \tilde{\boldsymbol{g}}(\boldsymbol{q}) = \boldsymbol{G}_C\boldsymbol{u}_C, \tag{4.114}$$

where

$$\begin{aligned}\tilde{\boldsymbol{M}} &= \boldsymbol{M}_O + \boldsymbol{G}_C\boldsymbol{J}_C^{-T}\boldsymbol{M}_F\boldsymbol{J}_C^{-1}\boldsymbol{G}_C^T\\ \tilde{\boldsymbol{C}} &= \boldsymbol{C}_O + \boldsymbol{G}_C\boldsymbol{J}_C^{-T}(\boldsymbol{C}_F\boldsymbol{J}_C^{-1}\boldsymbol{G}_C^T + \boldsymbol{M}_F\tfrac{d}{dt}(\boldsymbol{J}_C^{-1}\boldsymbol{G}_C^T))\\ \tilde{\boldsymbol{g}} &= \boldsymbol{g}_O + \boldsymbol{G}_C\boldsymbol{J}_C^{-T}\boldsymbol{g}_F\\ \boldsymbol{u}_C &= \boldsymbol{J}_C^{-T}\boldsymbol{\tau}_F.\end{aligned} \tag{4.115}$$

Equation (4.114) is the equation of motion of the object in which the constraint conditions have been deleted. This dynamic model has propositions similar to those seen in Propositions 4.5 and 4.6. Since it is assumed that the contact between the object and each finger is continuously maintained, and the contact force stays within the friction cone, it is necessary to confirm in the analysis that these conditions are satisfied.

4.9. Base Parameters of Robot Dynamic Model

The base parameters for a robot are defined as a minimum set of constant parameters that are necessary and sufficient to compute the robot dynamic model. Determining the base parameters will help reduce the computational cost of dynamic models and increase the robustness of the parameter identification process because the base parameters equal the identifiable parameters. Determination of the base parameters is also necessary for model-based adaptive control of robots.

There are two approaches for deriving the base parameters: the first is a numerical method that evaluates linear independency of the regressor numerically (Nakamura and Ghodoussi, 1989) and the second is symbolic formulation (Kawasaki and Nishimura, 1986; 1988; Kawasaki *et al.*, 1996; Mayeda *et al.*, 1990). Numerical results are not suitable for the design of mechanisms or for improving the computation algorithm of dynamic models. On the other hand, the results of symbolic formulation are suitable for the design and control of robots.

There are two approaches for deriving the symbolic formulation of base parameters. One is based on consideration of the geometrical structure of the robot and the direction of the gravitation. This approach was developed for serial-link and tree-structured robots. The other is based on a symbolic evaluation of the linear independency of the regressor. This approach is applicable for serial-link, tree-structured and closed-link robots. In general, a robotic hand with plural fingers is a tree-structured robot.

4.9.1. *Base Parameters and Regressor*

Let us present a symbolic formulation to derive the base parameters for robots. When $\dot{\boldsymbol{q}} = \dot{\boldsymbol{q}}_r$ and $\ddot{\boldsymbol{q}} = \ddot{\boldsymbol{q}}_r$ in Eq. (4.58), it can be represented by

$$\boldsymbol{\tau} = \boldsymbol{M}(\boldsymbol{q})\ddot{\boldsymbol{q}} + \boldsymbol{C}(\boldsymbol{q}, \dot{\boldsymbol{q}})\dot{\boldsymbol{q}} + \boldsymbol{g}(\boldsymbol{q}) = \boldsymbol{X}(\boldsymbol{q}, \dot{\boldsymbol{q}}, \ddot{\boldsymbol{q}})\boldsymbol{\sigma}, \tag{4.116}$$

where $\boldsymbol{\tau} \in \mathbb{R}^n$ is the generalized force vector, $\boldsymbol{q} \in \mathbb{R}^n$ is the generalized coordinate vector, $\boldsymbol{\sigma} \in \mathbb{R}^p$ is a dynamic parameter vector, and

$\boldsymbol{X}(\boldsymbol{q}, \dot{\boldsymbol{q}}, \ddot{\boldsymbol{q}}) \in \mathbb{R}^{n \times p}$, which does not include the dynamic parameters, is a regressor with respect to $\boldsymbol{\sigma}$. When $\boldsymbol{\sigma}$ represents the original dynamic parameters, its i-th sub-vector is given by Eq. (4.59). In this case, $p = 10n$. If the dynamic parameters are the base parameters, then column vectors of $\boldsymbol{X}(\boldsymbol{q}, \dot{\boldsymbol{q}}, \ddot{\boldsymbol{q}})$ should be linearly independent. If the column vectors of $\boldsymbol{X}(\boldsymbol{q}, \dot{\boldsymbol{q}}, \ddot{\boldsymbol{q}})$ are linearly dependent, then the dynamic parameters can be regrouped to obtain the base parameters as follows.

Let us define the i-th column vector of $\boldsymbol{X}$ as $\boldsymbol{x}_i$, and suppose that (i)$\boldsymbol{x}_1, \boldsymbol{x}_2, \ldots,$ and $\boldsymbol{x}_k$ are linearly independent, and (ii) $\boldsymbol{x}_1, \boldsymbol{x}_2, \ldots, \boldsymbol{x}_k$ and $\boldsymbol{x}_{k+1}$ are linearly dependent. There then exists a constant and non-zero vector $\boldsymbol{c} = [c_1, c_2, \ldots, c_k]^T$ satisfying the following equation:

$$\boldsymbol{x}_{k+1} = \sum_{i=1}^{k} c_i \boldsymbol{x}_i. \tag{4.117}$$

By substituting Eq. (4.117) into Eq. (4.116), it can be represented by

$$\boldsymbol{\tau} = \sum_{i=1}^{k} (\sigma_i + c_i \sigma_{k+1}) \boldsymbol{x}_i + \sum_{i=k+2}^{p} \sigma_i \boldsymbol{x}_i. \tag{4.118}$$

In Eq. (4.118), the vector $\boldsymbol{x}_{k+1}$ is eliminated, and the parameter σ_{k+1} is regrouped into σ_i(for $i = 1, \ldots, k$) as $\sigma_i + c_i \sigma_{k+1}$. Therefore, by removing $\boldsymbol{x}_{k+1}$ and σ_{k+1} from the regressor and the dynamic parameter vector, respectively, these can be renewed. The iteration of this process until all column vectors of the renewed regressor become linearly independent will lead to the base parameters. In general, it is not easy to obtain $\boldsymbol{c}$ for any given $\boldsymbol{x}_{k+1}$ directly, even if $\boldsymbol{c}$ exists, because $\boldsymbol{x}_i$ is a vector of multivariable functions.

4.9.2. *Fundamental Function of Regressor*

This section presents a symbolic analysis method to evaluate the set of base parameters for general robots systematically (Kawasaki *et al.*, 1996). The key idea is the derivation of the fundamental functions of the dynamic model.

First, we consider cases of serial-link and tree-structure robots. Then, the generalized coordinate vector $\boldsymbol{q}$ can be equated to the joint variable vector of the robot, and each element of $\boldsymbol{X}$ can be represented as a multivariable polynomial in $\boldsymbol{q}$, $\dot{\boldsymbol{q}}$ and $\ddot{\boldsymbol{q}}$. Linear independence of the column vectors of the regressor will depend on the motions of $\boldsymbol{q}$, $\dot{\boldsymbol{q}}$ and $\ddot{\boldsymbol{q}}$. We can declare, however, that these vectors are linear independent if they are not always linear dependent for all allowable joint motions. The term of the multivariable polynomial be can written by the product of a coefficient and a power product that is defined as a product of integer powers of variables given by

$$f_i = y_1^{n_1} y_2^{n_2} \cdots y_s^{n_s}, \tag{4.119}$$

where n_j is an integer and y_j is a variable that consists of $\boldsymbol{q}$, $\dot{\boldsymbol{q}}$, $\ddot{\boldsymbol{q}}$ and non-linear functions of $\boldsymbol{q}$. To evaluate the linear independency of the column vectors of the matrix $\boldsymbol{X} = [\boldsymbol{x}_1, \boldsymbol{x}_2, \ldots, \boldsymbol{x}_p]$, let us define fundamental functions of the matrix $\boldsymbol{X}$ as follows:

Definition 4.1. (Fundamental functions of $\boldsymbol{X}$)

Scalar functions $f_i (1 \leq i \leq r)$ satisfying the following conditions are called fundamental functions of $\boldsymbol{X}$:

(1) f_i is a scalar function of independent variables and their time derivative, and it does not include dependent variables.
(2) Any f_i cannot be represented by a linear combination of the other functions.
(3) All elements of $\boldsymbol{X}$ can be represented by a linear combination of $f_i (1 \leq i \leq r)$.

Next, we consider a closed-link robot that has a passive joint. A dynamic model of the closed-loop robot generally includes dependent joint variables. If the closed-link robot were transferred into a virtual open tree-structured robot by cutting a joint in the closed link, its dynamics would be represented by

$$\boldsymbol{\tau}_o = \boldsymbol{M}_o(\boldsymbol{q}_o)\ddot{\boldsymbol{q}}_o + \boldsymbol{C}_o(\boldsymbol{q}_o, \dot{\boldsymbol{q}}_o)\dot{\boldsymbol{q}}_o + \boldsymbol{g}_o(\boldsymbol{q}_o) = \boldsymbol{X}_o(\boldsymbol{q}_o, \dot{\boldsymbol{q}}_o, \ddot{\boldsymbol{q}}_o)\boldsymbol{\sigma}_o, \tag{4.120}$$

where subscript o indicates the object of the virtual tree-structured robot, $\boldsymbol{\tau}_o \in \mathbb{R}^{n_o}$ is a joint torque vector, $\boldsymbol{q}_o \in \mathbb{R}^{n_o}$ is a joint variable vector, $\boldsymbol{\sigma}_o \in \mathbb{R}^{p_o}$ is a dynamic parameter vector, and $\boldsymbol{X}_o \in \mathbb{R}^{n_o \times p_o}$ is a regressor with respect to $\boldsymbol{\sigma}_o$. Let us suppose that the motion of the virtual-structured robot is the same as that of the closed-link robot. When $\boldsymbol{q} \in \mathbb{R}^n$ is a generalized coordinate vector of the closed-loop link robot, a closed-loop kinematic constraint can be represented by

$$\boldsymbol{\phi}(\boldsymbol{q}, \boldsymbol{q}_o) = 0 \in \mathbb{R}^{n_o - n}. \tag{4.121}$$

This relation can be obtained from the homogeneous transformation of the closed loop. The generalized coordinates are selected as n independent variables from joint variables of the closed-link robot. The generalized force of the closed-link robot, $\boldsymbol{\tau} \in \mathbb{R}^n$, is given by

$$\boldsymbol{\tau} = \boldsymbol{J}^T \boldsymbol{\tau}_o, \tag{4.122}$$

where

$$\boldsymbol{J} = \frac{\partial \boldsymbol{q}_o}{\partial \boldsymbol{q}}. \tag{4.123}$$

From Eqs. (4.116), (4.120) and (4.122), we get

$$\boldsymbol{\tau} = \mathbf{X}\sigma, \tag{4.124}$$

where $\boldsymbol{X} = \boldsymbol{J}^T \boldsymbol{X}_o$ and $\boldsymbol{\sigma} = \boldsymbol{\sigma}_o$.

This shows that the dynamic model of a closed-link robot is linear with respect to the dynamic parameters of the virtual open tree-structured robot. Hence, it is possible to obtain the base parameters of the closed-link robot by evaluating the linear independency of $\boldsymbol{X}$. Each element of $\boldsymbol{X}$ is a function in $\boldsymbol{q}$, $\boldsymbol{q}_o$ and their time derivatives. $\boldsymbol{X}$ can be represented as a multivariate polynomial in $\boldsymbol{q}$, $\dot{\boldsymbol{q}}$ and $\ddot{\boldsymbol{q}}$ using relations of $\dot{\boldsymbol{q}}_o = \boldsymbol{J}\dot{\boldsymbol{q}}$ and $\ddot{\boldsymbol{q}}_o = \boldsymbol{J}\ddot{\boldsymbol{q}} + \dot{\boldsymbol{J}}\dot{q}$. There are two approaches for obtaining the fundamental functions by taking kinematic constraints into account.

Similarly, it is represented by

$$\boldsymbol{\tau} = \boldsymbol{S}^T \boldsymbol{\tau}_c, \tag{4.125}$$

where $\boldsymbol{\tau}_c$ is an active joint torque vector of the closed-link robot, and $\boldsymbol{S}$ is given by

$$\boldsymbol{S} = \frac{\partial \boldsymbol{q}_c}{\partial \boldsymbol{q}} \in \mathbb{R}^{n \times n}, \tag{4.126}$$

where $\boldsymbol{q}_c$ is an active joint variable vector of the closed-link robot. Now, we suppose that the closed-link robot is controllable; the joint torque of the closed-loop robot is then represented by

$$\boldsymbol{\tau}_c = (\boldsymbol{S}^T)^{-1}\mathbf{X}\sigma_o. \tag{4.127}$$

This shows that the base parameters of the closed-link robot do not depend on the selection of joint variables in regards to the active joint and passive joint if the selected joint variables are independent.

4.9.2.1. *Direct elimination method*

If the kinematic constraint given by Eq. (4.121) can be rewritten as

$$\boldsymbol{q}_o = \boldsymbol{q}_c(\boldsymbol{q}), \tag{4.128}$$

it is possible that each element of $\boldsymbol{X}$ is represented as a function of the independent joint variables and their time derivative only. That is, the dynamics can be represented by Eq. (4.116). If y_j is a nonlinear function, then restrictions may be imposed on n_j such that the n_j-th power of y_j may not be represented as a function of the other variables. For example, if $y_1 = \sin q_i, y_2 = \cos q_i$, where q_i is the i-th element of $\boldsymbol{q}$, and n_2 is an arbitrary integer, then n_1 must be less or equal to 1. This restriction can always be kept by using $y_1^2 = 1 - y_2^2$. If $y_1 = q_i$ and $y_2 = \sqrt{q_i}$, then n_2 must be less or equal to 1. This restriction can also be maintained by using $y_2^2 = y_1$. The power product becomes the fundamental function if the restrictions on the variables are maintained.

4.9.2.2. *Gröbner basis method*

A symbolic explicit form such as Eq. (4.128) is not always possible. However, it can be made possible in a modified sense by using a Gröbner basis. Let us describe the basic idea of the **Gröbner basis** briefly (Büchberger, 1988). A given set of polynomial equations is transformed into a certain standard form G_r, called a reduced Gröbner basis, which has the same solutions as the original set. A 'normal form' of a polynomial F on a set of polynomials R is defined as a polynomial that cannot be reduced any further by applying R. The normal form depends on a sequence of reduction on the element R; however, it is unique if the G_r is used. The term of the normal form is represented by the product of a coefficient and a power product. From the property of the Gröbner basis, the power products of the normal form are the fundamental functions. Note that the Gröbner basis method may require a lot of computation time to derive the base parameter sets.

Example 4.6. Let us derive the reduced form of the following polynomial over the variable $\{x, y, x\}$:

$$F = x^3 y^2 z, \tag{1}$$

with respect to the total degree ordering and a set of constrained polynomials:

$$R = \{x^2y^2 - z = 0 \ (2), x^2 + y^2 - 1 = 0 \ (3)\}. \tag{2}$$

We obtain the following equation using Eq. (2)

$$F = xz^2, \tag{3}$$

and $F = xy^2z - xy^4z$ using Eq. (3). This process is called the reduction. The two results are called the normal form, since they cannot be further reduced with respect to the total degree by applying R. Two normal forms are different. Now, calculating Eq. (3) $\times\ y^2$ – Eq. (2), we obtain,

$$y^4 - y^2 + z = 0. \tag{4}$$

Substituting Eq. (4) into $F = xy^2z - xy^4z$, we get the fully reduced form $F = xz^2$. This means that the original set is equivalent to the following set:

$$G = \{x^2y^2 - z = 0, x^2 + y^2 - 1 = 0, y^4 - y^2 + z = 0\}, \tag{5}$$

which is called the Gröbner basis that leads to the unique normal form. The G_r is defined as the minimal Gröbner basis, which is derived by the reduction of the element of G. In the example, it is

$$G_r = \{x^2 + y^2 - 1 = 0, y^4 - y^2 + z = 0\}. \tag{6}$$

Any polynomial is reduced to the unique normal form using the G_r. The G_r can be obtained by the **Buchberger's algorithm**.

4.9.3. *Regrouping Methods*

4.9.3.1. *Repetitive regrouping method*

Using the fundamental functions, an element of $\boldsymbol{x}_i$ can be represented as

$$x_{i_j} = \mathbf{a}_{ij}^T \boldsymbol{f}, \tag{4.129}$$

where x_{i_j} is a j-th element of $\boldsymbol{x}_i$, $\boldsymbol{f} = [f_1, f_2, \ldots, f_r]^T$ is a fundamental function vector, and $\mathbf{a}_{ij} \in \mathbb{R}^r$ is a constant coefficient vector. From Eq. (4.117) and Eq. (4.129), we obtain

$$\boldsymbol{a}_{k+1j}^T \boldsymbol{f} = \sum_{i=1}^{k} c_i \boldsymbol{a}_{i_j} \boldsymbol{f} \quad (j = 1, \ldots, n). \tag{4.130}$$

Since this equation should be satisfied for all values of the variables at any arbitrary time, we get

$$\boldsymbol{a}_{k+1j} = \sum_{i=1}^{k} c_i \boldsymbol{a}_{i_j} \quad (j = 1, \ldots, n). \tag{4.131}$$

Let us define the following vector and matrix as:

$$\boldsymbol{b}_i = \begin{bmatrix} \boldsymbol{a}_{i_1} \\ \vdots \\ \boldsymbol{a}_{i_n} \end{bmatrix} \in \mathbb{R}^{n*r}, \tag{4.132}$$

$$\boldsymbol{B} = [\boldsymbol{b}_1 \boldsymbol{b}_2 \cdots \boldsymbol{b}_{k+1}] \in \mathbb{R}^{(n*r)\times(k+1)}. \tag{4.133}$$

Then Eq. (4.131) is rewritten as,

$$\boldsymbol{B} \begin{bmatrix} \boldsymbol{c} \\ -1 \end{bmatrix} = 0. \tag{4.134}$$

If there exists a vector $\boldsymbol{c}$ satisfying Eq. (4.117), $\boldsymbol{x}_1, \boldsymbol{x}_2, \ldots, \boldsymbol{x}_k$ and $\boldsymbol{x}_{k+1}$ are linear dependent, otherwise these become linear independent at the appropriate joint motions. The Gauss–Jordan elimination method, consisting of row operations, can be used to examine the existence of $\boldsymbol{c}$. This transforms Eq. (4.134) into the Gauss–Jordan form, which is an upper triangle as follows:

$$\begin{bmatrix} \boldsymbol{E}_{k\times k} & \tilde{\boldsymbol{b}}^1_{k+1} \\ 0 & \tilde{\boldsymbol{b}}^2_{k+1} \end{bmatrix} \begin{bmatrix} \boldsymbol{c} \\ -1 \end{bmatrix} = 0, \tag{4.135}$$

where $\boldsymbol{E}_{k\times k} \in \mathbb{R}^{k\times k}$ is an identity matrix and $\tilde{\boldsymbol{b}}^1_{k+1} \in \mathbb{R}^k$ and $\tilde{\boldsymbol{b}}^2_{k+1} \in \mathbb{R}^{n*r-k}$ are sub-vectors of $\tilde{\boldsymbol{b}}_{k+1} \in \mathbb{R}^{n*r}$, which is a transformed vector of $\boldsymbol{b}_{k+1}$. Evidently, if $\tilde{\boldsymbol{b}}^2_{k+1} = 0$, then $\boldsymbol{c}$ exists and is given by:

$$\boldsymbol{c} = \tilde{\boldsymbol{b}}^1_{k+1}. \tag{4.136}$$

In this case the $k+1$-th element of the parameter vector $\boldsymbol{\sigma}$ can be regrouped into the others and eliminated as follows:

$$\begin{aligned} &\sigma_i \Leftarrow \sigma_i + c_i\sigma_{k+1} && \text{for } i = 1, \ldots, k \\ &\boldsymbol{x}_j \Leftarrow \boldsymbol{x}_{j+1} && \text{for } j = k+1, \ldots, r. \end{aligned} \tag{4.137}$$

As a result, one of the parameters and one of the column vectors are eliminated. If $\tilde{\boldsymbol{b}}^2_{k+1} \neq 0$, then $\boldsymbol{c}$ does not exist, and $\boldsymbol{x}_1, \boldsymbol{x}_2, \ldots, \boldsymbol{x}_k$ and $\boldsymbol{x}_{k+1}$ are linear independent. In this case, the $k+1$-th element of the parameter vector $\boldsymbol{\sigma}$ is one of the base parameters. By replacing k with $k+1$ when $\tilde{\boldsymbol{b}}^2_{k+1} \neq 0$ and repeating the above step, all the base parameters will be obtained.

4.9.3.2. *Lump regrouping method*

Recursive regrouping may not be an effective approach because the Gauss–Jordan elimination method is called $p-1$ times. This call can be reduced only once using the property of the Gauss–Jordan elimination. Now we define a matrix $\boldsymbol{B}$ for all column vectors of $\boldsymbol{X}$ from the beginning as follows:

$$\boldsymbol{B} = [\boldsymbol{b}_1\boldsymbol{b}_2 \cdots \boldsymbol{b}_p] \in \mathbb{R}^{(n*r)\times p}. \tag{4.138}$$

This matrix is then transformed into the Gauss–Jordan form $\tilde{\boldsymbol{B}}$ by the Gauss–Jordan elimination, consisting of row operations. For example, the obtained matrix is upper triangular such as:

$$\tilde{\boldsymbol{B}} = [\boldsymbol{e}_1\boldsymbol{e}_2 0 \tilde{\boldsymbol{b}}_4\boldsymbol{e}_3 \cdots], \tag{4.139}$$

where $\boldsymbol{e}_i$ is a unit vector in which the i-th element is 1 and the others are 0. In this example, the third column vector is a zero vector, which means that the third element of $\boldsymbol{\sigma}$ is not needed for the dynamic model. The fourth column vector corresponds to $\boldsymbol{c}$ given by Eq. (4.117) for the regrouping of the parameters. Therefore, we can get the element of regrouped parameter vector φ as follows:

$$\varphi_i = \begin{cases} \sigma_i + \sum\limits_{\substack{l=1 \\ \tilde{\boldsymbol{b}}_l \neq \boldsymbol{e}_l}}^{p} \tilde{b}_{l_i} \sigma_l & \text{if } \tilde{\boldsymbol{b}}_i = \boldsymbol{e}_i \\ 0 & \text{if } \tilde{\boldsymbol{b}}_i \neq \boldsymbol{e}_i \end{cases} \quad \text{for } i = 1, \ldots, p \tag{4.140}$$

where $\tilde{b}_{l_i}$ is an i-th element of $\tilde{\boldsymbol{b}}_l$. The base parameters are obtained by removing zero elements from φ and renumbering the non-zero elements. Similarly, by collecting the column vectors corresponding to the base parameters, the regressor of the base parameters is obtained.

Example 4.7. Using the direct elimination method, let us evaluate the base parameters of a two-DOF closed-link robot moving in the vertical plane, as shown in Fig. 4.7, in which $\boldsymbol{z}_i$ are joint axes. The joint axes are all parallel or perpendicular to each other. Joint $1'$ is a translational joint, and the others are rotational joints. The active joint variables are $d_{1'}$ and θ_2, and the passive joint variables

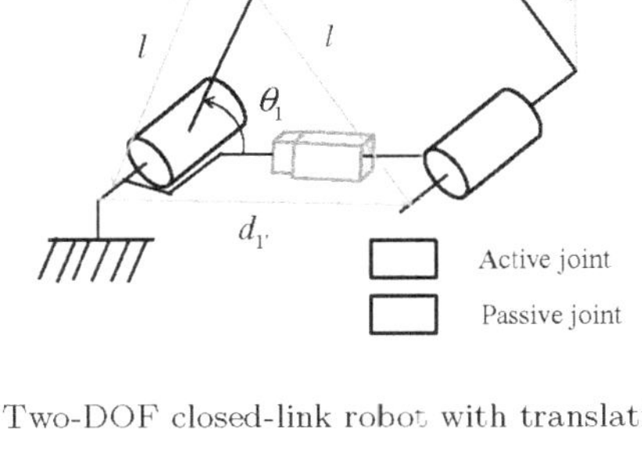

Fig. 4.7. Two-DOF closed-link robot with translational joint.

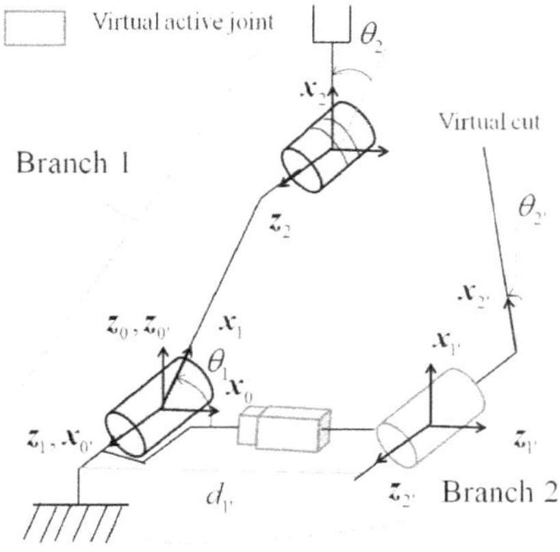

Fig. 4.8. Virtual tree-structured robot.

are θ_1 and $\theta_{2'}$. Joint $2'$ is cut to construct a virtual tree-structured mechanism, as shown in Fig. 4.8. Then $\boldsymbol{q}_c$ and $\boldsymbol{q}_o$ are given as:

$$\boldsymbol{q}_c = [d_{1'}, \theta_2]^T, \quad \boldsymbol{q}_o = [\theta_1, \theta_2, d_{1'}, \theta_{2'}]^T. \tag{1}$$

Let us define joint variables of branch 1 as the generalized coordinates $\boldsymbol{q}$, given by

$$\boldsymbol{q} = [\theta_1, \theta_2]^T. \tag{2}$$

From the kinematic constraints, we get

$$d_{1'} = 2l\cos(\theta_1), \quad \theta_{2'} = \frac{\pi}{2} - \theta_1. \tag{3}$$

Therefore $\boldsymbol{S}$ and $\boldsymbol{J}$ are given by

$$\boldsymbol{S} = \begin{bmatrix} -2l\sin\theta_1 & 0 \\ 0 & 1 \end{bmatrix}, \quad \boldsymbol{J} = \begin{bmatrix} 1 & 0 \\ 0 & 1 \\ -2l\sin\theta_1 & 0 \\ -1 & 0 \end{bmatrix}. \tag{4}$$

The original dynamic parameters of the virtual tree-structured robot can be represented by

$$\boldsymbol{\sigma}_o = [\boldsymbol{\mu}_1^T, \boldsymbol{\mu}_2^T, \boldsymbol{\mu}_{1'}^T, \boldsymbol{\mu}_{2'}^T]^T \in \mathbb{R}^{40}, \tag{5}$$

where $\boldsymbol{\mu}_i \in \mathbb{R}^{10}$ is a parameter vector of link$i$ given by Eq. (4.59). The generalized force is evaluated by the direct method. Then, the variables y_i consist of $\theta_i, \dot{\theta}_i, \sin\theta_i, \cos\theta_i$ where $i = 1, 2$. The fundamental function vector is given by

$$\begin{aligned}\boldsymbol{f} = [&c_1, s_1, \dot{\theta}_1, s_1 s_2, c_1 c_2, s_2\dot{\theta}_1\dot{\theta}_2, s_2\dot{\theta}_2^2, c_2\ddot{\theta}_1, c_2\ddot{\theta}_2, c_2 s_1, c_1 s_2,\\ &c_2\dot{\theta}_1\dot{\theta}_2, s_2\ddot{\theta}_2, s_2\ddot{\theta}_1, c_2\dot{\theta}_2^2, \ddot{\theta}_2, s_1 c_1\dot{\theta}_1^2, c_1\ddot{\theta}_1, c_1 s_1\ddot{\theta}_1, \dot{\theta}_1^2, c_1^2\dot{\theta}_1^2, s_2\dot{\theta}_1^2,\\ &c_2\dot{\theta}_1^2]^T \in \mathbb{R}^{23},\end{aligned} \tag{6}$$

where s_i and c_i are $\sin\theta_i$ and $\cos\theta_i$, respectively. It is easily shown that any element of $\boldsymbol{f}$ satisfies the condition of fundamental function. The matrix $\boldsymbol{X}$ is evaluated by

$$\begin{aligned}\boldsymbol{X} = &\left[\begin{matrix}0 & -gf_1 & gf_2 & 0\,0\,0\,0\,0\,0\,0 & \cdots & -glf_1 + l^2 f_3 & x_{112} & x_{113} & 0\,0\,0\,0\,0\,0 & f_3 + f_{16}\\ 0 & 0 & 0 & 0\,0\,0\,0\,0\,0\,0 & \cdots & 0 & x_{212} & x_{213} & 0\,0\,0\,0\,0\,0 & f_3 + f_{16}\end{matrix}\right.\\ &\left.\begin{matrix}4l^2(f_{13} + f_{17} - f_{18}) & 0\,0\,0\,0\,0\,0\,0\,0\,0 & \cdots & x_{141} & x_{142} & x_{143} & 0\,0\,0\,0\,0\,0 & f_3\\ 0 & 0\,0\,0\,0\,0\,0\,0\,0\,0 & \cdots & 0 & 0 & 0 & 0\,0\,0\,0\,0\,0 & 0\end{matrix}\right]\\ &\in \mathbb{R}^{2\times 40}\end{aligned} \tag{7}$$

where f_i is the i-th element of $\boldsymbol{f}$, g is a gravitation and

$$\begin{aligned}x_{112} &= g(f_4 - f_5) + l(-2f_6 - f_7 + 2f_8 + f_9)\\ x_{212} &= g(f_4 - f_5) + l(f_{18} + f_{22})\\ x_{113} &= g(f_{10} + f_{11}) - l(2f_{12} + f_{13} + 2f_{14} + f_{15})\\ x_{213} &= g(f_{10} + f_{11}) + l(-f_{14} + f_{23})\\ x_{141} &= 4l^2(f_{13} + f_{17} - f_{18})\\ x_{142} &= -gf_{11} - 4l(f_{13} + f_{17} - f_{18})\\ x_{143} &= -gf_2 + l(4f_{19} - 2f_{20} + 4f_{21}).\end{aligned} \tag{8}$$

From $\boldsymbol{X}$, we can get matrix $\boldsymbol{B}$. By the Gauss–Jordan elimination, $\boldsymbol{B}$ is transformed into the following matrix:

$$\begin{aligned}\tilde{\boldsymbol{B}} = [&\overset{1}{0}, \boldsymbol{e}_1, \boldsymbol{e}_2, 0, \ldots, 0, \overset{10}{\boldsymbol{e}_3}, \tilde{\boldsymbol{b}}_{11}, \boldsymbol{e}_4, \boldsymbol{e}_5, 0, \ldots, 0, \overset{20}{\boldsymbol{e}_6}, \boldsymbol{e}_7, 0, \ldots, \overset{30}{0},\\ &\tilde{\boldsymbol{b}}_{31}, \tilde{\boldsymbol{b}}_{32}, \boldsymbol{e}_8, 0, \ldots, 0, \overset{40}{\tilde{\boldsymbol{b}}_{40}}] \in \mathbb{R}^{46\times 40},\end{aligned} \tag{9}$$

where

$$\tilde{\boldsymbol{b}}_{11} = l\boldsymbol{e}_1 + l^2\boldsymbol{e}_3, \quad \tilde{\boldsymbol{b}}_{31} = \boldsymbol{e}_7, \tilde{\boldsymbol{b}}_{32} = \boldsymbol{e}_1 - (1/l)\boldsymbol{e}_7, \tilde{\boldsymbol{b}}_{40} = \boldsymbol{e}_3. \tag{10}$$

Regrouping original dynamic parameters we get,

$$\begin{aligned}\varphi = [&0, ms_{1x} + lm_2 + ms_{2'x}, ms_{1y}, 0, 0, 0, 0, 0, 0, I_{1zz} + l^2 m_2 + I_{2'zz},\\ &0, ms_{2x}, ms_{2y}, 0, 0, 0, 0, 0, 0, I_{2zz}\\ &m_{1'} + m_{2'} - ms_{2'x}/l, 0, 0, 0, 0, 0, 0, 0, 0, 0,\\ &0, 0, ms_{2'y}, 0, 0, 0, 0, 0, 0, 0]^T \in \mathbb{R}^{40}.\end{aligned} \tag{11}$$

Non-zero elements of φ equal the base parameters of the closed-loop robot; that is:

$$\begin{aligned}\boldsymbol{\sigma} = [&ms_{1x} + lm_2 + ms_{2'x}, ms_{1y}, I_{1zz} + l^2 m_2 + I_{2'zz}, ms_{2x},\\ &ms_{2y}, I_{2zz}, m_{1'} + m_{2'} - ms_{2'x}/l, ms_{2'y}]^T \in \mathbb{R}^8.\end{aligned} \tag{12}$$

The result shows that the number of base parameters for the closed-link robot is significantly lower than that of the original dynamic parameters.

Exercises

4.1. Solve Eq. (4.11).

4.2. Prove that C_{ij} given by Eq. (4.52) satisfies Eq. (4.47).

4.3. Using the Lagrange's method, find the equation of motion of the finger with two joints in the plane, as shown in Fig. 4.9.

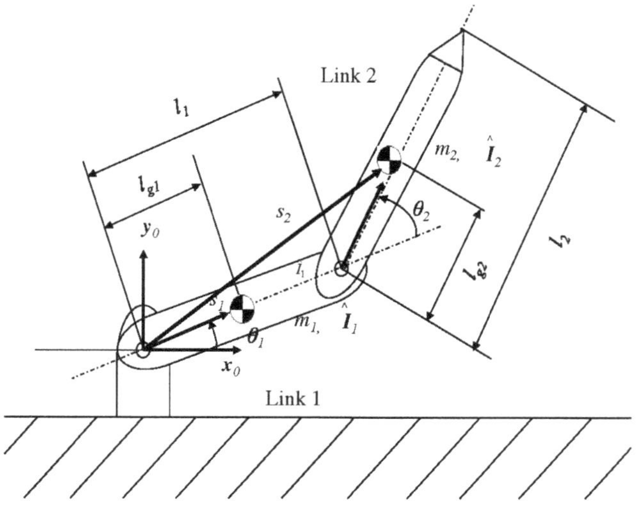

Fig. 4.9. A two-joint finger.

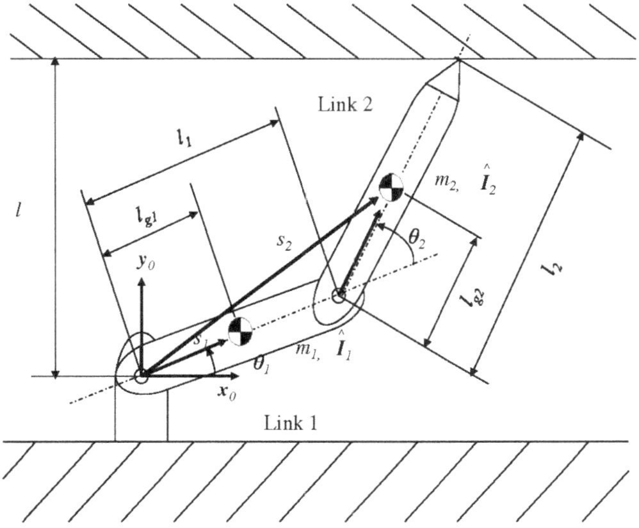

Fig. 4.10. A two-joint finger with constraints.

4.4. Find the equation of motion of a planar finger with two joints when it contacts the horizontal wall, as shown in Fig. 4.10.

4.5. In Fig. 4.8, find the low-dimension model when assuming θ_1 as an independent variable, and θ_2 as a dependent variable.

4.6. In the dynamic model in Eq. (4.114), prove that (1) $\boldsymbol{M}$ is a symmetric and positive matrix, and (2) $\dot{\boldsymbol{M}} - 2\boldsymbol{C}$ is a skew-symmetric matrix.

4.7. Prove that a finger with three joints, as shown in Fig. 4.3, has the properties of 4.3 and 4.5.

Chapter 5

Stability Theory of Non-Linear Systems

Robotic hands are mechanical **non-linear systems**. This chapter outlines the basics of non-linear systems and Lyapunov's methods for stability analyses. Lyapunov theorem is useful in analyzing the stability of non-linear control systems. Proofs of theorems presented are outlined; more details may be found in the references at the back of the book (Slotine and Li, 1991; La Salle and Leffschtz, 1961; Bacciotti and Rosier, 2005; Imura, 2000; Queiros *et al.*, 2000).

5.1. Basics of Non-Linear Systems

5.1.1. *Existence and Uniqueness of Solution*

In non-linear systems, the existence, uniqueness, global property, and continuousness of the solution are basic concepts. For example, a solution of:

$$\dot{x} = \sqrt{x}(x(t_0) = 0)$$

is given by $x = (t - t_0)^2/4$, but $x = 0$ is also its solution as shown in Fig. 5.1. Furthermore, $x = 0$ until time t_1 and $x = (t - t_1)^2/4$ subsequently is also one of solutions. This shows that there exist an infinite number of solutions, and that the solution is not unique.

A non-linear dynamic system can usually be represented by a set of non-linear differential equations in the form of:

$$\dot{\boldsymbol{x}} = \boldsymbol{f}(\boldsymbol{x}, t), \quad t \geq t_0, \quad \boldsymbol{x}(t_0) = \boldsymbol{x}_0 \tag{5.1}$$

where $\boldsymbol{x} \in \mathbb{R}^n$ is an n-dimensional state vector, $\boldsymbol{f}$ is an n-dimensional non-linear vector function, and $\boldsymbol{x}_0$ is an initial state at $t = t_0$. We

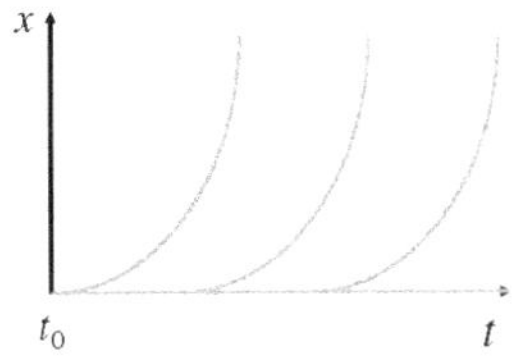

Fig. 5.1. Example of non-unique solutions ($\dot{x} = \sqrt{x}(x(t_0) = 0)$).

first describe a simple and sufficient condition for the existence of solutions, and explain the conditions necessary for this solution to be unique. Hereafter, let us use a notation of set $A = \{x : P(x)\}$, which means the set of all x for which $P(x)$ is true and the word *region* to refer to a subset of real space.

Theorem 5.1. (Local existence)
Consider Eq. (5.1) defined on the finite closed region:

$$\Omega(\boldsymbol{x}_0, t_0) = \{(\boldsymbol{x}, t) : \|\boldsymbol{x} - \boldsymbol{x}_0\| \leq b, |t - t_0| \leq c, b, c > 0\},$$

where b and c are strictly positive constants. If the function $\boldsymbol{f}(\boldsymbol{x}, t)$ is continuous with respect to $\boldsymbol{x}$ and t on the Ω and the partial derivative $\partial \boldsymbol{f}(\boldsymbol{x}, t)/\partial \boldsymbol{x}$ is also continuous on the Ω, then the equation has at least one solution $\boldsymbol{x}(t)$ which satisfies the initial condition and is continuous over a finite interval $[t_0, t_1]$, where $t_0 < t_1$.

This theorem demonstrates the existence of solutions if $\boldsymbol{f}(\boldsymbol{x}, t)$ is continuous with respect to $\boldsymbol{x}$ and t on a finite closed region; it does not, however, show the uniqueness of the solution. The sufficient condition for the existence of a unique solution is given by the following theorem.

Theorem 5.2. (Local uniqueness)
If the function $\boldsymbol{f}(\boldsymbol{x}, t)$ is continuous with respect to $\boldsymbol{x}$ and t on the finite closed region Ω, and if there exist a positive constant K such that,

$$\|\boldsymbol{f}(\boldsymbol{x}_1, t) - \boldsymbol{f}(\boldsymbol{x}_2, t)\| \leq K\|\boldsymbol{x}_1 - \boldsymbol{x}_2\|, \tag{5.2}$$

For $\boldsymbol{x}_1$ and $\boldsymbol{x}_2$ on the Ω, then Eq. (5.1) has a unique solution $\boldsymbol{x}(t)$, that satisfies the initial condition on the Ω.

The condition of Eq. (5.2) is called the **Lipschitz condition** and the constant K is referred to as a **Lipschitz constant**. This condition means, intuitively, that 'small' changes in input result in 'small' changes in output. If the function $\boldsymbol{f}(\boldsymbol{x}, t)$ satisfies the Lipschitz condition in the neighborhood of $\boldsymbol{x}_0$, this function is called the **Lipschitz continuous at $\boldsymbol{x}_0$**. It should be noted that this Lipschitz condition shows the sufficient condition of a unique solution. Moreover, this Lipschitz condition is described in the neighborhood of $\boldsymbol{x}_0$, and global existence and uniqueness of solution globally are different issues. As an example, consider following differential equation:

$$\dot{x} = x^2(x(t_0) = x_0).$$

From theorem 5.1 and 5.2, this has a unique solution in a region of Ω and the solution is given by

$$x = -1/(t - t_0 - 1/x_0).$$

As shown in Fig. 5.2, if $x_0 > 0$, then the solution is discontinuous at time $t = t_0 + 1/x_0$ and the Lipschitz condition is not satisfied at the time. When $\|\boldsymbol{x}\| \to \infty$ at a certain time, the time is called the **finite escape time**.

If Eq. (5.2) is satisfied for all times and for any $\boldsymbol{x}_1$ and $\boldsymbol{x}_2$, then a finite escape time does not exist, and a unique solution exists globally. The sufficient condition for the global existence of a unique solution is given by following theorem.

Theorem 5.3. (Global existence and uniqueness)
If the function $\boldsymbol{f}(\boldsymbol{x}, t)$ is continuous with respect to $\boldsymbol{x}$ and t, and if there exists a positive constant K such that:

$$\|\boldsymbol{f}(\boldsymbol{x}_1, t) - \boldsymbol{f}(\boldsymbol{x}_2, t)\| \leq K\|\boldsymbol{x}_1 - \boldsymbol{x}_2\|, \tag{5.3}$$

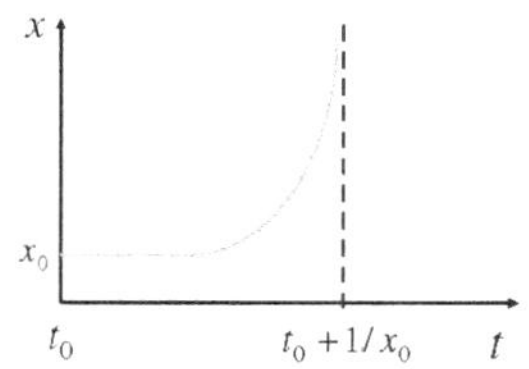

Fig. 5.2. Example of finite escape ($\dot{x} = x^2(x(t_0) = x_0)$).

for all $\boldsymbol{x}_1$ *and* $\boldsymbol{x}_2$, *then Eq.* (5.1) *has a unique solution* $\boldsymbol{x}(t)$ *with an initial value* $\boldsymbol{x}_0$ *for all* $t \geq t_0$.

The condition given in Theorem 5.3 is called the **globally Lipschitz condition**. This is a sufficient condition for the existence of a unique solution, but it is not a necessary condition. A differential equation $\dot{x} = x(x(t_0) = x_0)$ also satisfies the globally Lipschitz condition, and it has the solution $x(t) = x(t_0)e^{t-t_0}$. Thus, it is evident that $x \to \infty$ as $t \to \infty$. This means that the boundedness of the solution is not shown even if the solution exists globally.

5.1.2. *Autonomous and Non-Autonomous Systems*

The non-linear dynamic system described by Eq. (5.1) is said to be **autonomous** if $\boldsymbol{f}$ does not depend explicitly on time, i.e., if the non-linear dynamic system can be written by

$$\dot{\boldsymbol{x}} = \boldsymbol{f}(\boldsymbol{x}). \tag{5.4}$$

Otherwise, the non-linear dynamic system is called **non-autonomous**.

It is generally known that dynamical control system is controlled by the control input. Hence, non-linear dynamic systems with control input can be represented by

$$\dot{\boldsymbol{x}} = \boldsymbol{f}(\boldsymbol{x}, \mathbf{u}) \tag{5.5}$$

where $\boldsymbol{u}$ is an m-dimensional control input vector. When the control input does not depend explicitly on time, the control system becomes autonomous. It becomes non-autonomous, however, when the control input includes time explicitly as in $\boldsymbol{u} = \boldsymbol{u}(\boldsymbol{x}, t)$.

The basic difference between autonomous and non-autonomous non-linear dynamic systems is that the state trajectory of autonomous systems depends on the initial state $\boldsymbol{x}_0$ and does not depend on initial time t_0, in contrast, that of non-autonomous systems depends on both the initial state and the initial time. Hence, initial time should be considered in stability analyses of non-autonomous systems, and a more complex stability analysis will be necessary.

5.2. Non-Linear Autonomous Systems

5.2.1. *Concepts of Stability*

In this section, we discuss the stability of non-linear autonomous systems, described by Eq. (5.4). Assume that $\boldsymbol{f}(\boldsymbol{x})$ and its first-order time derivative is continuous in a neighborhood region of the origin Ω. A state $\boldsymbol{x}_e$ is called an **equilibrium point** if

$$\boldsymbol{f}(\boldsymbol{x}_e) = 0. \tag{5.6}$$

The state vector $\boldsymbol{x}(t)$ stays on $\boldsymbol{x}_e$ for all $t \geq t_0$ if $\boldsymbol{x}(t_0) = \boldsymbol{x}_e$. The equilibrium points are in general not unique and multiple. For example, the equilibrium points of $\dot{x} = -\sin(x)$ are given by $x_e = \pm 2k\pi (k = 0, 1, 2, \ldots)$ and their number is infinite. Therefore, the stability described here focuses on the behavior of solutions at a single equilibrium point. When we discuss the stability around equilibrium point $\boldsymbol{x}_e$, we can suppose that the equilibrium point is the origin, i.e., $\boldsymbol{x}_e = 0$. This is because, if a new state vector $\boldsymbol{z} = \boldsymbol{x} - \boldsymbol{x}_e$ is introduced into Eq. (5.6), it can be represented by $\dot{\boldsymbol{z}} = \boldsymbol{f}(\boldsymbol{z})$, with an equilibrium point 0. Therefore, without loss of generality, it is supposed hereafter that the equilibrium point is the origin. Now, we define stability around the origin, asymptotic stability, exponential stability, global asymptotic stability, and global exponential stability. Some mathematical notations are introduced. Let B_r denote a spherical region defined by $\|\boldsymbol{x}\| < r$, and S_r denote a sphere defined by $\|\boldsymbol{x}\| = r$. We use three spherical regions $B_R \supseteq B_\varepsilon \supseteq B_r$, where $R \geq \varepsilon \geq r$.

Definition 5.1. (Stability)

The equilibrium point of Eq. (5.4) is said to be stable if, for arbitrary $\varepsilon > 0$, there exists a $r = r(\varepsilon) > 0$, such that if $\|\boldsymbol{x}(0)\| < r$, then $\|\boldsymbol{x}(t)\| < \varepsilon$ for all $t \geq 0$.

There are other definitions of stability, and the above definition is called ***stability in the sense of Lyapunov*** or ***Lyapunov stability***. Lyapunov stability means that solutions starting close enough to the equilibrium within a distance $r = r(\varepsilon) > 0$ remain close enough forever within a distance $\varepsilon > 0$ from the origin. Note that this must

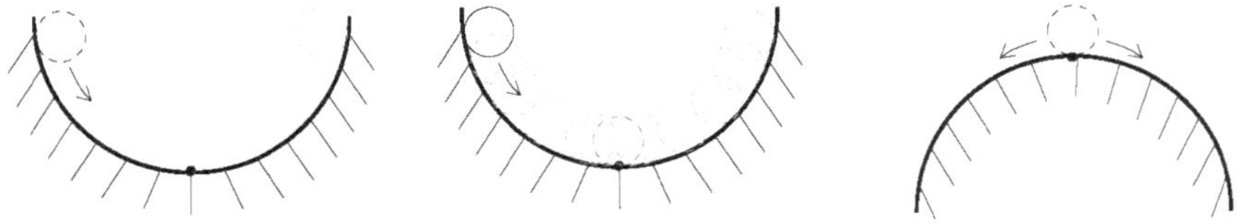

(a) Stable equilibrium position (b) Asymptotic equilibrium position (c) Unstable equilibrium position

Fig. 5.3. Classification of equilibrium positions.

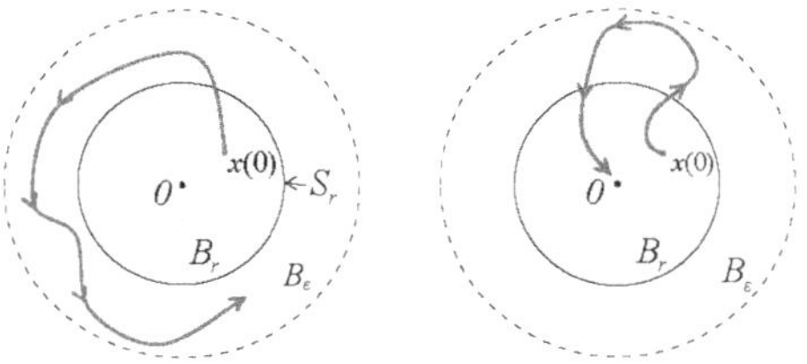

(a) Stability (b) Asymptotic stability

Fig. 5.4. Stability and asymptotic stability.

be true for any $\varepsilon > 0$ that one may choose. If the equilibrium point is not stable, it is called **unstable**.

Definition 5.2. (Asymptotic stability)

The equilibrium point of Eq. (5.4) is said to be asymptotically stable if it is stable, and if there exists some $r > 0$ such that $\|\boldsymbol{x}(0)\| < r$ implies that $\boldsymbol{x}(t) \to 0$ as $t \to \infty$.

Asymptotic stability means that solutions that start from a finite region with $\|\boldsymbol{x}(0)\| < r$ converge to the origin. The concepts of stability and asymptotic stability are shown in Fig. 5.4. When a system is stable, a solution starting from an initial point on region B_r stays on region B_ε subsequently. When a system is asymptotically stable, solutions starting from initial points on region B_r converge to the origin with time. Region B_r is called the **domain of attraction** of the equilibrium point. Figure 5.3 presents examples of stable, asymptotically stable, and unstable positions. In the case of Fig. 5.3(a), a ball rotates on a concave sphere without friction and is stable with un-damping motion; In Fig. 5.3(b), a ball rotates on a concave sphere with damping and it is asymptotically stable because it converges to the bottom; finally, in Fig. 5.3(c), a ball rotates on a convex curve and is unstable because it can fall off the top.

Example 5.1. Consider the system:

$$\begin{aligned} \dot{x}_1(t) &= x_2(t) \\ \dot{x}_2(t) &= -x_1(t). \end{aligned} \tag{1}$$

A solution of the system is given by

$$\begin{aligned} x_1(t) &= c_1 \sin(t) + c_2 \cos(t) \\ x_2(t) &= c_1(t) \cos(t) - c_2 \sin(t), \end{aligned} \tag{2}$$

where c_1 and c_2 are constant parameters satisfying an initial condition. Then, the norm of the state vector is:

$$\|\boldsymbol{x}(t)\| = (x_1^2(t) + x_2^2(t))^{1/2} = (c_1^2 + c_2^2)^{1/2}. \tag{3}$$

The right side of Eq. (3) is determined from the initial conditions. Hence, the system is stable because there exits an initial condition with $(c_1^2 + c_2^2)^{1/2} < \varepsilon$ such that $\|\boldsymbol{x}(t)\| < \varepsilon$ for any $\varepsilon > 0$.

Example 5.2. Consider the system:

$$\begin{aligned} \dot{x}_1(t) &= x_2(t) \\ \dot{x}_2(t) &= -x_1(t) + (1 - x_1^2(t))x_2(t). \end{aligned} \tag{1}$$

This is well known as the **Van der Pol equation**. The solution from arbitrary initial value except for the origin converges to a limit cycle as shown in Fig. 5.5. The system is unstable, because an initial condition such that $\|\boldsymbol{x}(t)\| < \varepsilon$ for an arbitrary $\varepsilon > 0$ does not exist.

From the perspective of many engineering applications, knowing that a system converges to the origin after an infinite time period is not sufficient. What is needed is an estimate of how fast the state of the dynamic system converges to the origin. The concept of exponential stability will be used for this purpose.

Definition 5.3. (Exponential stability)

The equilibrium point of Eq. (5.4) is said to be exponentially stable if it is asymptotically stable and if there exist three positive numbers

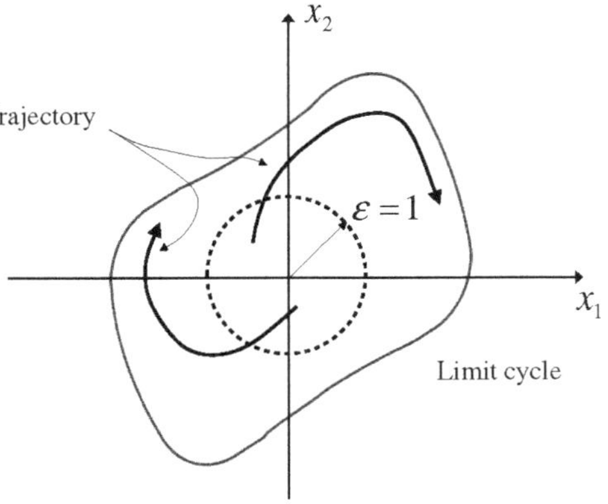

Fig. 5.5. Unstable origin of van der Pol equation.

a, b, r such that if $\|\boldsymbol{x}(0)\| < r$, then $\|\boldsymbol{x}(t)\| \leq a\|\boldsymbol{x}(0)\|e^{-bt}$ for all $t \geq 0$.

Exponential stability means that solutions not only converge, but converge to the origin faster than the exponential function $a\|\boldsymbol{x}(0)\|e^{-bt}$. The positive number b is often called the **rate of exponential convergence**. The stabilities mentioned above are defined on a neighborhood around an equilibrium point. In some applications, global properties are important. There are two definitions of stability for the arbitrary initial state on $\mathbb{R}^n$.

Definition 5.4. (Global asymptotic stability)

The equilibrium point of Eq. (5.4) is said to be globally asymptotically stable if it is asymptotically stable for arbitrary initial states.

Definition 5.5. (Global exponential stability)

The equilibrium point of Eq. (5.4) is said to be globally exponentially stable if it is exponentially stable for arbitrary initial states.

Example 5.3. Consider the system

$$\dot{x} = -(1 + \cos^2 x)x. \tag{1}$$

A solution of the system is given by

$$x(t) = x(0) \exp\left(-\int_0^t (1 + \cos^2 x(\tau))d\tau\right). \tag{2}$$

Then, we get

$$|x(t)| = |x(0)| \exp\left(-\int_0^t (1 + \cos^2 x(\tau))d\tau\right) \leq |x(0)|e^{-t}. \quad (3)$$

Equation (3) holds for arbitrary initial state $x(0)$. This shows that the origin is globally exponentially stable.

5.2.2. *Linearization of a Non-Linear System*

It is conceivable that a non-linear system could behave similarly to its linearized approximation in a small range around a linearized point. The Taylor expansion of the non-linear system Eq. (5.4) with the first-order term is represented by

$$\dot{\boldsymbol{x}} = \left.\frac{\partial \boldsymbol{f}}{\partial \boldsymbol{x}}\right|_{\boldsymbol{x}=0} \boldsymbol{x} + \boldsymbol{f}_{h.o.t.}(\boldsymbol{x}), \quad (5.7)$$

Where $\boldsymbol{f}_{h.o.t.}$ stands for higher-order terms in $\boldsymbol{x}$. Let us denote the Jacobian matrix $\boldsymbol{A} \in R^{n\times n}$ of $\boldsymbol{f}$ with respect to $\boldsymbol{x}$ at $\boldsymbol{x} = 0$ as:

$$\boldsymbol{A} = \left.\frac{\partial \boldsymbol{f}}{\partial \boldsymbol{x}}\right|_{\boldsymbol{x}=0}. \quad (5.8)$$

Then, the system

$$\dot{\boldsymbol{x}} = \mathbf{Ax}, \quad (5.9)$$

is called the **linearization** or **linearized system** of the original non-linear system at the origin. The matrix $\boldsymbol{A}$ is called the **system matrix** of the linear system. The origin of the linear system is said to be **strictly stable** if all eigenvalues of $\boldsymbol{A}$ have negative real parts; it is said to be **marginally stable** if some but not all of the eigenvalues of, have negative real parts and the rest have zero real parts; finally, it is said to be **unstable** if one of eigenvalues of $\boldsymbol{A}$ has positive real part. The following theorem makes precise the relationship between the stability of the original non-linear system Eq. (5.4) and that of the linearized system Eq. (5.9).

Theorem 5.4. (Lyapunov's linearization method)

(1) *If the linearized system is strictly stable, then the equilibrium point of the original non-linear system is asymptotically stable.*

(2) *If the linearized system is unstable, then the equilibrium point of the original non-linear system is unstable.*
(3) *If the linearized system is marginally stable, then one cannot conclude anything from the linearized system. The equilibrium point of the original non-linear system may be asymptotically stable, stable, or unstable.*

The proof of this theorem is outlined intuitively as follows. If the linearized system is strictly stable or unstable, then; the non-linear system is locally stable or locally unstable since the approximation is valid near the equilibrium. However, if the linearized system is marginally stable, the higher-order terms in Eq. (5.7) can have a decisive effect on whether the non-linear system is stable or unstable.

Example 5.4. Consider the non-linear system:

$$\begin{aligned} \dot{x}_1 &= -x_2^2 - x_1 \cos x_2 \\ \dot{x}_2 &= (1 + x_1)x_2 - 3x_2 \end{aligned} \tag{1}$$

Evidently, the origin is the equilibrium point. The Taylor expansion with the first-order term of Eq. (1) at $\boldsymbol{x} = [x_1, x_2]^T = 0$ is represented by

$$\begin{aligned} \dot{x}_1 &= \left.\frac{\partial(-x_2^2 - x_1 \cos x_2)}{\partial x_1}\right|_{x=0} x_1 \\ &\quad + \left.\frac{\partial(-x_2^2 - x_1 \cos x_2)}{\partial x_2}\right|_{x=0} x_2 = -x_1 \\ \dot{x}_2 &= \left.\frac{\partial((1 + x_1)x_2 - 3x_2)}{\partial x_1}\right|_{x=0} x_1 \\ &\quad + \left.\frac{\partial((1 + x_1)x_2 - 3x_2)}{\partial x_2}\right|_{x=0} x_2 = -2x_2. \end{aligned} \tag{2}$$

Then, the linearized system is:

$$\dot{\boldsymbol{x}} = \begin{bmatrix} -1 & 0 \\ 0 & -2 \end{bmatrix} \boldsymbol{x}. \tag{3}$$

The eigenvalues of the system matrix are $\lambda = -1, -2$. All real parts are negative. This derivation indicates that the original non-linear system is asymptotically stable at the origin.

Example 5.5. Consider the non-linear system:

$$\dot{x} = -x^3. \tag{1}$$

Evidently, the origin is the equilibrium point. The linearized system of Eq. (1) at $x = 0$ is represented by

$$\dot{x} = 0. \tag{2}$$

Since the eigenvalue of the linearized system is $\lambda = 0$, the linearized system is marginal stable. This means that the linearization method fails, but the Lyapunov's direct method can easily solve this problem as shown in Example 5.7.

5.2.3. *Lyapunov's Direct Method*

First, we define positive definite, negative definite, positive semi-definite, and negative semi-definite for scalar functions.

Definition 5.6. (Positive (negative) definite)
A scalar continuous function $V(\boldsymbol{x})$ is said to be positive (negative) definite in a neighborhood B_R of the origin if:

- $V(0) = 0$.
- $V(\boldsymbol{x}) > 0$ ($V(\boldsymbol{x}) < 0$) for all non-zero $\boldsymbol{x} \in B_R$.

Definition 5.7. (Positive (negative) semi-definite)
A scalar continuous function $V(\boldsymbol{x})$ is said to be positive (negative) semi-definite in a neighborhood B_R of the origin if:

- $V(0) = 0$.
- $V(\boldsymbol{x}) \geq 0$ ($V(\boldsymbol{x}) \leq 0$) for all non-zero $\boldsymbol{x} \in B_R$.

These definitions are characterized in a neighborhood of the origin. Hence, they are often referred to as locally positive (negative) definite or locally positive (negative) semi-definite. The definitions can be extended over the whole state space; i.e., if $V(0) = 0$ and $V(\boldsymbol{x}) > 0$ ($V(\boldsymbol{x}) < 0$) for all $\boldsymbol{x}$ over the whole state space, then $V(\boldsymbol{x})$ is said to be globally positive (negative) definite. Globally positive (negative) semi-definite is similarly defined.

A positive definite function is minimum at the origin. For $\boldsymbol{x} = [x_1, x_2]^T \in \mathbb{R}^2$, $V(\boldsymbol{x}) = x_1^2 + x_2^2$ is positive definite, $V(\boldsymbol{x}) = x_1^2$ is positive semi-definite because of $V(\boldsymbol{x}) = 0$ at $x_1 = 0, x_2 \neq 0$, and $V(\boldsymbol{x}) = x_1^2 - x_2^2$ is neither positive definite nor positive semi-definite. Next, we introduce the definition of **Lyapunov function**.

Definition 5.8. (Lyapunov function)

If a function $V(\boldsymbol{x})$ is positive definite in a neighborhood B_R of the origin, has continuous partial derivatives $\partial V(\boldsymbol{x})/\partial \boldsymbol{x}$, and its time derivative along any state trajectory of the system Eq. (5.4) is negative semi-definite, i.e.,

$$\dot{V}(\boldsymbol{x}) = \left(\frac{\partial V(\boldsymbol{x})}{\partial \boldsymbol{x}}\right)^T \dot{\boldsymbol{x}} = \left(\frac{\partial V(\boldsymbol{x})}{\partial \boldsymbol{x}}\right)^T \boldsymbol{f}(\boldsymbol{x}) \leq 0 \quad \text{for all } \boldsymbol{x} \in B_R, \tag{5.10}$$

then, $V(\boldsymbol{x})$ is said to be a Lyapunov function for the system.

Using the Lyapunov function, following local stability theorems are presented as **Lyapunov's direct method**.

Theorem 5.5. (Local stability)

If there exists a Lyapunov function $V(\boldsymbol{x})$ for the system Eq. (5.4) in a neighborhood B_R of the origin, then the origin is stable.

Theorem 5.6. (Local asymptotically stability)

If $V(\boldsymbol{x})$ is a Lyapunov function and $\dot{V}(\boldsymbol{x})$ is negative definite in a neighborhood B_R of the origin for all non-zero $\boldsymbol{x} \in B_R$, then the equilibrium point is asymptotically stable.

Proof. The proofs of these theorems are derived using the geometric interpretation of a Lyapunov function as shown in Fig. 5.6 for the case $n = 2$. The function $V(\boldsymbol{x})$ is a parabolic curve downward and is minimum at the origin. The state trajectory of the system in Eq. (5.4) is plotted in the x_1x_2 plane, and projected on the function $V(\boldsymbol{x})$.

First, we consider a case in which the equilibrium point is stable and $V(\boldsymbol{x})$ is a Lyapunov function. To show stability, we must show that for any given number $r > 0$, there exists a number $\varepsilon > 0$ such that any trajectory starting inside region B_r remains inside region $B_\varepsilon (\subseteq B_R)$ for all $t \geq t_0$. Let m be the minimum of $V(\boldsymbol{x})$ on the

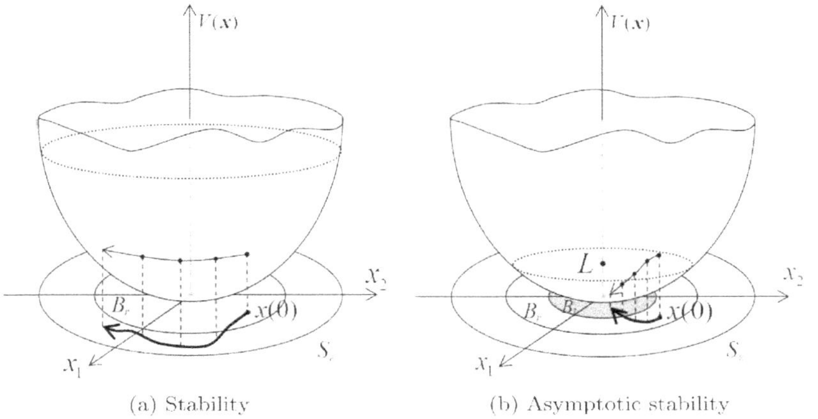

Fig. 5.6. Stability and asymptotic stability of Lyapunov function.

sphere S_ε. Since $V(\boldsymbol{x})$ is positive definite, m exists. Furthermore, since $V(0) = 0$, there exists a region B_r around the origin such that $V(\boldsymbol{x}) < m$ for any $\boldsymbol{x}$ inside the region. Consider now a trajectory whose initial state $\boldsymbol{x}(0)$ is within region B_r. Since $V(\boldsymbol{x})$ is non-increasing along system trajectories, $V(\boldsymbol{x})$ remains strictly smaller than m, and therefore the trajectory cannot possibly cross the outside sphere S_ε. Thus, any trajectory starting inside region B_r remains inside the ball B_ε, and therefore local stability is guaranteed.

Next, let us show that the equilibrium is asymptotically stable if $\dot{V}(\boldsymbol{x}) < 0$, by contradiction. A trajectory starting at $\boldsymbol{x}(0) \in B_r$ remains in region B_ε for all future time. Since $V(\boldsymbol{x})$ is lower bounded and decreases continually, $V(\boldsymbol{x})$ tends towards a lower limit L. It is assumed that this limit is not zero, i.e., $L > 0$. Since $V(\boldsymbol{x})$ is continuous and $V(0) = 0$, there thus exists a region B_L that the system trajectory never enters. Since $-\dot{V}(\boldsymbol{x})$ is also continuous and positive definite, however, and since B_ε is bounded, $-\dot{V}(\boldsymbol{x})$ must remain larger than some positive number L_0. This is a contradiction, because it would imply that $V(t)$ decreases from its initial value $V(0)$ to a value smaller than L in a finite time. Hence, all trajectories staring in B_r asymptotically converge to the origin.

The Lyapunov stability theorem presents the sufficient condition. Therefore, it should be noted that one cannot conclude instability in a system when a Lyapunov function cannot be found. Furthermore, Lyapunov stability is defined in a local region, following a global stability theorem.

Theorem 5.7. (Global asymptotic stability)

If over the whole state space there exists a Lyapunov function $V(\boldsymbol{x})$ *for the system in Eq.* (5.4) *and* $\dot{V}(\boldsymbol{x})$ *is negative definite, and if* $V(\boldsymbol{x}) \to \infty$ *as* $\|\boldsymbol{x}\| \to \infty$, *then the origin is globally asymptotically stable.*

The proof is omitted because it is almost same as in the local case. In global asymptotic stability, $V(\boldsymbol{x})$ must satisfy the condition that $V(\boldsymbol{x}) \to \infty$ as $\|\boldsymbol{x}\| \to \infty$. This is said to be **radially unbounded** condition.

Example 5.6. Consider the system:

$$\dot{x} = -x \sin^2(x). \tag{1}$$

Let us define a Lyapunov function candidate as

$$V(x) = x^2. \tag{2}$$

It is evident that $V(0) = 0$ and $V(x) > 0$ $(x \neq 0)$. A time derivative along the trajectory of system is given by

$$\dot{V}(x) = 2x\dot{x} = -2x^2 \sin^2(x) \leq 0. \tag{3}$$

Therefore, $V(x)$ is a Lyapunov function of the system. Since $V(x) \to \infty$ as $|x\| \to \infty$, the origin is globally stable. Moreover, since $V(x) \to 0$ in a range $-\pi < x < \pi$, the origin is locally asymptotically stable in the range.

Example 5.7. Consider the system:

$$\dot{x} = -x^3. \tag{1}$$

This system is the same as that in Example 5.5. The linearized method could not show the stability of the system. Let us define a Lyapunov function candidate as:

$$V(x) = x^2. \tag{2}$$

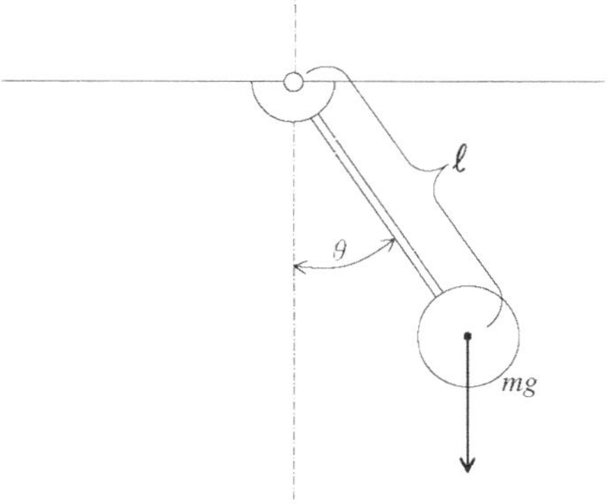

Fig. 5.7. Single pendulum.

It is evident that $V(0) = 0$ and $V(x) > 0$ $(x \neq 0)$. A time derivative along the trajectory of system is given by

$$\dot{V}(x) = 2x\dot{x} = -x^4 < 0. \tag{3}$$

Therefore, $V(x)$ is negative definite. Since $V \to \infty$ as $x \to \infty$, the origin is globally asymptotically stable.

Example 5.8. Let us consider a single pendulum motion, as shown in Fig. 5.7. The dynamic equation is

$$ml^2\ddot{\theta} + b\dot{\theta} + mgl\sin\theta = 0 \tag{1}$$

where m is the mass, l is the length of the pendulum, b is the coefficient of viscosity, and g is the gravitational acceleration. To convert this to a first-order simultaneous differential equation, we set $x_1 = \theta$ and $x_2 = \dot{\theta}$. We then get,

$$\begin{bmatrix} \dot{x}_1 \\ \dot{x}_2 \end{bmatrix} = \begin{bmatrix} x_2 \\ -\dfrac{b}{ml^2}x_2 - \dfrac{g}{l}\sin(x_1) \end{bmatrix}. \tag{2}$$

It is assumed that a motion range of the pendulum is $-\pi < \theta < \pi$. Then, because the equilibrium is

$$x_2 = 0, \quad x_1 = 0,$$

the origin is the equilibrium. Let us define a Lyapunov function candidate as the total energy, consisting of both kinematic and potential energy; i.e.,

$$V(\boldsymbol{x}) = \frac{1}{2}m(lx_2)^2 + mg(1 - \cos x_1) \geq 0. \tag{3}$$

A time derivative along the trajectory of the system is given by

$$\dot{V}(\boldsymbol{x}) = ml^2 x_2 \dot{x}_2 + mgx_2 \sin x_1 = -bx_2^2 \leq 0. \tag{4}$$

Therefore, $V(\boldsymbol{x})$ is a Lyapunov function and the origin is locally stable. From physical insight, the pendulum is evidently a damping system that converges to a minimum point. We cannot, however, derive the asymptotic stability from the Lyapunov theory. This will be shown by the invariant set theorem described in Sec. 5.2.4.

5.2.4. *Invariant Set Theorems*

It is often difficult to demonstrate the asymptotic stability of a nonlinear system because the time derivative of the Lyapunov function candidate is only negative semi-definite, as was seen in Example 5.8. Asymptotic stability for control systems is a very important property. In this kind of situation, it is still possible to derive concussions on asymptotic stability with the help of the **invariant set theorems**, attributed to **La Salle**.

Definition 5.9. (Invariant set)

A set of states $\Omega \subseteq \mathbb{R}^n$ is an invariant set for the system if every state trajectory $\boldsymbol{x}(t)$ that starts from an arbitral $\boldsymbol{x}_0 \in \Omega$ remains in Ω at all future times.

It is evident that if a set Ω equals the whole state space, (i.e.,$\Omega = \mathbb{R}^n$) then it is an invariant set. Any equilibrium point is an invariant set, and the domain of attraction of an equilibrium point is also an invariant set. Limit cycles that are closed curves in the phase plane are also invariant sets. A Lyapunov theory is extended by the invariant set theorem so as to describe convergence to dynamic behaviors more general than equilibrium.

Theorem 5.8. (Local invariant set theorem [La Salle's theorem])

Consider an autonomous system $\dot{\boldsymbol{x}} = \boldsymbol{f}(\boldsymbol{x})$ with $\boldsymbol{f}$ continuous. Let $V(\boldsymbol{x})$ be a scalar function with continuous first partial derivatives.

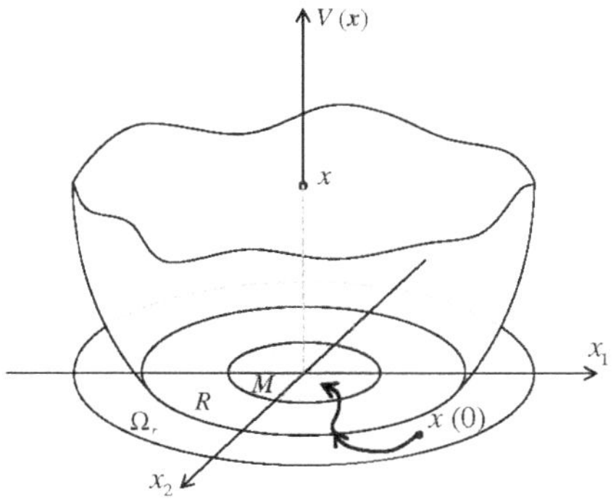

Fig. 5.8. Maximum invariant set.

Assume that:

- *For some $r > 0$, the region Ω_r defined by $V(\boldsymbol{x}) < r$ is bounded.*
- *$V(\boldsymbol{x}) > 0$ for all non-zero $\boldsymbol{x}$ in Ω_r.*
- *$\dot{V}(\boldsymbol{x}) \leq 0$ for all $\boldsymbol{x}$ in Ω_r.*

Let R be the set of all points within Ω_r where $\dot{V}(\boldsymbol{x}) = 0$, and let M be the largest invariant set in R. Then, every solution $\boldsymbol{x}(t)$ in Ω_r tends to M as $t \to \infty$.

Figure 5.8 shows the geometrical meaning of the theorem, where a trajectory starting from within the bounded region Ω_r is seen to converge to the largest invariant set M. If set R is an invariant set, then $M = R$. We give only an outline of the proof because detailed proof involves a number of mathematical concepts.

Proof. Consider a state trajectory, $\boldsymbol{x}(t)$, which starts from an arbitrary point $\boldsymbol{x}_0$ at $t = t_0$ in Ω_r. The trajectory stays in Ω_r for $t \geq t_0$, because $\dot{V}(\boldsymbol{x}) \leq 0$ implies $V(\boldsymbol{x}(t)) \leq V(\boldsymbol{x}_0)$ for $t \geq t_0$. Since $V(\boldsymbol{x})$ is continuous in $\boldsymbol{x}$ over the bounded region Ω_r, it is lower bounded in the region. Since $\boldsymbol{f}$ is continuous, $V(\boldsymbol{x})$ has continuous partial derivatives, and the region Ω_r is bounded, then $\dot{V}(x)$ is uniform continuous. Therefore, from Barbalat's lemma (see Sec. 5.3.3), it is derived that $\dot{V}(\boldsymbol{x}) \to 0$ as $t \to \infty$. This implies that all state trajectories staring from within Ω_r converge to set R. Furthermore, they converge to the largest invariant set M in set R. This is because, if they do not converge to the largest invariant set M, then they converge to a set in set R, but outside of the largest invariant set M. This contradicts the notion that M is the largest invariant set.

This theorem makes mention of the convergence of the trajectory to the maximum invariant set, but no mention of the asymptotic stability of the trajectory. Note that the asymptotic stability result in the local Lyapunov theorem can be viewed as a special case of the invariant set theorem, where the set M is only the origin.

Corollary 5.1. (Slotine and Li)

Consider a system $\dot{\boldsymbol{x}} = \boldsymbol{f}(\boldsymbol{x})$ with $\boldsymbol{f}$ continuous, and let $V(\boldsymbol{x})$ be a scalar function with continuous partial derivatives. Assume that within a neighborhood Ω of the origin,

- *$V(\boldsymbol{x})$ is locally positive definite.*
- *$\dot{V}(\boldsymbol{x})$ is negative semi-definite.*
- *The set R defined by $\dot{V}(\boldsymbol{x}) = 0$ contains no trajectories of $\dot{\boldsymbol{x}} = \boldsymbol{f}(\boldsymbol{x})$ other than the trivial trajectory $\boldsymbol{x} = 0$.*

Then, the origin is asymptotically stable. Furthermore, the largest connected region of the form Ω_r within Ω is a domain of attraction of the equilibrium point.

The above invariant set theorem can be extended to the **global invariant set theorem** by enlarging the involved region to include the whole space.

Theorem 5.9. (Global invariant set theorem)

Consider an autonomous system $\dot{\boldsymbol{x}} = \boldsymbol{f}(\boldsymbol{x})$ with $\boldsymbol{f}$ continuous. Let $V(\boldsymbol{x})$ be a scalar positive definite function with continuous first partial derivatives. Assume that:

- *$\dot{V}(\boldsymbol{x}) \leq 0$ over the whole state space.*
- *$V(\boldsymbol{x}) \to \infty$ as $\|\boldsymbol{x}\| \to \infty$.*

Let R be the set of all points within Ω_r where $\dot{V}(\boldsymbol{x}) = 0$, and M be the largest invariant set in R. Then, every solution globally asymptotically converges to M as $t \to \infty$.

Example 5.9. Consider the single pendulum in Example 5.7. The time derivative along the trajectory of the Lyapunov function given by Eq. (3) in Example 5.7 is

$$\dot{V}(\boldsymbol{x}) = -bx_2^2 \leq 0. \tag{1}$$

A set R, defined as $\dot{V}(\boldsymbol{x}) = 0$, is represented by $R = \{\boldsymbol{x} : x_2 = 0\}$. Then, the system is represented by

$$ml^2\dot{x}_2 + mgl\sin x_1 = 0. \tag{2}$$

A set satisfying Eq. (2) is represented by

$$M = \{\boldsymbol{x} : x_1 = 0, x_2 = 0\}, \tag{3}$$

and this is the maximum invariant set. Therefore, $\boldsymbol{x}(t)$ converges to the origin. It is noted that the convergence is not global but local, because the condition such that $V(\boldsymbol{x}) \to \infty$ as $\|\boldsymbol{x}\| \to \infty$ is not satisfied.

5.2.5. *Stability of Robot Position Control*

In general, the dynamic equation of an n-DOF (degree of freedom) robot can be represented by

$$\boldsymbol{M}(\boldsymbol{q})\ddot{\boldsymbol{q}} + \boldsymbol{C}(\boldsymbol{q}, \dot{\boldsymbol{q}})\dot{\boldsymbol{q}} + \boldsymbol{g}(\boldsymbol{q}) = \boldsymbol{\tau}, \tag{5.11}$$

where $\boldsymbol{q} \in \mathbb{R}^n$ is the joint variable, $\boldsymbol{\tau} \in \mathbb{R}^n$ is joint torque, $\boldsymbol{M} \in \mathbb{R}^{n\times n}$ is the inertia matrix, $\boldsymbol{C}\dot{\boldsymbol{q}}$ is the velocity square term consisting of centrifugal and Coriolis forces, and $\boldsymbol{g}$ is the gravitation term. The proportional and derivative (PD) position control with gravity compensator, which drives the joint variable $\boldsymbol{q}(t)$ to a desired constant joint position $\boldsymbol{q}_d$ in the joint space, is represented by

$$\boldsymbol{\tau} = \boldsymbol{K}_v\dot{\boldsymbol{e}} + \boldsymbol{K}_p\boldsymbol{e} + \boldsymbol{g}(\boldsymbol{q}), \tag{5.12}$$

where $\boldsymbol{e}$ is the joint position error defined by $\boldsymbol{e} = \boldsymbol{q}_d - \boldsymbol{q}$, $\boldsymbol{K}_v > 0$ is the velocity feedback gain and $\boldsymbol{K}_p > 0$ is the position feedback gain (Arimoto and Miyazaki, 1984).

Let us consider the asymptotic convergence to the desired position by this control input. Let a Lyapunov function candidate be:

$$V(t) = \frac{1}{2}(\dot{\boldsymbol{e}}^T\boldsymbol{M}(\boldsymbol{q})\dot{\boldsymbol{e}} + \boldsymbol{e}^T\boldsymbol{K}_p\boldsymbol{e}). \tag{5.13}$$

Evidently, $V(t)$ is positive definite. The time derivative along the trajectory of the system given by Eq. (5.11) is

$$\dot{V}(t) = \dot{\boldsymbol{e}}^T \left(\boldsymbol{M}(\boldsymbol{q})\ddot{e} + \frac{1}{2}\dot{\boldsymbol{M}}(\boldsymbol{q})\dot{e} + \boldsymbol{K}_p \boldsymbol{e} \right),$$

where by defining $\boldsymbol{C}$ suitably, $\dot{\boldsymbol{M}} - 2\boldsymbol{C}$ is a skew-symmetric matrix (refer Proposition 4.3 in Sec. 4.4 of Chapter 4). Using this property and the relations of $\ddot{\boldsymbol{e}} = -\ddot{\boldsymbol{q}}$ and $\dot{\boldsymbol{e}} = -\dot{\boldsymbol{q}}$, we can get;

$$\dot{V}(t) = -\dot{\boldsymbol{e}}^T \boldsymbol{K}_v \dot{\boldsymbol{e}} \leq 0. \tag{5.14}$$

Therefore, $V(t)$ is the Lyapunov function. A set defined by $\dot{V}(t) = 0$ is $\{\dot{\boldsymbol{e}} : \dot{\boldsymbol{e}} = 0\}$. In this region, the system dynamics can be represented by $\boldsymbol{M}(\boldsymbol{q})\ddot{e} + \boldsymbol{K}_p \boldsymbol{e} = 0$. Therefore, the invariant set is represented by $\{(\boldsymbol{e}, \dot{\boldsymbol{e}}) : \boldsymbol{e} = 0, \dot{\boldsymbol{e}} = 0\}$. This implies that the error converges to the origin as $t \to \infty$, i.e., $\boldsymbol{q}(t) \to \boldsymbol{q}_d$ and $\dot{\boldsymbol{q}}(t) \to 0$ as $\|\boldsymbol{e}\| \to \infty$. Furthermore, since $V(t) \to \infty$ as $\|\boldsymbol{e}\| \to \infty$, the system is globally asymptotically stable.

When the kinematic Jacobian $\boldsymbol{J}$, which makes a correlation between joint velocity and end-effector velocity, is non-singular, the dynamic equation of an n-DOF robot in Cartesian space is represented by

$$\boldsymbol{M}_r(\boldsymbol{q})\ddot{\boldsymbol{r}} + \boldsymbol{C}_r(\boldsymbol{q}, \dot{\boldsymbol{q}})\dot{\boldsymbol{r}} + \boldsymbol{g}_r(\boldsymbol{q}) = \boldsymbol{J}(\boldsymbol{q})^{-T}\boldsymbol{\tau}, \tag{5.15}$$

where $\boldsymbol{r}$ is the position and orientation vector of the end-effector, $\boldsymbol{M}_r = \boldsymbol{J}^{-T}\mathrm{MJ}^{-1}$, $\boldsymbol{C}_r = \boldsymbol{J}^{-T}(\boldsymbol{C} - \boldsymbol{M}_r\dot{\mathrm{J}})\boldsymbol{J}^{-1}$, and $\boldsymbol{g}_r = \boldsymbol{J}^{-T}\boldsymbol{g}$. If $\boldsymbol{C}$ is defined suitably, such that $\dot{\boldsymbol{M}} - 2\boldsymbol{C}$ is skew-symmetric, then $\dot{\boldsymbol{M}}_r - 2\boldsymbol{C}_r$ is also skew-symmetric. Therefore, let us define $\boldsymbol{F} = \boldsymbol{J}(\boldsymbol{q})^{-T}\boldsymbol{\tau}$ and consider a PD control input with gravity compensator in Cartesian space, which is similar to the PD control given by Eq. (5.12), and represented by

$$\boldsymbol{F} = \boldsymbol{K}_v \dot{\boldsymbol{e}}_r + \boldsymbol{K}_p \boldsymbol{e}_r + \boldsymbol{g}_r(\boldsymbol{q}),$$

where $\boldsymbol{e}_r$ is the position error in Cartesian space defined by $\boldsymbol{e}_r = \boldsymbol{r}_d - \boldsymbol{r}$. Then, the control input is given by

$$\boldsymbol{\tau} = \boldsymbol{J}^T(\boldsymbol{K}_v \dot{\boldsymbol{e}}_r + \boldsymbol{K}_p \boldsymbol{e}_r) + \boldsymbol{g}(\boldsymbol{q}). \tag{5.16}$$

Since a motion range of end-effector is not the whole Cartesian space, it can be proved that the origin is locally asymptotically stable.

5.3. Non-Autonomous Systems

5.3.1. *Concepts of Stability for Non-Autonomous Systems*

Consider the non-autonomous system,

$$\dot{\boldsymbol{x}} = \boldsymbol{f}(\boldsymbol{x}, t) \tag{5.17}$$

where it is assumed that $\boldsymbol{f}(\boldsymbol{x}, t)$ is continuous with respect to x and t. The equilibrium points $\boldsymbol{x}^*$ for the non-autonomous system are defined by

$$\boldsymbol{f}(\boldsymbol{x}^*, t) = 0 \quad \text{for all } t \geq t_0. \tag{5.18}$$

Note that this equation should be satisfied for arbitrary $t \geq t_0$. This implies that the system should be able to stay at point $\boldsymbol{x}^*$ all the time. As with autonomous systems, we can assume without loss of generality that the equilibrium point is the origin. For non-autonomous systems, stability, asymptotic stability, and exponential stability are defined by extending the definitions for autonomous systems as follows.

Definition 5.10. (Stability for non-autonomous systems)

The equilibrium point of the system Eq. (5.17) is said to be stable at t_0 if for any $\varepsilon > 0$, there exists a scalar $r(\varepsilon, t_0) > 0$ such that if $\|\boldsymbol{x}(t_0)\| < r(t_0)$ then $\|\boldsymbol{x}(t)\| < \varepsilon$ for all $t \geq t_0$.

If the equilibrium point is not stable, then it is said to be **unstable**. The definition of stability for non-autonomous systems differs from that of autonomous systems in that the radius $r(\varepsilon, t_0) > 0$ of the initial region may depend on the initial time t_0.

Definition 5.11. (Asymptotic stability for non-autonomous systems)

The equilibrium point of Eq. (5.17) is said to be asymptotically stable at t_0 if:

- It is stable.
- There exists a scalar $r(t_0) > 0$ such that if $\|\boldsymbol{x}(t_0)\| < r(t_C)$ then $\|\boldsymbol{x}(t)\| \to 0$ as $t \to \infty$.

This definition means that asymptotic stability requires that there exists an attractive region for every initial time t_0.

Definition 5.12. (Exponential stability for non-autonomous systems)

The equilibrium point of Eq. (5.17) is said to be exponentially stable at t_0 if there exists two positive scalars a and b, such that, for sufficiently small $\boldsymbol{x}(t_0)$, $\|\boldsymbol{x}(t)\| \leq a\|\boldsymbol{x}(t_0)\|e^{-b(t-t_0)}$ for all $t \geq t_0$.

Definition 5.13. (Global asymptotic stability for non-autonomous systems) The equilibrium point of Eq. (5.17) is said to be globally asymptotically stable at t_0 if $\boldsymbol{x}(t) \rightarrow 0$ as $t \rightarrow \infty$ for arbitral $\boldsymbol{x}(t_0)$.

These definitions relate to initial time t_0. Most applications of non-autonomous systems, however, need properties that do not depend on initial time. Hence, there are two concepts — **uniformly stable** and **uniformly asymptotically stable** — to analyze such properties.

Definition 5.14. (Uniform stability)

The equilibrium point of Eq. (5.17) is said to be uniformly stable if the scalar $\varepsilon > 0$ in Definition 5.10 can be chosen independently of t_0.

Definition 5.15. (Uniform asymptotic stability)

The equilibrium point of Eq. (5.17) is said to be uniformly asymptotically stable if:

- It is uniformly stable.
- There exists a scalar $r > 0$, which is independent of t_0, such that any trajectory $\boldsymbol{x}(t)$ with initial states $\boldsymbol{x}(t_0)$ of $\|\boldsymbol{x}(t_0)\| < r$ converges to the origin uniformly in t_0.

The uniform convergence in terms of t_0 means that for $0 < r_2 < r_1 < r$, any trajectory starting from $\boldsymbol{x}(t_0)$ with the condition of $\|\boldsymbol{x}(t_0)\| < r_1$ converges into a smaller range $\|\boldsymbol{x}(t)\| < r_2$ after a time period $T(r_1, r_2)$ which is independent of t_0; i.e., if $\|\boldsymbol{x}(t_0)\| < r_1$ then $\|\boldsymbol{x}(t)\| < r_2$ for arbitrary $t \geq t_0 + T(r_1, r_2)$.

Example 5.10. Consider the non-autonomous system:

$$\dot{x}(t) = (t\cos t - 1)x(t), x(t_0) = x_0. \tag{1}$$

The general solution of this system is:

$$x(t) = \frac{x_0 e^{\cos(t)+t\sin(t)-t}}{e^{\cos(t_0)+t_0\sin(t_0)-t_0}}. \tag{2}$$

From a relation of $e^{\cos(t)+t\sin(t)-t} = e^{\cos(t)}e^{t\sin(t)-t} \leq e$, we get

$$|x(t)| \leq \frac{x_0 e}{e^{\cos(t_0)+t_0\sin(t_0)-t_0}} \quad \text{for all } t \geq t_0 \geq 0. \tag{3}$$

The origin, then, is stable. But a solution at $t = (2n+1)\pi$ with $t_0 = 2n\pi$ is represented by

$$x((2n+1)\pi) = x_0 e^{2n\pi-1}. \tag{4}$$

This shows that $x((2n+1)\pi) \to \infty$ as $n \to \infty$. Therefore, the origin is not uniformly stable because $r > 0$ such that $\|\boldsymbol{x}(t_0)\| < r$ depends on initial time t_0.

Example 5.11. Consider the system:

$$\dot{x}(t) = (\cos t - 1)x(t), x(t_0) = x_0. \tag{1}$$

The general solution of this system is:

$$x(t) = \frac{x_0 e^{\sin(t)-t}}{e^{\sin(t_0)-t_0}} \tag{2}$$

From a relation of $e^{\sin(t)-t}/e^{\sin(t_0)-t_0} = e^{\sin(t)-\sin(t_0)-t+t_0} \leq e^2$, we get

$$|x(t)| \leq |x_0|e^2 \quad \text{for all } t \geq t_0 \geq 0 \tag{3}$$

This shows that the origin is uniformly stable. Furthermore, since the trajectory staring from arbitrary initial point x_0 converges to the origin uniformly, it is uniformly asymptotically stable.

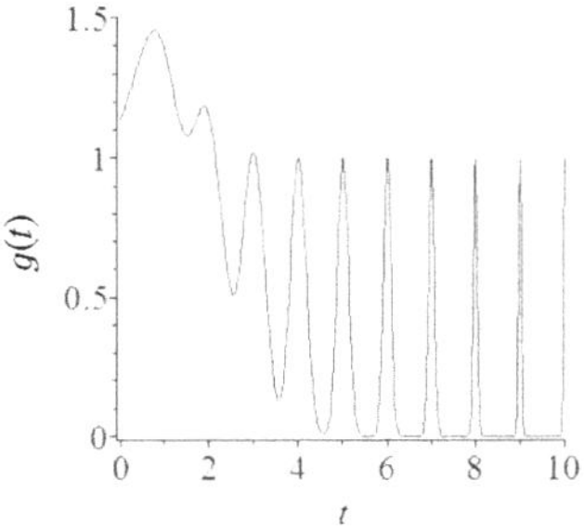

(a) $g(t)$

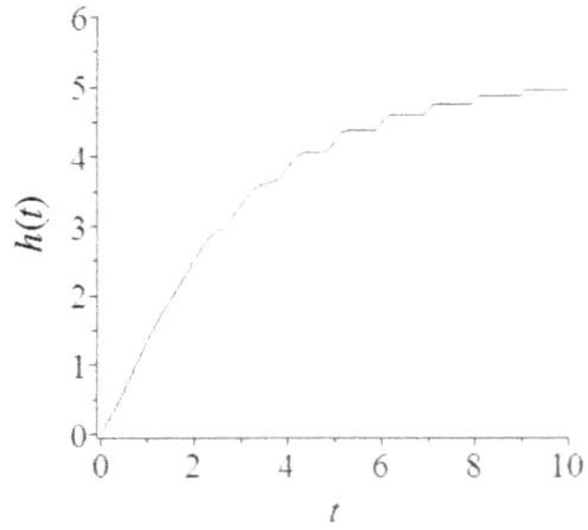

(b) $h(t)=\int_0^t g(t)dt$

Fig. 5.9. Pulse train functions $g(t)$.

5.3.2. *Stability Theorems for Non-Autonomous Systems*

The stability analyses of non-autonomous systems are derived by extending those for autonomous systems. We cannot, however, apply the invariant set theorem to non-autonomous systems. Let us consider this using the following example.

Example 5.12. Consider the continuous function:

$$g(t) = \sum_{n=0}^{\infty} e^{-2^n (t-n)^2} > 0. \tag{1}$$

This function is a pulse train, as shown in Fig. 5.9(a), and it has an upper limit a such that:

$$g(t) < a. \tag{2}$$

The time integral of Eq. (1) is shown in Fig. 5.9 (b) and has a finite limit b such that:

$$h(t) = \int_0^t g(t)dt < b. \tag{3}$$

Consider the stability of the system:

$$\dot{x}(t) = \frac{\dot{g}(t)}{g(t)} x(t). \tag{4}$$

Let us consider as a Lyapunov function candidate:

$$V(x,t) = \frac{x(t)^2}{g(t)^2}(1+b-h(t)). \tag{5}$$

This is positive definite because

$$V(x,t) > \frac{x(t)^2}{a^2}(1+b-h(t)) > \frac{x(t)^2}{a^2} > 0$$

The time derivative along the trajectory of Eq. (4) is

$$\dot{V}(x,t) = \frac{2x(\dot{x}g - x\dot{g})}{g^3}(1+b-h) - \frac{x^2}{g^2}h' = -\frac{x^2}{g} < -\frac{x^2}{a} < 0, \tag{6}$$

which is negative definite. If the invariant set theorem can be applied to the stability analysis for the system, the origin is asymptotically stable. However, the general solution of Eq. (4) is:

$$\dot{x}(t) = \frac{g(t)}{g(t_0)}x_n(t_0). \tag{7}$$

It is thus clear that the origin is not asymptotically stable. This shows that the invariant set theorem cannot be applied to the non-autonomous system.

When the Lyapunov function of a non-autonomous system $V(\boldsymbol{x}, t)$ is a time varying function, the time derivative along the trajectory of Eq. (5.17) is represented by

$$\dot{V}(\boldsymbol{x},t) = \frac{\partial V}{\partial t} + \frac{\partial V}{\partial \boldsymbol{x}}\dot{\boldsymbol{x}} = \frac{\partial V}{\partial t} + \frac{\partial V}{\partial \boldsymbol{x}}\boldsymbol{f}(\boldsymbol{x},t). \tag{5.19}$$

To analyze this function, let us define **positive definite** and **decrescent** for time varying functions.

Definition 5.16. (Positive definite for time varying functions) A scalar time varying function $V(\boldsymbol{x}, t)$ is said to be locally positive definite if:

- $V(0,t) = 0$.
- There exists a time invariant positive definite function $V_0(\boldsymbol{x})$ such that $V(\boldsymbol{x},t) > V_0(\boldsymbol{x}) > 0$ for all non-zero $\boldsymbol{x}$.

In the above definition, a function $V(\boldsymbol{x}, t)$ is said to be locally **positive semi-definite** if $V_0(\boldsymbol{x}) \geq 0$. **Negative definite** is also defined similarly for scalar time varying functions. When a time varying function $f(\boldsymbol{x}, t)$ is positive definite (positive semi-definite) or negative definite (negative semi-definite), it is described by $f(\boldsymbol{x}, t) > 0$ ($f(\boldsymbol{x}, t) \geq 0$) or $f(\boldsymbol{x}, t) < 0$ ($f(\boldsymbol{x}, t) \leq 0$).

Definition 5.17. (Decrescent)

A scalar time varying function $V(\boldsymbol{x}, t)$ is said to be decrescent if:

- $V(0, t) = 0$.
- There exists a time invariant positive definite function $V_1(\boldsymbol{x})$ such that $V(\boldsymbol{x}, t) \leq V_1(\boldsymbol{x})$ for arbitrary $t \geq t_0$.

Example 5.13. Consider a scalar time varying function:

$$V(\boldsymbol{x}, t) = (x_1^2 + x_2^2)(1 + \sin^2 t), \boldsymbol{x} = [x_1, x_2]^T.$$

Since $V(0, t) = 0$ and $V(\boldsymbol{x}, t) \leq 2(x_1^2 + x_2^2)$, this function is decrescent.

The Lyapunov stability for non-autonomous systems is summarized as the following theorem.

Theorem 5.10. (Lyapunov theorem for non-autonomous systems)

(1) *Stability*: *The origin is stable in the sense of Lyapunov if in a region B_r around the origin, there exists a scalar time varying function $V(\boldsymbol{x}, t)$ with continuous partial derivatives such that:*

- $V(\boldsymbol{x}, t) > 0$.
- $\dot{V}(\boldsymbol{x}, t) \leq 0$.

(2) *Uniform stability*: *The origin is uniformly stable if:*

- *It is stable in the sense of Lyapunov.*
- *$V(\boldsymbol{x}, t)$is decrescent.*

(3) ***Uniform asymptotic stability***: *The origin is uniformly asymptotically stable if*:

- *It is uniformly stable.*
- $\dot{V}(\boldsymbol{x}, t) < 0$.

(4) ***Globally uniform asymptotic stable***: *The origin is globally uniformly asymptotically stable if*:

- *It is uniformly asymptotically stable.*
- $V(\boldsymbol{x}, t) \to \infty$ *as* $\|\boldsymbol{x}\| \to \infty$.

As in the case of autonomous systems, if $V(\boldsymbol{x}, t) > 0$ and $\dot{V}(\boldsymbol{x}, t) \leq 0$, then $V(\boldsymbol{x}, t)$ is said to be a Lyapunov function. In order to prove this theorem, the concept of **class *K*** is introduced.

Definition 5.18. (Class-*K*)

A continuous positive definite function $\alpha(s)$ is said to be of class K if:

- $\alpha(0) = 0$.
- $\alpha(s)$ is non-decreasing for s.

Following theorem is derived using the concept of class K.

Theorem 5.11.

(1) ***Stability***: *The origin is stable in the sense of Lyapunov if there exists a scalar function* $V(\boldsymbol{x}, t)$ *with continuous first order derivatives and a class K function* $\alpha(\|\boldsymbol{x}\|)$ *such that* $V(\boldsymbol{x}, t) \geq \alpha(\|\boldsymbol{x}\|) > 0$ *and* $\dot{V}(\boldsymbol{x}, t) \leq 0$ *for arbitrary* $\boldsymbol{x} \neq 0$.

(2) ***Uniform stability***: *The origin is uniformly stable if it is stable in the sense of Lyapunov and if there exists a scalar class K function* $\beta(|\boldsymbol{x}\|)$ *such that* $V(\boldsymbol{x}, t) \leq \beta(\|\boldsymbol{x}\|)$.

(3) ***Uniform asymptotic stability***: *The origin is uniformly asymptotically stable if it is uniformly stable and if there exists a scalar class K function* $\gamma(\|\boldsymbol{x}\|)$ *such that* $\dot{V}(\boldsymbol{x}, t) \leq -\gamma(\|\boldsymbol{x}\|) < 0$.

(4) ***Globally uniform asymptotic stability***: *The origin is globally uniformly asymptotically stable if it is uniformly asymptotically stable in the whole state space and if* $\alpha(\|\boldsymbol{x}\|) \to \infty$ *as* $\|\boldsymbol{x}\| \to \infty$.

The proof of this theorem is omitted (refer to Slotine and Li, 1991). It is noted that for autonomous systems, the origin is guaranteed to be asymptotically stable if $V > 0$ and $\dot{V} \leq 0$. However, the origin of non-autonomous systems is not guaranteed to be asymptotically stable even if these are satisfied, as shown in Example 5.12. A condition that $V(\boldsymbol{x}, t)$ is decrescent is required to guarantee asymptotic stability.

The following exponential stability theorem is useful because it permits us to estimate convergence rates in some nonlinear systems.

Theorem 5.12. (Exponential stability of non-autonomous systems)

The origin of Eq. (5.17) *is said to be exponentially stable if* $\boldsymbol{f}(\boldsymbol{x}, t)$ *has continuous, bounded, first partial derivatives with respect to* $\boldsymbol{x}$ *and* t, *for all* $\boldsymbol{x}$ *in a region* B_r *around the origin, and for all* $t \geq 0$; *in addition, if there exist* $V(\boldsymbol{x}, t)$ *and four positive constants* $\alpha_1, \alpha_2, \alpha_3, \alpha_4 > 0$ *such that* $\alpha_1 \|\boldsymbol{x}\|^2 \leq V(\boldsymbol{x}, t) \leq \alpha_2 \|\boldsymbol{x}\|^2, \dot{V}(\boldsymbol{x}, t) \leq -\alpha_3 \|\boldsymbol{x}\|^2$, *and* $\|\partial V(\boldsymbol{x}, t)/\partial \boldsymbol{x}\| \leq \alpha_4 \|\boldsymbol{x}\|$ *for all* $\boldsymbol{x}$ *in a region* B_r *for arbitrary* $t \geq 0$.

Example 5.14. Consider the system:

$$\begin{aligned} \dot{x}_1(t) &= -x_1(t) + e^{-t} x_2(t) \\ \dot{x}_2(t) &= -x_1(t) - x_2(t). \end{aligned} \tag{1}$$

Let us consider a Lyapunov function candidate:

$$V(x, t) = x_1(t)^2 + (1 + e^{-t}) x_2(t)^2. \tag{2}$$

This function is positive definite and decrescent because of:

$$x_1^2 + 2x_2^2 \geq V(x, t) \geq x_1^2 + x_2^2 > 0.$$

A time derivative along the trajectory is:

$$\begin{aligned} \dot{V}(x, t) &= -2x_1^2 - 2x_1 x_2 - (2 + 3e^{-t}) x_2^2 \\ &= -x_1^2 - (x_1 + x_2)^2 - (1 + 3e^{-t}) x_2^2 < 0 \end{aligned} \tag{3}$$

Since $\dot{V}(x, t)$ is negative definite and decrescent, and $\alpha(\|\boldsymbol{x}\|) = x_1^2 + x_2^2 \to \infty$ as $\|\boldsymbol{x}\| \to \infty$, the origin is globally uniformly asymptotically stable.

5.3.3. *Barbalat's Lemma*

Usually it is hard to confirm the asymptotic stability of time varying systems because it is very difficult to find Liapunov functions with a negative definite derivative. In the case of autonomous systems, if $\dot{V}$ is negative semi-definite, then it is possible to know the asymptotic behaviors through the use of invariant set theorems. However, invariant set theorems are not available for time varying systems. **Babarlat's lemma**, which indicates asymptotical properties of functions and their derivative, is then useful. The followings are important as properties of functions and their derivatives.

(1) $\dot{f} \to 0$ does not imply $f \to$ constant as $t \to \infty$.

E.g., consider a function $f(t) = \cos(\log t)$. Its time derivative is $\dot{f}(t) = -\sin(\log t)/t$ and $\dot{f} \to 0$ as $t \to \infty$. However, $f(t)$ keeps vibrating in the interval $[-1, 1]$.

(2) $f \to$ constant does not imply $\dot{f} \to 0$ as $t \to \infty$.

E.g., consider a function $f(t) = e^{-t}\cos(e^t)$. Its time derivative is $\dot{f}(t) = -\sin(\log t)/t$ and $f \to 0$ as $t \to \infty$. However, $\dot{f}(t)$ keeps vibrating in the interval $[-1, 1]$.

Let us define **continuity** and **uniform continuity** of functions. Here, we use a mathematical symbol '$\Rightarrow$', which means 'implies that'.

Definition 5.19. (Continuity of functions)

A function $f(t)$ is said to be continuous on $[0, \infty)$ if:

- There exists $r(M, t_0) > 0$ for any $t_0 \geq 0$ and any $M > 0$.
- $|t - t_0| < r$ for all $t \geq 0 \quad \Rightarrow \quad |f(t) - f(t_0)| < M$.

Definition 5.20. (Uniform continuity of functions)

A function $f(t)$ is said to be uniform continuous on $[0, \infty)$ if:

- There exists $r(M) > 0$ for any $M > 0$.
- $|t - t_0| < r$ for any $t_0 \geq 0$, $t \geq 0 \quad \Rightarrow |f(t) - f(t_0)| < M$.

The difference between continuity and uniform continuity is that the r of the former depends on the time t_0 and that of later does not depend on it. From the mean-value theorem for a differentiable function $f(t)$, in the interval $[t_0, t]$, there exits t_1 such that $f(t) -$

$f(t_0) = \dot{f}(t_1)(t-t_0)$. When $M_1 > 0$ is an upper bound on the function $|\dot{f}(t)|$, we can derive $|f(t) - f(t_0)| < M$ by putting $r = M/M_1$. This implies that a differentiable function is uniformly continuous if its derivative is bounded.

Lemma 5.1. (Barbalat's lemma)

If the differentiable function $f(t)$ has a finite limit as $t \to \infty$, and if $\dot{f}(t)$ is uniformly continuous, then $\dot{f}(t) \to 0$ as $t \to \infty$.

Proof. This will be proven by contradiction. Suppose that $\dot{f}$ does not converge to zero as $t \to \infty$. Then, for an arbitrary $T > 0$, there exists $t \geq T$ and $\varepsilon > 0$ such that $|\dot{f}(t)| \geq \varepsilon$. Since $\dot{f}(t)$ is uniformly continuous, there exits r such that $|\dot{f}(t_1) - \dot{f}(t_2)| < \varepsilon/2$ for any t_1 and t_2 satisfying $|t_1 - t_2| < r$. This implies that, for all t_i,

$$\left| \int_{t_i-r}^{t_i+r} \dot{f}(\tau)d\tau \right| = \int_{t_i-r}^{t_i+r} |\dot{f}(\tau)|d\tau \geq \frac{\varepsilon}{2}2r = \varepsilon r,$$

where the left equality comes from the fact $\dot{f}$ keeps a constant sign over the integration interval because $\dot{f}$ is continuous and lower bounded as shown by:

$$\begin{aligned} |\dot{f}(t)| &= |f(t_2) - (f(t_2) - f(t))| \\ &\geq |f(t_2)| - |f(t_2) - f(t)| \geq \varepsilon - \frac{\varepsilon}{2} = \frac{\varepsilon}{2}. \end{aligned}$$

This would mean that $\int_0^t \dot{f}(\tau)d\tau \to \infty$ as $t \to \infty$. At the same time, since $f(t)$ is bounded, $\int_0^t \dot{f}(\tau)d\tau$ has a limit of $f(\infty) - f(0)$ as $t \to \infty$. This is a contradiction. Therefore, $\dot{f} \to 0$ as $t \to \infty$.

An immediate and practical corollary of Barbalat's lemma can be stated as the following corollary.

Corollary 5.2. *If the differentiable function $f(t)$ has a finite limit as $t \to \infty$, and $\ddot{f}(t)$ exits and is bounded, then $\dot{f}(t) \to 0$ as $t \to \infty$.*

This corollary is derived by the replacement uniform continuity of $\dot{f}(t)$ in Barbalet's lemma to the existence and boundedness of $\ddot{f}(t)$. When the Barbalat's lemma is applied to the analysis of dynamic systems, a Lyapunov-like lemma is derived.

Lemma 5.2. (Lyapunov-like lemma)

If a scalar function $V(\boldsymbol{x}, t)$ satisfies the following conditions:

- $V(\boldsymbol{x}, t)$ *is lower bounded.*
- $\dot{V}(\boldsymbol{x}, t) \leq 0$.
- $\dot{V}(\boldsymbol{x}, t)$ *is uniformly continuous in time.*

Then $\dot{V}(\boldsymbol{x}, t) \to 0$ *as* $t \to \infty$.

Notice that the first two assumptions imply that $V(\boldsymbol{x}, t)$ tends to a limit as $t \to \infty$. This lemma is derived from Barbalat's lemma.

Example 5.15. Assume that $a(t) > 0$ is bounded and it's time derivative is also bounded. Consider the uniform stability of the following system:

$$\begin{aligned} \dot{x}_1(t) &= -x_1(t) + a(t)x_2(t) \\ \dot{x}_2(t) &= -a(t)x_1(t). \end{aligned} \tag{1}$$

Now, let us define a lower bounded function

$$V(\boldsymbol{x}, t) = x_1(t)^2 + x_2(t)^2 \geq 0. \tag{2}$$

Its time derivative is represented by:

$$\begin{aligned} \dot{V}(\boldsymbol{x}, t) &= 2x_1\dot{x}_1 + 2x_2\dot{x}_2 \\ &= 2x_1(-x_1 + ax_2) + 2x_2(-ax_1) = -2x_1^2 \leq 0. \end{aligned} \tag{3}$$

This implies $V(\boldsymbol{x}, t) \leq V(\boldsymbol{x}, 0)$, and x_1 and x_2 are bounded. It should be noted that, since the system is a non-autonomous system, we cannot conclude that x_1 converges to zero from the invariant set theorem. Now, the time derivative of $\dot{V}(\boldsymbol{x}, t)$ along the trajectory is:

$$\ddot{V}(\boldsymbol{x}, t) = -4x_1(-x_1 + ax_2). \tag{4}$$

Since $a(t)$ is bounded from the assumption, $\ddot{V}(\boldsymbol{x}, t)$ is also bounded. Therefore, $\dot{V}(\boldsymbol{x}, t)$ is uniformly continuous. From this, Lyapunov-like lemma, $\dot{V}(\boldsymbol{x}, t) \to 0$ as $t \to \infty$, that is, $x_1 \to 0$ as $t \to \infty$. Then, $\dot{x}_2 \to 0$ as $t \to \infty$. The asymptotic convergence to the origin is not guaranteed because we cannot conclude that $x_2 \to 0$ as $t \to \infty$.

5.3.4. *Stability of Adaptive Control for Robots*

The analysis of robot dynamic systems require dynamic parameters such as the inertial tensor of link, the mass of link, and the mass center mass of link, but it is not easy to measure these precisely. Furthermore, the parameters of the end links vary depending on target objects. Adaptive control based on the dynamics models, in which the parameters are estimated in real time and the control input is generated based on these estimations, is one way to achieve effective robot controls.

Assume that the desired joint variable $\boldsymbol{q}_d(t)$, desired joint velocity $\dot{\boldsymbol{q}}_d(t)$, and desired joint acceleration $\ddot{\boldsymbol{q}}_d(t)$ are continuous and bounded. The robot dynamics given by Eq. (5.11) is linear with respect to the dynamic parameters that are chosen appropriately (see Sec. 4.4 of Chapter 4), and it can be represented using the reference velocity by

$$\begin{aligned} &\boldsymbol{M}(\boldsymbol{q}(t))\ddot{\boldsymbol{q}}_r(t) + \boldsymbol{C}(\boldsymbol{q}(t), \dot{\boldsymbol{q}}(t))\dot{\boldsymbol{q}}_r(t) + \boldsymbol{g}(\boldsymbol{q}(t)) \\ &\quad = \mathbf{Y}(\boldsymbol{q}(t), \dot{\boldsymbol{q}}(t), \dot{\boldsymbol{q}}_r(t), \ddot{\boldsymbol{q}}_r(t))\sigma, \end{aligned} \tag{5.20}$$

where $\dot{\boldsymbol{q}}_r \in \mathbb{R}^n$ is the reference velocity vector, $\boldsymbol{\sigma} \in \mathbb{R}^p$ is the dynamic parameter vector, and $\boldsymbol{Y}(\boldsymbol{q}, \dot{\boldsymbol{q}}, \dot{\boldsymbol{q}}_r, \ddot{\boldsymbol{q}}_r) \in \mathbb{R}^{n\times p}$ is the regressor with respect to $\boldsymbol{\sigma}$. Let us define the trajectory error for the desired joint variable by

$$\boldsymbol{e}(t) = \boldsymbol{q}(t) - \boldsymbol{q}_d(t); \tag{5.21}$$

$\dot{\boldsymbol{q}}_r$is defined by

$$\dot{\boldsymbol{q}}_r(t) = \dot{\boldsymbol{q}}_d(t) + \boldsymbol{\Lambda}\boldsymbol{e}(t) \tag{5.22}$$

where $\boldsymbol{\Lambda} \in \mathbb{R}^{n\times n}$ is a positive symmetric matrix. Consider an adaptive control whose control law and adaptive law are given by

$$\boldsymbol{\tau}(t) = \boldsymbol{Y}(\boldsymbol{q}(t), \dot{\boldsymbol{q}}(t), \dot{\boldsymbol{q}}_r(t), \ddot{\boldsymbol{q}}_r(t))\hat{\boldsymbol{\sigma}}(t) - \mathbf{Ks}(t), \tag{5.23}$$

$$\dot{\hat{\boldsymbol{\sigma}}}(t) = -\boldsymbol{\Gamma}\boldsymbol{Y}(\boldsymbol{q}(t), \dot{\boldsymbol{q}}(t), \dot{\boldsymbol{q}}_r(t), \ddot{\boldsymbol{q}}_r(t))\mathbf{s}(t), \tag{5.24}$$

where $\hat{\boldsymbol{\sigma}}$ is an estimate of $\boldsymbol{\sigma}$, $\boldsymbol{\Gamma}, \boldsymbol{K} \in \mathbb{R}^{n\times n}$ are positive symmetric matrices, and $\mathbf{s}(t)$ is defined by

$$\boldsymbol{s}(t) = \dot{\boldsymbol{q}}(t) - \boldsymbol{q}_r(t). \tag{5.25}$$

This system is a non-autonomous because the desired variables are given by the time functions. The first term of the right side of the control law is a dynamic compensation term based on the parameter estimate, and the second term is a PD control term. Then, we can derive the following theorem by applying the Lyapunov-like lemma.

Theorem 5.13. (Adaptive trajectory control for robots)
For the robot dynamic systems represented by Eq. (5.11), *apply the control law given by Eq.* (5.23) *and the adaptation law given by Eq.* (5.24), *where the desired* $\boldsymbol{q}_d(t)$, $\dot{\boldsymbol{q}}_d(t)$, *and* $\ddot{\boldsymbol{q}}_d(t)$ *are time continuous and bounded. Then,* $\boldsymbol{q}(t) \to \boldsymbol{q}_d(t)$ *and* $\dot{\boldsymbol{q}}(t) \to \dot{\boldsymbol{q}}_d(t)$ *as* $t \to \infty$.

Proof. By substituting Eq. (5.23) into Eq. (5.11), the robot dynamics can be rewritten as

$$\boldsymbol{M}(\boldsymbol{q})\dot{\boldsymbol{s}}(t) + \boldsymbol{C}(\boldsymbol{q}, \dot{\boldsymbol{q}})\boldsymbol{s}(t) = \boldsymbol{Y}(\boldsymbol{q}, \dot{\boldsymbol{q}}, \dot{\boldsymbol{q}}_r, \ddot{\boldsymbol{q}}_r)\tilde{\boldsymbol{\sigma}}(t) - \boldsymbol{K}\boldsymbol{s}(t), \tag{1}$$

where $\tilde{\boldsymbol{\sigma}}(t) = \hat{\boldsymbol{\sigma}}(t) - \boldsymbol{\sigma}$ is the estimate error of parameters. This implies that the state vector of the dynamics consists of $\boldsymbol{s}(t)$ and $\tilde{\boldsymbol{\sigma}}(t)$. Then, consider a Lyapunov function candidate given by:

$$V(t) = \frac{1}{2}\left(\boldsymbol{s}(t)^T\boldsymbol{M}(t)\boldsymbol{s}(t) + \tilde{\boldsymbol{\sigma}}(t)^T\boldsymbol{\Gamma}\tilde{\boldsymbol{\sigma}}(t)\right) \geq 0. \tag{2}$$

The time derivative of $V(t)$ is represented by:

$$\dot{V}(t) = \frac{1}{2}\boldsymbol{s}^T(\dot{\boldsymbol{M}} + 2\boldsymbol{M}\boldsymbol{s}) + \tilde{\boldsymbol{\sigma}}^T\boldsymbol{\Gamma}\dot{\tilde{\boldsymbol{\sigma}}}.$$

Since $\dot{\boldsymbol{M}} - 2\boldsymbol{C}$ is skew-symmetric, by substituting the estimation law into the above equation, we get:

$$\dot{V}(t) = -\boldsymbol{s}^T\mathbf{K}\mathbf{s} \leq 0. \tag{3}$$

Therefore, $V(t)$ is a Lyapunov function, and $\boldsymbol{s}$ and $\tilde{\boldsymbol{\sigma}}$ are bounded. From $\boldsymbol{s} = -\dot{\boldsymbol{e}} - \boldsymbol{\Lambda}\boldsymbol{e}$, $\boldsymbol{e}$ and $\dot{\boldsymbol{e}}$ are bounded. Using this fact, we can show sequentially that $\boldsymbol{q}, \dot{\boldsymbol{q}}, \dot{\boldsymbol{q}}_r$ and $\ddot{\boldsymbol{q}}_r$ are bounded. Since $\boldsymbol{M}(\boldsymbol{q})$ is the non-singular matrix, Eq. (1) shows that $\dot{\boldsymbol{s}}$ is bounded.

The time derivative of $\dot{V}(t)$ is given by

$$\ddot{V}(t) = -2\boldsymbol{s}^T\boldsymbol{K}\dot{s}. \tag{4}$$

Evidently, $\ddot{V}(t)$ is bounded. This implies that $\dot{V}(t)$ is uniformly continuous. From the Lyapunov-like lemma, $\dot{V}(t) \to 0$ as $t \to \infty$.

Therefore, $\boldsymbol{s} \to 0$ as $t \to \infty$, and $\boldsymbol{e} \to 0$ and $\dot{\boldsymbol{e}} \to 0$ as $t \to \infty$. This means $\boldsymbol{q}(t) \to \boldsymbol{q}_d(t)$ and $\dot{\boldsymbol{q}}(t) \to \dot{\boldsymbol{q}}_d(t)$ as $t \to \infty$.

The convergence to the desired trajectory does not guarantee that the estimate parameter $\hat{\boldsymbol{\sigma}}(t)$ converges to the true parameter $\boldsymbol{\sigma}$, but the estimate parameter converges to a constant $\boldsymbol{\sigma}(\infty)$ asymptotically. If $\boldsymbol{Y}$ satisfies the '**persistently exciting condition**' and is uniformly continuous, the estimate parameters converge to the true value asymptotically (see Morgan and Narendra, 1977; Boyd and Sastry, 1986). By a persistently exciting condition on $\boldsymbol{Y}$, we mean that there exist strictly positive constants α and T such that for any $t > 0$,

$$\int_t^{t+T} \boldsymbol{Y}^T \boldsymbol{Y} d\tau \geq \alpha \mathbf{I}. \tag{5.26}$$

The proof of this is outlined as follows: $\hat{\boldsymbol{\sigma}} \to$ constant and $\boldsymbol{s} \to 0$ as $t \to \infty$; $\dot{\boldsymbol{s}}$ is uniformly continuous, which is derived from Eq. (1). From Barbalat's lemma, $\dot{\boldsymbol{s}} \to 0$ as $t \to \infty$. By substituting these into Eq. (1), we get:

$$\boldsymbol{Y}\tilde{\sigma} \to 0$$

as $t \to \infty$. By multiplying $\boldsymbol{Y}^T$ from the left side to the above equation and integrating the equation for a period of time T, we get:

$$\int_t^{t+T} \boldsymbol{Y}^T \boldsymbol{Y} dt \tilde{\sigma} \to 0.$$

If Eq. (5.26) is satisfied, then $\tilde{\boldsymbol{\sigma}} \to 0$ as $t \to \infty$ and $\hat{\boldsymbol{\sigma}} \to \boldsymbol{\sigma}$ as $t \to \infty$.

Exercises

5.1. Discuss the stability around the equilibrium point $x = 1$ of the system: $\dot{x} = x(1 - x)$.

5.2. Discuss the stability for the system: $\ddot{x}(t) + (1+t)\dot{x}(t) + x(t) = 0$.

5.3. Assume that a closed-loop error system, including an estimation law for an unknown parameter is given by

$$\dot{e} = -2e + \sigma \sin(t)$$
$$\dot{\sigma} = -e \sin(t),$$

where e is the error and σ is the parameter estimate. Then, prove that $e \to 0$ as $t \to \infty$.

5.4. Prove that, for any symmetric matrix $\boldsymbol{M}(t)$, a time derivative of $\boldsymbol{x}(t)^T \boldsymbol{M}(t)\boldsymbol{x}(t)$, which is the quadratic form in $\boldsymbol{x}(t)$, is:

$$\frac{d}{dt}(\boldsymbol{x}^T \boldsymbol{M}\boldsymbol{x}) = 2\dot{\boldsymbol{x}}^T \boldsymbol{M}\boldsymbol{x} + \boldsymbol{x}^T \dot{\boldsymbol{M}} x.$$

5.5. Prove by applying the invariant set theorem that if the linear system

$$\dot{\boldsymbol{x}} = \boldsymbol{A}\boldsymbol{x}, \quad \boldsymbol{y} = \boldsymbol{c}^T \boldsymbol{x},$$

is observable, and there exists a positive symmetric matrix **P** such that

$$\boldsymbol{A}^T \boldsymbol{P} + \boldsymbol{P}\boldsymbol{A} = -\boldsymbol{c}\boldsymbol{c}^T,$$

then the linear system is asymptotically stable.

5.6. Prove that the system adopting the control law given by Eq. (5.16) is asymptotically stable.

Chapter 6

Robot Hand Control

The cooperative control of multiple fingers on a robotic hand used to perform an operation by grasping a target object is a basic technique of dexterous manipulation. For many cooperative controls, widely used control laws are based on compliance control (Cutkosky and Kao, 1989), impedance control (Kosuge *et al.*, 1995), control of underactuated fingers (Birglen *et al.*, 2010), dynamic control (Zheng *et al.*, 2000), together with position and force laws based on hybrid control (Hayati, 1986; Yoshikawa and Zheng, 1990). In these control laws, it is assumed that the dynamics of the system control is known. However, it is often difficult to find the exact dynamics of the system. In addition, the dynamic parameters of the system will vary depending on the object being manipulated. Cooperative control using a neural network (Hwang and Toda, 1994), learning control (Naniwa and Arimoto, 1995), control without object sensing (Ozawa *et al.*, 2005), adaptive sliding control (Su and Stepanenco, 1995; Ueki *et al.*, 2006) and adaptive coordinated control (Kawasaki *et al.*, 2006) have been proposed to accommodate cases where the model parameters of the system are either unknown or fluctuating. This chapter describes impedance control for manipulating an object with multiple fingers, computed torque control, and adaptive control. Introduction of robot hand control may be found in the references at the back of the book (Mason and Salisbury, 1985).

6.1. Impedance Control

This section discusses **impedance control** in the dynamic model of Eq. (4.114) of Chapter 4, which summarizes the motion of the finger

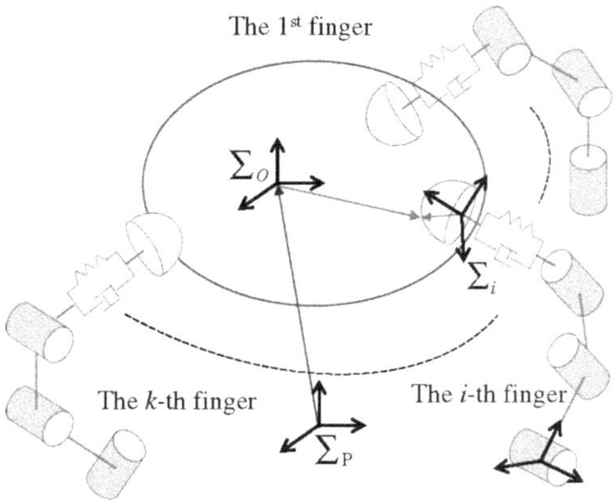

Fig. 6.1. Multi-fingered robotic hand equipped with spring-damper mechanisms at the fingertips.

and the target object.

$$\tilde{\boldsymbol{M}}\dot{\boldsymbol{v}}_O + \tilde{\boldsymbol{C}}\boldsymbol{v}_O + \tilde{\boldsymbol{g}} = \boldsymbol{G}_C\boldsymbol{u}, \tag{6.1}$$

where $\boldsymbol{u} = \boldsymbol{J}_C^{-T}\boldsymbol{\tau}_F$. It is assumed that the integrated system of the finger and the target object includes a spring-damper mechanism at each fingertip virtually, as shown in Fig. 6.1. Consider the impedance control that generates the following force due to the effect of this mechanism.

$$\boldsymbol{F}_m = \tilde{\boldsymbol{M}}_d\dot{\boldsymbol{v}}_O + \tilde{\boldsymbol{D}}_d(\dot{\boldsymbol{r}}_O - \dot{\boldsymbol{r}}_O^d) + \tilde{\boldsymbol{K}}_d(\boldsymbol{r}_O - \boldsymbol{r}_O^d), \tag{6.2}$$

where $\boldsymbol{r}_O^d$ is the desired position and orientation of the target object, and $\tilde{\boldsymbol{M}}_d$, $\tilde{\boldsymbol{D}}_d$, and $\tilde{\boldsymbol{K}}_d$ represent a inertia matrix, a damping coefficient matrix, and a stiffness matrix of the virtual impedance of the object, respectively. The force and moment applied to the object by the contact force at the fingertips are $\mathbf{G}_C\boldsymbol{u}_C$. This indicates that the vector in the null space of $\boldsymbol{G}_C$ is the internal force that does not affect the motion of the object. Thus, the desired internal force at the grasping of the object can be provided by

$$\boldsymbol{F}_{int}^d = (\boldsymbol{I}_{6k\times 6k} - \boldsymbol{G}_C^+\boldsymbol{G}_C)\boldsymbol{f}_{int}^d, \tag{6.3}$$

where $\boldsymbol{f}_{int}^d = [(\boldsymbol{f}_{int_1}^d)^T, \ldots, (\boldsymbol{f}_{int_k}^d)^T]^T$ is an arbitrary force, and $\boldsymbol{G}_C^+$ is a pseudo-inverse matrix of $\boldsymbol{G}_C$ given by

$$\boldsymbol{G}_C^+ = \boldsymbol{G}_C^T\left(\boldsymbol{G}_C\boldsymbol{G}_C^T\right)^{-1}. \tag{6.4}$$

Clearly, $\boldsymbol{G}_C \boldsymbol{F}_{int}^d$ is zero, regardless of $\boldsymbol{f}_{int}^d$, which does not affect the motion of the target object. By adding gravity compensation, the control input is given by

$$\boldsymbol{u} = \boldsymbol{G}_C^+(\boldsymbol{F}_m + \tilde{\boldsymbol{g}}) + \boldsymbol{F}_{int}^d. \tag{6.5}$$

The first term on the right-hand side of the above equation represents the impedance control term by the feedback and feed-forward compensation terms of gravity, and the second term represents the feed-forward compensation term of the internal force. By substituting Eq. (6.5) into Eq. (6.1), it can be represented by

$$(\tilde{\boldsymbol{M}}_d - \tilde{\boldsymbol{M}})\dot{\boldsymbol{v}}_O + \tilde{\boldsymbol{D}}_d(\dot{\boldsymbol{r}}_O - \dot{\boldsymbol{r}}_O^d) + \tilde{\boldsymbol{K}}_d(\boldsymbol{r}_O - \boldsymbol{r}_O^d) - \tilde{\mathbf{C}}_O \boldsymbol{v}_O = 0. \tag{6.6}$$

If it is assumed that the target object velocity is low, and its quadratic velocity term is negligible, it can be expressed by

$$(\tilde{\boldsymbol{M}}_d - \tilde{\boldsymbol{M}})\dot{\boldsymbol{v}}_O + \tilde{\boldsymbol{D}}_d(\dot{\boldsymbol{r}}_O - \dot{\boldsymbol{r}}_O^d) + \tilde{\boldsymbol{K}}_d(\boldsymbol{r}_O - \boldsymbol{r}_O^d) = 0. \tag{6.7}$$

By setting the impedance characteristics of the system properly, the system will respond in a stable manner. For instance, when $\tilde{\boldsymbol{M}}_d = \tilde{\boldsymbol{M}}$, and the quadratic velocity term is negligible because of the low velocity, the system response can be represented as

$$\tilde{\boldsymbol{D}}_d(\dot{\boldsymbol{r}}_O - \dot{\boldsymbol{r}}_O^d) + \tilde{\boldsymbol{K}}_d(\boldsymbol{r}_O - \boldsymbol{r}_O^d) = 0. \tag{6.8}$$

When $\tilde{\boldsymbol{D}}_d$ and $\tilde{\boldsymbol{K}}_d$ are set to positive diagonal matrices, the system is asymptotically stable. Now, let it be defined as

$$\boldsymbol{G}_{C_i}^* = \boldsymbol{G}_{C_i}^T \left(\sum_{j=1}^{k} \boldsymbol{G}_{C_j} \boldsymbol{G}_{C_j}^T \right)^{-1}; \tag{6.9}$$

then the control input of the i-th finger can be represented by

$$\boldsymbol{\tau}_{F_i} = \boldsymbol{J}_{C_i}(\boldsymbol{G}_{C_i}^*(\boldsymbol{F}_m + \tilde{\boldsymbol{g}}) + \boldsymbol{F}_{int_i}^d), \tag{6.10}$$

where

$$\boldsymbol{F}_{int_i}^d = \left(\boldsymbol{I}_{6\times 6} - \boldsymbol{G}_{C_i}^* \boldsymbol{G}_{C_i}\right) \boldsymbol{f}_{int_i}^d. \tag{6.11}$$

Equation (6.10) indicates that the control input of each finger consists of impedance characteristics given by Eq. (6.2), the gravity

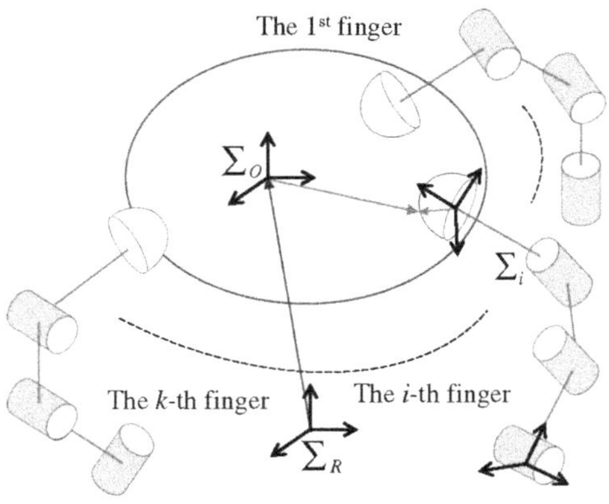

Fig. 6.2. Multi-fingered robotic hand.

compensation, and the internal force and grasp matrix. While this control is relatively easy, trajectory errors will increase when the object is dynamically manipulated. Furthermore, the grasp is not guaranteed due to lack of control of the internal force. When $\tilde{\boldsymbol{M}}_d = \tilde{\boldsymbol{M}}$ is given, the calculation of $\tilde{\boldsymbol{M}}_d$ given by Eq. (4.115) of Chapter 4 becomes very complicated.

6.2. Computed Torque Control

When a robotic hand with k fingers grasps and manipulates a rigid body, as shown in Fig. 6.2, the dynamics of the integrated system of hand and object, which is presented in Sec. 4.6 of Chapter 4, can be written as

$$\begin{aligned} &\boldsymbol{M}_O(\boldsymbol{r}_O)\dot{\boldsymbol{v}}_O + \boldsymbol{C}_O(\boldsymbol{r}_O, \dot{\boldsymbol{r}}_O)\boldsymbol{v}_O + \mathbf{g}_O(\boldsymbol{r}_O) \\ &\quad + \boldsymbol{G}_C\boldsymbol{J}_C^{-T}(\boldsymbol{M}_F\ddot{\boldsymbol{q}}_F + \boldsymbol{C}_F(\boldsymbol{g}_F, \dot{\boldsymbol{q}}_F)\dot{\boldsymbol{q}}_F + \boldsymbol{g}_F(\boldsymbol{g}_F)) = \boldsymbol{G}_C\boldsymbol{J}_C^{-T}\boldsymbol{\tau}_F. \end{aligned} \tag{6.12}$$

This dynamic model does not explicitly include forces at the contact points, suggesting that the object trajectory can be converged to the desired trajectory without using a force sensor. In general, force sensors are expensive and susceptible to noise, and can thus prevent the control system from being highly reliable. If a stable controller is configurable without a force sensor, the control law increases the usefulness of the control system. Thus, **computed torque control** with and without a force sensor will be discussed. Computed torque control cancels the non-linear terms of the finger dynamics by feed-forward compensation, and gives the feedback compensation generated by object trajectory errors. This is one of the possible dynamic control methods. Computed torque control can be designed using the following three steps:

(i) Calculate the desired external force, $\boldsymbol{F}_O^d$, to be applied to the object using a reference model. The reference model of the object is generated from the desired trajectory $\boldsymbol{r}_O^d, \dot{\boldsymbol{r}}_O^d$, and $\ddot{\boldsymbol{r}}_O^d$. It is assumed that the desired values are bounded and continuous.
(ii) Compute the desired contact force at the contact point using the desired external force $\boldsymbol{F}_C^d$.
(iii) Compute the control law of the i-th finger from the reference model of the i-th finger.

It should be noted that the control law described here does not explicitly ask for the desired position or velocity of each fingertip.

6.2.1. *Computed Torque Control without Using a Force Sensor*

First, the reference velocity of the object $\boldsymbol{v}_{Or} \in \mathbb{R}^6$ is defined as

$$\boldsymbol{v}_{Or} = \boldsymbol{T}_O(\dot{\boldsymbol{r}}_O^d + \boldsymbol{\Lambda}\boldsymbol{e}_O), \tag{6.13}$$

where $\boldsymbol{e}_O = \boldsymbol{r}_O^d - \boldsymbol{r}_O$ is the position and orientation error vector of the object, $\boldsymbol{T}_O \in \mathbb{R}^{6\times 6}$ is the matrix that converts the velocity of the Euler angles into angular velocity [see Eq. (4.13) of Chapter 4], and $\boldsymbol{\Lambda} \in \mathbb{R}^{6\times 6}$ is a weighting matrix that is positive and diagonal. Moreover, the residual velocity of the object $\boldsymbol{s}_O \in \mathbb{R}^6$ is defined as

$$\boldsymbol{s}_O = \boldsymbol{v}_O - \boldsymbol{v}_{Or}. \tag{6.14}$$

Then, the relation of $\boldsymbol{s}_O = -\boldsymbol{T}_O(\dot{\boldsymbol{e}}_O + \boldsymbol{\Lambda}\boldsymbol{e}_O)$ is obtained. The desired external force to be applied to the object, $\boldsymbol{F}_O^d$, is generated from the reference model of the object using the following equation:

$$\begin{aligned} \boldsymbol{F}_O^d &= \boldsymbol{M}_O(\boldsymbol{r}_O)\,\dot{\boldsymbol{v}}_{Or} + \boldsymbol{C}_O(\boldsymbol{r}_O, \dot{\boldsymbol{r}}_O)\,\boldsymbol{v}_{Or} + \boldsymbol{g}_O(\boldsymbol{r}_O) \\ &= \boldsymbol{Y}_O(\boldsymbol{r}_O, \dot{\boldsymbol{r}}_O, \boldsymbol{v}_{Or}, \dot{\boldsymbol{v}}_{Or})\,\boldsymbol{\sigma}_O, \end{aligned} \tag{6.15}$$

where $\boldsymbol{\sigma}_O$ is a dynamic parameter vector of the object. It is assumed that forces and moments at the contact point are in equilibrium with the external force. In this case, the force at the contact point (i.e., where it is in equilibrium with the desired external force) should satisfy the relationship of $\boldsymbol{F}_O = \boldsymbol{G}_C\boldsymbol{F}_C$. Furthermore, forces and moments at the contact point generate the internal force acting within the object, which is required for a stable grasp. Therefore, the desired contact force at the contact point, $\boldsymbol{F}_C^d = [(\boldsymbol{F}_{C_1}^d)^T, \ldots, (\boldsymbol{F}_{C_k}^d)^T]^T$, is provided by

$$\boldsymbol{F}_C^d = \boldsymbol{G}_C^+\boldsymbol{F}_O^d + \left(\boldsymbol{I}_{6k\times 6k} - \boldsymbol{G}_C^+\boldsymbol{G}_C\right)\boldsymbol{f}_{int}^d. \tag{6.16}$$

The first term of Eq. (6.16) is the external force that has an effect on the object motion, and the second term is the internal force that does not have such an effect. The internal force that is bounded and uniformly continuous is a vector in the null space of $\boldsymbol{G}_C$. Therefore, as long as it meets the contact conditions, the way of setting is unlimited. The joint torque that can balance the desired force is $\boldsymbol{J}_C^T\boldsymbol{F}_C^d$.

The following relationship is established in the joint velocity of the i-th finger and the object velocity, as was seen in Eq. (2.42) of Chapter 2:

$$\dot{\boldsymbol{q}}_{F_i} = \boldsymbol{J}_{C_i}^{-1}\boldsymbol{G}_{C_i}^T\boldsymbol{v}_O. \tag{6.17}$$

From this relationship, the reference joint velocity of the i-th finger $\dot{\boldsymbol{q}}_{Fr_i}$ is given by

$$\dot{\boldsymbol{q}}_{Fr_i} = \boldsymbol{J}_{C_i}^{-1}\boldsymbol{G}_{C_i}^T\boldsymbol{v}_{Or}. \tag{6.18}$$

Let us define the reference model of the i-th finger as

$$\begin{aligned} &\boldsymbol{M}_{F_i}\ddot{\boldsymbol{q}}_{Fr_i} + \boldsymbol{C}_{F_i}(\boldsymbol{g}_{F_i}, \dot{\boldsymbol{q}}_{F_i})\dot{\boldsymbol{q}}_{Fr_i} + \boldsymbol{g}_{F_i}(\boldsymbol{g}_{F_i}) \\ &\quad = \boldsymbol{Y}_{F_i}(\boldsymbol{g}_{F_i}, \dot{\boldsymbol{q}}_{F_i}, \dot{\boldsymbol{q}}_{Fr_i}, \ddot{\boldsymbol{q}}_{Fr_i})\boldsymbol{\sigma}_{F_i}; \end{aligned} \tag{6.19}$$

the control law of the i-th finger is then given by

$$\boldsymbol{\tau}_{F_i} = \boldsymbol{Y}_{F_i}(\boldsymbol{q}_{F_i}, \dot{\boldsymbol{q}}_{F_i}, \dot{\boldsymbol{q}}_{Fr_i}, \ddot{\boldsymbol{q}}_{Fr_i})\boldsymbol{\sigma}_{F_i} + \boldsymbol{J}_{C_i}^T \boldsymbol{F}_{C_i}^d - \boldsymbol{K}_i \boldsymbol{s}_i, \tag{6.20}$$

where $\boldsymbol{\sigma}_{F_i}$ is the dynamic parameter vector of the i-th finger, $\boldsymbol{K}_i > 0$ is the symmetric gain matrix, and $\boldsymbol{s}_i$ is the residual between the reference joint velocity and the actual joint velocity and is given by

$$\boldsymbol{s}_i = \dot{\boldsymbol{q}}_{F_i} - \dot{\boldsymbol{q}}_{Fr_i}. \tag{6.21}$$

Then, the relation of $\boldsymbol{s}_i = \boldsymbol{J}_{C_i}^{-1} \boldsymbol{G}_{C_i}^T \boldsymbol{s}_O$ is obtained. In the right-hand side of Eq. (6.20), the first term is the feed-forward input based on the reference model of the finger, the second term is the feed-forward input that is balanced with the desired external force acting on the object, and the third term is the feedback input corresponding to the trajectory error of the joint velocity and joint displacement.

The control law that integrates the first finger with the k-th finger can thus be represented as

$$\boldsymbol{\tau}_F = \boldsymbol{Y}_F(\boldsymbol{q}_F, \dot{\boldsymbol{q}}_F, \dot{\boldsymbol{q}}_{Fr}, \ddot{\boldsymbol{q}}_{Fr})\boldsymbol{\sigma}_F + \boldsymbol{J}_C^T \boldsymbol{F}_C^d - \boldsymbol{K}\boldsymbol{s}, \tag{6.22}$$

where $\boldsymbol{K} = block\ diag[\boldsymbol{K}_1, \ldots, \boldsymbol{K}_k]$, $\boldsymbol{q}_{Fr} = [\boldsymbol{q}_{Fr_1}^T \cdots \boldsymbol{q}_{Fr_k}^T]^T$, $\boldsymbol{s} = [\boldsymbol{s}_1^T \ldots \boldsymbol{s}_k^T]^T$, and $\boldsymbol{\sigma} = [\boldsymbol{\sigma}_1^T \ldots \boldsymbol{\sigma}_k^T]^T$. The following theorem is proved if the assumptions of grasping conditions in Sec. 4.8 of Chapter 4: A1–A4 are made.

Theorem 6.1. *Consider a rigid body grasped by k fingers. For the system in Eq.* (6.12), *given that the desired external force of the object is Eq.* (6.15), *the desired force at the contact point is Eq.* (6.16), *the residual is Eq.* (6.21), *and the control law is Eq.* (6.22). *The closed-loop system is then asymptotically stable in the following sense:*

(1) $\boldsymbol{r}_O \to \boldsymbol{r}_O^d$ *and* $\dot{\boldsymbol{r}}_O \to \dot{\boldsymbol{r}}_O^d$ *as* $t \to \infty$.
(2) *The force error at the contact point,* $\boldsymbol{F}_C - \boldsymbol{F}_C^d$, *is bounded.*

Proof. Consider a candidate Lyapunov function given by

$$V = \frac{1}{2}\left(\boldsymbol{s}_O^T \boldsymbol{M}_O \boldsymbol{s}_O + \boldsymbol{s}^T \boldsymbol{M}_F \boldsymbol{s}\right). \tag{1}$$

Since $\dot{\boldsymbol{M}}_O - 2\boldsymbol{C}_O$ and $\dot{\boldsymbol{M}}_F - 2\boldsymbol{C}_F$ are skew-symmetric, a time derivative of V along the trajectory using these properties is given by

$$\begin{aligned}\dot{V} &= \frac{1}{2}\left(\boldsymbol{s}_O^T \dot{\boldsymbol{M}}_O \boldsymbol{s}_O + 2\boldsymbol{s}_O^T \boldsymbol{M}_O \dot{\mathbf{s}}_O + \boldsymbol{s}^T \dot{\boldsymbol{M}}_F \boldsymbol{s} + 2\boldsymbol{s}^T \boldsymbol{M}_F \dot{\mathbf{s}}\right)\\ &= \boldsymbol{s}_O^T(\boldsymbol{M}_O \dot{\mathbf{s}}_O + \boldsymbol{C}_O \boldsymbol{s}_O) + \boldsymbol{s}^T(\boldsymbol{M}_F \dot{\mathbf{s}} + \boldsymbol{C}_F \boldsymbol{s}).\end{aligned} \tag{2}$$

By substituting Eq. (6.14), Eq. (6.15) , Eqs. (4.109) and (4.111) of Chapter 4 into $\boldsymbol{M}_O \dot{\boldsymbol{s}}_O + \boldsymbol{C}_O \boldsymbol{s}_O$, in order, and defining $\Delta \boldsymbol{F}_O = \boldsymbol{F}_O^d - \boldsymbol{F}_O$, it can be represented as

$$\boldsymbol{M}_O \dot{\boldsymbol{s}}_O + \boldsymbol{C}_O \boldsymbol{s}_O = -\Delta \boldsymbol{F}_O. \tag{3}$$

In the same way, by substituting Eq. (6.21), Eq. (4.110) of Chapter 4, Eq. (6.19), and Eq. (6.20) into $\boldsymbol{M}_F \dot{\boldsymbol{s}} + \boldsymbol{C}_F \boldsymbol{s}$ in order, and defining $\Delta \boldsymbol{F}_C = \boldsymbol{F}_C^d - \boldsymbol{F}_C$, it is represented as

$$\boldsymbol{M}_F \dot{\boldsymbol{s}} + \boldsymbol{C}_F \boldsymbol{s} = \boldsymbol{J}_C^T \Delta \boldsymbol{F}_C - \boldsymbol{K}\boldsymbol{s}. \tag{4}$$

Now, by substituting Eq. (3) and Eq. (4) into Eq. (2) and using the relationship of $\boldsymbol{G}_C \boldsymbol{\Delta} \boldsymbol{F}_C = \Delta \boldsymbol{F}_O$, Eq. (2) is represented by

$$\dot{V} = -\boldsymbol{s}^T \boldsymbol{K} \boldsymbol{s} \leq 0. \tag{5}$$

Therefore, V is a Lyapunov function and $\boldsymbol{s}_O$ and $\boldsymbol{s}$ are bounded. From Eq. (6.14), $\boldsymbol{v}_O, \boldsymbol{v}_{Or}, \dot{\boldsymbol{e}}_O$, and $\boldsymbol{e}_O$ are also bounded, and so, $\dot{\boldsymbol{r}}_O, \boldsymbol{r}_O, \boldsymbol{G}_C$ and $\boldsymbol{J}_O$ are bounded. $\dot{\boldsymbol{v}}_{Or}$ is bounded from the time differentiation of Eq. (6.13), and $\boldsymbol{F}_O^d$ and $\boldsymbol{Y}_O$ are bounded from Eq. (6.15). In addition, $\boldsymbol{F}_C^d$ is bounded from Eq. (6.16), $\dot{\boldsymbol{q}}_{Fr}$ is bounded from Eq. (6.18), and $\dot{\boldsymbol{q}}_F$ is bounded from Eq. (6.17).

Next, the following is derived from Eq. (3):

$$\dot{\boldsymbol{s}}_O = -\boldsymbol{M}_O^{-1}(\boldsymbol{G}_C \Delta \boldsymbol{F}_C + (\boldsymbol{K}_O + \boldsymbol{C}_O)\boldsymbol{s}_O). \tag{6}$$

Then the time derivative of Eq. (6.21) is given as

$$\begin{aligned}\dot{\boldsymbol{s}} &= \dot{\boldsymbol{J}}_C^{-1} \boldsymbol{G}_C^T \boldsymbol{s}_O + \boldsymbol{J}_C^{-1}(\dot{\boldsymbol{G}}_C^T \boldsymbol{s}_O + \boldsymbol{G}_C^T \dot{\boldsymbol{s}}_O)\\ &= -\boldsymbol{J}_C^{-1} \boldsymbol{G}_C^T \boldsymbol{M}_O^{-1} \boldsymbol{G}_C \Delta \boldsymbol{F}_C + \mathrm{b},\end{aligned} \tag{7}$$

where $\boldsymbol{b}$ is a bounded vector given by

$$\boldsymbol{b} = \dot{\boldsymbol{J}}_C^{-1} \boldsymbol{G}_C^T \boldsymbol{s}_O + \boldsymbol{J}_C^{-1}(\dot{\boldsymbol{G}}_C^T \boldsymbol{s}_O - \boldsymbol{G}_C^T(\boldsymbol{K}_O + \boldsymbol{C}_O)\boldsymbol{s}_O). \tag{8}$$

By substituting Eq. (7) into Eq. (4), it shows

$$(\boldsymbol{J}_C^T + \boldsymbol{M}_F \boldsymbol{J}_C^{-1} \boldsymbol{G}_C^T \boldsymbol{M}_O^{-1} \boldsymbol{G}_C)\Delta \boldsymbol{F}_C = \boldsymbol{M}_F \boldsymbol{b} + (\boldsymbol{C}_F + \boldsymbol{K})\boldsymbol{s}. \quad (9)$$

By multiplying $\boldsymbol{J}_C \boldsymbol{M}_F^{-1}$ from left to this equation, it is represented as

$$\boldsymbol{U}\Delta \boldsymbol{F}_C = \boldsymbol{J}_C(\boldsymbol{b} + \boldsymbol{M}_F^{-1}(\boldsymbol{C}_F + \boldsymbol{K})\boldsymbol{s}), \quad (10)$$

where

$$\boldsymbol{U} = \boldsymbol{J}_C \boldsymbol{M}_F^{-1} \boldsymbol{J}_C^T + \boldsymbol{G}_C^T \boldsymbol{M}_O^{-1} \boldsymbol{G}_C. \quad (11)$$

Obviously, $\boldsymbol{U}$ is a positive and symmetric matrix. Therefore, $\Delta \boldsymbol{F}_C$ is bounded. As a result, $\Delta \boldsymbol{F}_O$ is bounded, $\dot{\boldsymbol{s}}_O$ is bounded from Eq. (6), and $\dot{\boldsymbol{s}}$ is bounded from Eq. (4). From these relations, the time derivative of Eq. (5) is given as

$$\ddot{V} = -2\boldsymbol{s}^T \mathbf{K}\dot{\mathbf{s}}, \quad (12)$$

and it is also bounded. Hence, $\dot{V}$ is uniformly continuous. According to the Lyapunov-like Lemma, $\dot{V} \to 0$ as $t \to \infty$. Therefore, $\boldsymbol{s}_O \to 0$ and $\boldsymbol{s} \to 0$ as $t \to \infty$. As a result, from Eq. (6.14), $\dot{\boldsymbol{e}}_O$ and $\boldsymbol{e}_O$ converge to zero, and the object trajectory converges to the desired trajectory as $t \to \infty$. Also, $\boldsymbol{b} \to 0$ as $t \to \infty$. From Eq. (10), $\boldsymbol{F}_C \to \boldsymbol{F}_C^d$ and $\boldsymbol{F}_O \to \boldsymbol{F}_O^d$ as $t \to \infty$.

It is noted that the explicit trajectory of each fingertip in this control is not required. Moreover, there is the assumption that the object is grasped and the internal force is sufficiently large to keep the grasp on the object.

6.2.2. *Computed Torque Control Using a Force Sensor*

Consider a control law that uses a force sensor in which the contact force converges to the desired contact force asymptotically. The definitions of the reference velocity, $\boldsymbol{v}_{Or}$, and the residual velocity, $\boldsymbol{s}_O$, of the object are used for Eq. (6.13) and Eq. (6.14), respectively. The vector of contact force error,

$$\Delta \boldsymbol{F}_{C_i} = \boldsymbol{F}_{C_i}^d - \boldsymbol{F}_{C_i}, \quad (6.23)$$

is passed through a low-pass filter given by

$$\dot{\boldsymbol{v}}_{C_i} + \alpha \boldsymbol{v}_{C_i} = \alpha \Delta \boldsymbol{F}_{C_i}, \quad (6.24)$$

where $\alpha > 0$ is a design parameter of the filter, $\boldsymbol{v}_{C_i}$ is an output signal passed through the filter, and its continuity is guaranteed. The reference model of the object is given as follows on behalf of Eq. (6.15):

$$\begin{aligned} \boldsymbol{F}_O^d &= \boldsymbol{M}_O(\boldsymbol{r}_O)\dot{\boldsymbol{v}}_{Or} + \boldsymbol{C}_O(\boldsymbol{r}_O, \dot{\boldsymbol{r}}_O)\boldsymbol{v}_{Or} + \boldsymbol{g}_O(\boldsymbol{r}_O) - \boldsymbol{K}_O\boldsymbol{s}_O \\ &= \boldsymbol{Y}_O(\boldsymbol{r}_O, \dot{\boldsymbol{r}}_O, \boldsymbol{v}_{Or}, \dot{\boldsymbol{v}}_{Or})\boldsymbol{\sigma}_O - \boldsymbol{K}_O\boldsymbol{s}_O, \end{aligned} \tag{6.25}$$

where $\boldsymbol{K}_O > 0$ is a symmetric feedback gain matrix. Furthermore, the reference angular velocity of the i-th finger, $\dot{\boldsymbol{q}}_{Fr_i}$, is defined as the following formula, instead of as Eq. (6.18), and feedback terms of the force error and integral force errors are added.

$$\dot{\boldsymbol{q}}_{Fr_i} = \boldsymbol{J}_{C_i}^{-1}(\boldsymbol{G}_{C_i}^T\boldsymbol{v}_{Or} + \boldsymbol{\Omega}_i\boldsymbol{v}_{C_i} + \boldsymbol{\Psi}_i\boldsymbol{\eta}_{C_i}), \tag{6.26}$$

where $\boldsymbol{\eta}_{C_i}$ is an integral of the contact force error defined by

$$\boldsymbol{\eta}_{C_i} = \int_0^t \Delta\boldsymbol{F}_{C_i}dt. \tag{6.27}$$

In addition, $\boldsymbol{\Omega}_i > 0$ and $\boldsymbol{\Psi}_i > 0$ are the symmetric feedback gain matrices of force error and integral force error, respectively. Then, the residuals defined by Eq. (6.21) are represented as

$$\boldsymbol{s}_i = \boldsymbol{J}_{C_i}^{-1}(\boldsymbol{G}_{C_i}^T\boldsymbol{s}_O - \boldsymbol{\Omega}_i\boldsymbol{v}_{C_i} - \boldsymbol{\Psi}_i\boldsymbol{\eta}_{C_i}). \tag{6.28}$$

The control law of the i-th finger is same as Eq. (6.20), which is given by

$$\boldsymbol{\tau}_{F_i} = \boldsymbol{Y}_{F_i}(\boldsymbol{g}_{F_i}, \dot{\boldsymbol{q}}_{F_i}, \dot{\boldsymbol{q}}_{Fr_i}, \ddot{\boldsymbol{q}}_{Fr_i})\boldsymbol{\sigma}_{F_i} + \boldsymbol{J}_{C_i}^T\boldsymbol{F}_{C_i}^d - \boldsymbol{K}_i\boldsymbol{s}_i. \tag{6.29}$$

It is noted that $\boldsymbol{s}_i$ contains a feedback term of the contact force error. The control law and estimation law that are integrated from the i-th to the k-th finger are represented in the same expression as Eq. (6.22), as

$$\boldsymbol{\tau}_F = \boldsymbol{Y}_F(\boldsymbol{q}_F, \dot{\boldsymbol{q}}_F, \dot{\boldsymbol{q}}_{Fr}, \ddot{\boldsymbol{q}}_{Fr})\boldsymbol{\sigma}_F + \boldsymbol{J}_C^T\boldsymbol{F}_C^d - \boldsymbol{K}\boldsymbol{s}. \tag{6.30}$$

The following theorem is established if the grasping conditions A1–A4 as shown in Sec. 4.8 of Chapter 4 are assumed.

Theorem 6.2. *Consider a rigid body grasped by k fingers. For the system in Eq.* (6.12), *given that the desired external force is Eq.* (6.25), *the desired force at the contact point is Eq.* (6.16), *the* *residue is Eq. (6.28), and the control law is Eq. (6.30). The closed-loop system is then asymptotically stable in the following sense.*

(1) $\boldsymbol{r}_O \to \boldsymbol{r}_O^d$ *and* $\dot{\boldsymbol{r}}_O \to \dot{\boldsymbol{r}}_O^d$ *as* $t \to \infty$.
(2) $\boldsymbol{F}_C \to \boldsymbol{F}_C^d$ *as* $t \to \infty$.

This theorem can be proved by defining $\boldsymbol{v}_C = [\boldsymbol{v}_{C_1}^T, \ldots, \boldsymbol{v}_{C_k}^T]^T$, $\boldsymbol{\eta}_C = [\boldsymbol{\eta}_{C_1}^T, \ldots, \boldsymbol{\eta}_{C_k}^T]^T, \boldsymbol{\Omega} = block\ diag[\boldsymbol{\Omega}_1, \ldots, \boldsymbol{\Omega}_k] > 0$, and $\boldsymbol{\Psi} = block\ diag[\boldsymbol{\Psi}_1, \ldots, \boldsymbol{\Psi}_k] > 0$, and giving a candidate for a Lyapunov function by (see Exercise 6.1)

$$V = \frac{1}{2}\left(\boldsymbol{s}_O^T \boldsymbol{M}_O \boldsymbol{s}_O + \boldsymbol{s}^T \boldsymbol{M}_F \boldsymbol{s} + \frac{1}{\alpha}\boldsymbol{v}_C^T \boldsymbol{\Omega} v_C + \boldsymbol{\eta}_C^T \boldsymbol{\Psi} \eta_C\right).$$

It is noted that the asymptotic stability of the computed torque control is proved in cases of either fixed contact or rolling contact under grasping conditions of A1–A4 as shown in Sec. 4.8 of Chapter 4.

6.3. Adaptive Control

Dynamic models of the fingers and object include dynamic parameters such as the mass and inertia tensors of the object's and fingers' links. When these dynamic parameters are unknown, it is difficult to precisely control the object motion with computed torque control. The purpose of control here is to find a control law such that the object motion follows the desired trajectory asymptotically. Like computed torque control, the **adaptive control** introduced below is designed in three steps; the reference model of the object is calculated using an estimate of the dynamic parameters.

6.3.1. *Adaptive Control without Using a Force Sensor*

Using the estimate $\hat{\boldsymbol{\sigma}}_O$ of the dynamic parameter vector of the object $\boldsymbol{\sigma}_O$, the desired external force acting on the object is generated from an estimated reference model of the object given by the following equation, instead of Eq. (6.15):

$$\begin{aligned} \boldsymbol{F}_O^d &= \hat{\boldsymbol{M}}_O(\boldsymbol{r}_O)\,\dot{\boldsymbol{v}}_{Or} + \hat{\boldsymbol{C}}_O(\boldsymbol{r}_O, \dot{\boldsymbol{r}}_O)\,\boldsymbol{v}_{Or} + \hat{\boldsymbol{g}}_O(\boldsymbol{r}_O) \\ &= \boldsymbol{Y}_O(\boldsymbol{r}_O, \dot{\boldsymbol{r}}_O, \boldsymbol{v}_{Or}, \dot{\boldsymbol{v}}_{Or})\,\hat{\boldsymbol{\sigma}}_O, \end{aligned} \tag{6.31}$$

where $\boldsymbol{v}_{Or}$ is the reference velocity defined by Eq. (6.13) and $\hat{\boldsymbol{M}}_O$, $\hat{\boldsymbol{C}}_O$ and $\hat{\boldsymbol{g}}_O$ are the estimates of $\boldsymbol{M}_O$, $\boldsymbol{C}_O$ and $\boldsymbol{g}_O$ using $\hat{\boldsymbol{\sigma}}_O$. Now, an estimated law of dynamics parameter vector of the object is given by

$$\dot{\hat{\boldsymbol{\sigma}}}_O = -\boldsymbol{\Gamma}_O \boldsymbol{Y}_O^T(\boldsymbol{r}_O, \dot{\boldsymbol{r}}_O, \boldsymbol{v}_{Or}\dot{\boldsymbol{v}}_{Or})\boldsymbol{s}_O, \tag{6.32}$$

where $\boldsymbol{\Gamma}_O > 0$ is an adaptive gain matrix that is symmetric, and $\boldsymbol{s}_O$ is the residual given by Eq. (6.14), which is calculated as

$$\boldsymbol{s}_O = -\boldsymbol{T}_O(\dot{\boldsymbol{e}}_O + \boldsymbol{\Lambda}\boldsymbol{e}_O). \tag{6.33}$$

The desired force of the object is sequentially updated based on the estimation of the dynamics parameters of the object. Forces and moments at the contact points are in equilibrium with the external force. Therefore, the desired force at the contact point $\boldsymbol{F}_c^d$ is provided in the following equation, the same as for computed torque control:

$$\boldsymbol{F}_C^d = \boldsymbol{G}_C^+\boldsymbol{F}_O^d + (\boldsymbol{I}_{6k\times 6k} - \boldsymbol{G}_C^+\boldsymbol{G}_C)\boldsymbol{f}_{\text{int}}^d. \tag{6.34}$$

The difference from computed torque control is that $\boldsymbol{F}_O^d$ is calculated from Eq. (6.31) using the estimated value $\hat{\boldsymbol{\sigma}}_O$.

By using the estimate $\hat{\boldsymbol{\sigma}}_{F_i}$ of the dynamic parameter vector of the i-th finger $\boldsymbol{\sigma}_{F_i}$, the reference model of the i-th finger is given by

$$\begin{aligned}\hat{\boldsymbol{M}}_{F_i}\ddot{\boldsymbol{q}}_{Fr_i} + \hat{\boldsymbol{C}}_{F_i}(\boldsymbol{q}_{F_i}, \dot{\boldsymbol{q}}_{F_i})\dot{\boldsymbol{q}}_{Fr_i} + \hat{\boldsymbol{q}}_{F_i}(\boldsymbol{g}_{F_i})\\ = \boldsymbol{Y}_{F_i}(\boldsymbol{g}_{F_i}, \boldsymbol{q}_{F_i}, \dot{\boldsymbol{q}}_{Fr_i}, \ddot{\boldsymbol{q}}_{Fr_i})\hat{\boldsymbol{\sigma}}_{F_i},\end{aligned} \tag{6.35}$$

where $\hat{\boldsymbol{M}}_{F_i}$, $\hat{\boldsymbol{C}}_{F_i}$ and $\hat{\boldsymbol{g}}_{F_i}$ are estimates of each $\boldsymbol{M}_{F_i}$, $\boldsymbol{C}_{F_i}$ and $\boldsymbol{g}_{F_i}$ using $\hat{\boldsymbol{\sigma}}_{F_i}$, and $\dot{\boldsymbol{q}}_{Fr_i}$ is given by Eq. (6.18). Instead of Eq. (6.20), the control law of the i-th finger is given as

$$\boldsymbol{\tau}_{F_i} = \boldsymbol{Y}_{F_i}(\boldsymbol{q}_{F_i}, \dot{\boldsymbol{q}}_{F_i}, \dot{\boldsymbol{q}}_{Fr_i}, \ddot{\boldsymbol{q}}_{Fr_i})\hat{\boldsymbol{\sigma}}_{F_i} + \boldsymbol{J}_{C_i}^T\boldsymbol{F}_{C_i}^d - \boldsymbol{K}_i\boldsymbol{s}_i, \tag{6.36}$$

where $\boldsymbol{K}_i > 0 \in \mathbb{R}^{6\times 6}$ is a feedback gain matrix, and $\boldsymbol{s}_i$ is the residual between the actual velocity and the reference velocity defined by Eq. (6.21) and is represented by

$$\boldsymbol{s}_i = \boldsymbol{J}_{C_i}^{-1}\boldsymbol{G}_{C_i}^T\boldsymbol{s}_O. \tag{6.37}$$

The adaptive law of the parameter estimate of the i-th finger is given by

$$\dot{\hat{\boldsymbol{\sigma}}}_{F_i} = -\boldsymbol{\Gamma}_i \boldsymbol{Y}_{F_i}^T(\boldsymbol{q}_{F_i}, \dot{\boldsymbol{q}}_{F_i}, \dot{\boldsymbol{q}}_{Fr_i}, \ddot{\boldsymbol{q}}_{Fr_i})\boldsymbol{s}_i, \tag{6.38}$$

where $\boldsymbol{\Gamma}_i > 0$ is an adaptive gain matrix that is symmetric. In the right-hand side of Eq. (6.36), the first term is a feed-forward input term based on the estimated reference model, the second term is a feed-forward input term corresponding to the desired force at the contact point, and the third term is a feedback input term based on the object trajectory errors. It should be noted that the adaptive control does not require measurements of forces and moments at each contact point. The integrated control law and adaptive law can be written as follows:

$$\boldsymbol{\tau}_F = \boldsymbol{Y}_F(\boldsymbol{q}_F, \dot{\boldsymbol{q}}_F, \dot{\boldsymbol{q}}_{Fr}, \ddot{\boldsymbol{q}}_{Fr})\hat{\boldsymbol{\sigma}}_F + \boldsymbol{J}_C^T \boldsymbol{F}_C^d - \boldsymbol{K}\boldsymbol{s} \tag{6.39}$$

$$\dot{\hat{\boldsymbol{\sigma}}}_F = -\boldsymbol{\Gamma} \boldsymbol{Y}_F^T(\boldsymbol{g}_F, \dot{\boldsymbol{q}}_F, \dot{\boldsymbol{q}}_{Fr}, \ddot{\boldsymbol{q}}_{Fr})\boldsymbol{s}, \tag{6.40}$$

where $\hat{\boldsymbol{\sigma}}_F = [\hat{\boldsymbol{\sigma}}_{F_1}^T \cdots \hat{\boldsymbol{\sigma}}_{F_k}^T]^T$ and $\boldsymbol{\Gamma} = block\ diag[\boldsymbol{\Gamma}_1, \ldots, \boldsymbol{\Gamma}_k]$. If the grasping conditions A1–A4 as described in Sec. 4.8 of Chapter 4 are assumed, the following theorem is proved for adaptive control without use of a force sensor (Kawasaki *et al.*, 2006).

Theorem 6.3. *Consider a rigid body that is grasped by k fingers. For the system in Eq. (6.12), given that Eq. (6.32) is the parameter estimation law of the object, Eq. (6.31) is the desired external force of the object, Eq. (6.34) is the desired force at the contact point, Eq. (6.37) is the residual, Eq. (6.40) is the parameter estimate law of the finger, and Eq. (6.39) is the control law. The closed-loop system is then asymptotically stable in the following sense:*

(1) $\boldsymbol{r}_O \to \boldsymbol{r}_O^d$ *and* $\dot{\boldsymbol{r}}_O \to \dot{\boldsymbol{r}}_O^d$ *as* $t \to \infty$.
(2) *The force error at the contact point* $\boldsymbol{F}_C - \boldsymbol{F}_C^d$ *is bounded.*

Theorem 6.3 can be proved by using a candidate Lyapunov function given by (see Exercise 6.2)

$$V = \frac{1}{2}\left(\boldsymbol{s}_O^T \boldsymbol{M}_O \boldsymbol{s}_O + \Delta\boldsymbol{\sigma}_O^T \boldsymbol{\Gamma}_O \Delta\boldsymbol{\sigma}_O + \boldsymbol{s}^T \boldsymbol{M}_F \boldsymbol{s} + \Delta\boldsymbol{\sigma}_F^T \boldsymbol{\Gamma} \Delta\boldsymbol{\sigma}_F\right), \tag{6.41}$$

where $\Delta\boldsymbol{\sigma}_O = \hat{\boldsymbol{\sigma}}_O - \boldsymbol{\sigma}_O$ and $\Delta\boldsymbol{\sigma}_F = \hat{\boldsymbol{\sigma}}_F - \boldsymbol{\sigma}_F$.

6.3.2. *Adaptive Control Using a Force Sensor*

Consider adaptive control when each fingertip has a force sensor and the dynamic parameters of the object and fingers are unknown. Using the estimate $\hat{\boldsymbol{\sigma}}_O$ of the dynamic parameter vector of object $\boldsymbol{\sigma}_O$, the desired external force acting on the object is generated from the estimated reference model of the object given by the following equation instead of Eq. (6.25):

$$\begin{aligned}\boldsymbol{F}_O^d &= \hat{\boldsymbol{M}}_O(\boldsymbol{r}_O)\,\dot{\boldsymbol{v}}_{Or} + \hat{\boldsymbol{C}}_O(\boldsymbol{r}_O, \dot{\boldsymbol{r}}_O)\,\boldsymbol{v}_{Or} + \hat{\boldsymbol{g}}_O(\boldsymbol{r}_O) - \boldsymbol{K}_O\boldsymbol{s}_O \\ &= \boldsymbol{Y}_O(\boldsymbol{r}_O, \dot{\boldsymbol{r}}_O, \boldsymbol{v}_{Or}, \dot{\boldsymbol{v}}_{Or})\,\hat{\boldsymbol{\sigma}}_O - \boldsymbol{K}_O\boldsymbol{s}_O, \end{aligned} \tag{6.42}$$

where $\boldsymbol{v}_{Or}$ is the reference velocity defined by Eq. (6.13), $\boldsymbol{s}_O$ is the residual defined in Eq. (6.14), and the estimation law of the estimate $\hat{\boldsymbol{\sigma}}_O$ is same as in Eq. (6.32); therefore,

$$\dot{\hat{\boldsymbol{\sigma}}}_O = -\boldsymbol{\Gamma}_O\boldsymbol{Y}_O^T(\boldsymbol{r}_O, \dot{\boldsymbol{r}}_O, \boldsymbol{v}_{Or}\dot{\boldsymbol{v}}_{Or})\,\boldsymbol{s}_O, \tag{6.43}$$

In addition, as in the case of computed torque control using a force sensor, the reference velocity of the i-th finger is given by Eq. (6.26), using the filtered output $\boldsymbol{v}_{C_i}$ through a low-pass filter on the contact force error given by

$$\dot{\boldsymbol{v}}_{C_i} + \alpha\boldsymbol{v}_{C_i} = \alpha\Delta\boldsymbol{F}_{C_i}, \tag{6.44}$$

i.e.,

$$\dot{\boldsymbol{q}}_{Fr_i} = \boldsymbol{J}_{C_i}^{-1}(\boldsymbol{G}_{C_i}^T\boldsymbol{v}_{Or} + \boldsymbol{\Omega}_i\boldsymbol{v}_{C_i} + \boldsymbol{\Psi}_i\boldsymbol{\eta}_{C_i}). \tag{6.45}$$

In this case, the residual velocity of the i-th finger is defined in the same way as Eq. (6.21) and obtained by

$$\boldsymbol{s}_i = \boldsymbol{J}_{C_i}^{-1}(\boldsymbol{G}_{C_i}^T\boldsymbol{s}_O - \boldsymbol{\Omega}_i\boldsymbol{v}_{C_i} - \boldsymbol{\Psi}_i\boldsymbol{\eta}_{C_i}). \tag{6.46}$$

The control law and estimated law are given by

$$\boldsymbol{\tau}_{F_i} = \boldsymbol{Y}_{F_i}(\boldsymbol{q}_{F_i}, \dot{\boldsymbol{q}}_{F_i}, \dot{\boldsymbol{q}}_{Fr_i}, \ddot{\boldsymbol{q}}_{Fr_i})\hat{\boldsymbol{\sigma}}_{F_i} + \boldsymbol{J}_{C_i}^T\boldsymbol{F}_{C_i}^d - \boldsymbol{K}_i\boldsymbol{s}_i \tag{6.47}$$

$$\dot{\hat{\boldsymbol{\sigma}}}_{F_i} = -\boldsymbol{\Gamma}_i\boldsymbol{Y}_{F_i}^T(\boldsymbol{q}_{F_i}, \dot{\boldsymbol{q}}_{F_i}, \dot{\boldsymbol{q}}_{Fr_i}, \ddot{\boldsymbol{q}}_{Fr_i})\boldsymbol{s}_i. \tag{6.48}$$

The control law and estimated law that integrate the 1st to the k-th finger are represented by

$$\boldsymbol{\tau}_F = \boldsymbol{Y}_F(\boldsymbol{q}_F, \dot{\boldsymbol{q}}_F, \dot{\boldsymbol{q}}_{Fr}, \ddot{\boldsymbol{q}}_{Fr})\hat{\boldsymbol{\sigma}}_F + \boldsymbol{J}_C^T\boldsymbol{F}_C^d - \boldsymbol{K}\boldsymbol{s} \tag{6.49}$$

$$\dot{\hat{\boldsymbol{\sigma}}}_F = -\boldsymbol{\Gamma}\boldsymbol{Y}_F^T(\boldsymbol{q}_F, \dot{\boldsymbol{q}}_F, \dot{\boldsymbol{q}}_{Fr}, \ddot{\boldsymbol{q}}_{Fr})\boldsymbol{s}. \tag{6.50}$$

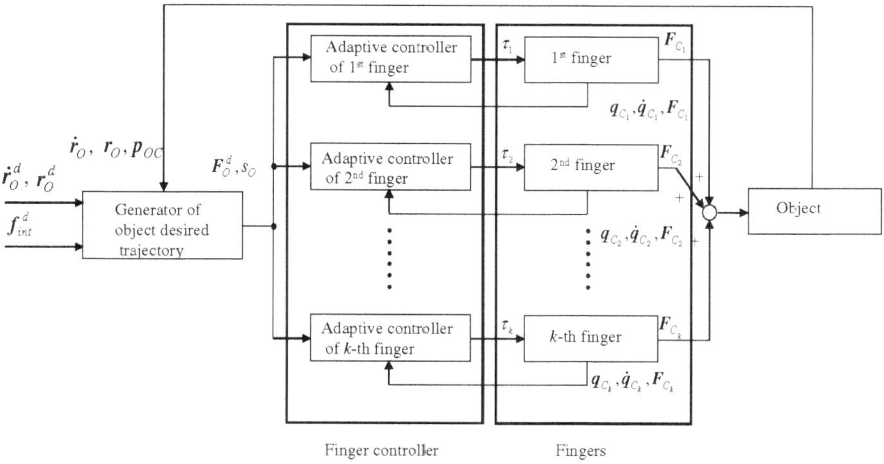

Fig. 6.3. Adaptive control system.

Figure 6.3 shows the structure of this adaptive control. The control input of each finger is calculated from the trajectory error of the object and the desired internal force. Each finger is adaptively controlled independently. Communication between fingers is not necessary. Adaptive coordination control is performed by providing the generator of the desired object trajectory at the top of the controller of each finger. If the grasping conditions A1–A4 as described in Sec. 4.8 of Chapter 4 are assumed, the following theorem is proved for adaptive control using a force sensor (Ueki *et al.*, 2009).

Theorem 6.4. *Consider a rigid body grasped by k fingers. For the system in Eq.* (6.12), *given that Eq.* (6.43)*is the parameter estimate law of the object, Eq.* (6.42) *is the desired external force of the object, Eq.* (6.34) *is the desired force at the contact point, Eq.* (6.50) *is the parameter estimate law of the finger, and Eq.* (6.49) *is the control law. The closed-loop system is then asymptotically stable in the following conditions*:

(1) $\boldsymbol{r}_O \rightarrow \boldsymbol{r}_O^d$, and $\dot{\boldsymbol{r}}_O \rightarrow \dot{\boldsymbol{r}}_O^d$ *as* $t \rightarrow \infty$.
(2) $\boldsymbol{F}_C \rightarrow \boldsymbol{F}_C^d$ *as* $t \rightarrow \infty$.

Outline of proof. To prove this theorem, a candidate Lyapunov function is given by

$$V = \frac{1}{2}\left(\boldsymbol{s}_O^T \boldsymbol{M}_O \boldsymbol{s}_O + \Delta\boldsymbol{\sigma}_O^T \boldsymbol{\Gamma}_O^{-1} \Delta\boldsymbol{\sigma}_O + \frac{1}{\alpha}\boldsymbol{v}_C^T \boldsymbol{\Omega} v_C + \boldsymbol{\eta}_C^T \boldsymbol{\Psi} \eta_C \right.$$
$$\left. + \ \boldsymbol{s}^T \boldsymbol{M}_F \boldsymbol{s} + \Delta\boldsymbol{\sigma}_F^T \boldsymbol{\Gamma}^{-1} \Delta\boldsymbol{\sigma}_F\right) \tag{6.51}$$

Its time derivative is represented by

$$\dot{V} = -\boldsymbol{s}_O^T \boldsymbol{K}_O \boldsymbol{s}_O - \boldsymbol{s}^T \boldsymbol{K} \boldsymbol{s} - \boldsymbol{v}_C^T \boldsymbol{\Omega} v_C \leq 0. \tag{6.52}$$

The asymptotic stability can be shown by the Lyapunov-like Lemma (Exercise 6.4). Although it is proved that $\boldsymbol{Y}_F \Delta\boldsymbol{\sigma}_F \to 0$ and $\boldsymbol{Y}_O\ \Delta\boldsymbol{\sigma}_O \to 0$ as $t \to \infty$, parameter convergence to the true values is not guaranteed. If $\boldsymbol{Y}_i (i = O, F)$ satisfy the **persistently exciting condition** — that is, there exists parameter $\gamma > 0$ and T such that $^\forall T \geq 0$

$$\int_t^{t+T} \boldsymbol{Y}_i^T \boldsymbol{Y}_i dt \geq \gamma \boldsymbol{I}_i, \quad (i = O, F), \tag{6.53}$$

where $\boldsymbol{I}_i (i = O, F)$ is the identity matrix of appropriate size — then the parameter error $\Delta\boldsymbol{\sigma}_i (i = O, F)$ will converge to zero exponentially (Anderson, 1977).

6.4. Adaptive Control at a Sliding Contact

Consider control when the fingers are in point contact with an object; some fingers are in slide rolling contact on the object surface, while the rest are in twist rolling contact. In the following given k fingers, the number of fingers in sliding contact is s; it is assumed that the degree of freedom (DOF) of each finger is 3, where the conditions $m = 3k - 2s > 6$ and the grasping conditions A1–A4 presented in Sec. 4.8 of Chapter 4 are satisfied. Consider that the control of the finger without sliding contact is the same as that with adaptive control using a force sensor in Sec. 6.3.2, and that control of the finger(s) with sliding contact is such that the desired velocity and position of the finger are expressed with the curved surface coordinate.

First, consider the relationship between the sliding velocity and the velocity with the surface coordinate variable. From Eqs. (2.15) and (2.35) of Chapter 2, sliding velocity at the contact point of the i-th finger $\boldsymbol{v}_{s_i} = [\dot{\boldsymbol{p}}_{s_i}^T, \boldsymbol{\omega}_{s_i}^T]^T$ is represented by

$$\boldsymbol{v}_{s_i} = \boldsymbol{J}_{FC_i}\dot{\boldsymbol{q}}_{F_i} - \boldsymbol{W}_{OC_i}^T\boldsymbol{v}_O = \boldsymbol{J}_{FC_i}\dot{\boldsymbol{q}}_{F_i} - \begin{bmatrix} \boldsymbol{D}_{OC_i} & \vdots & 0 \\ & \vdots & \boldsymbol{I}_{3\times 3} \end{bmatrix}^T \boldsymbol{v}_O, \tag{6.54}$$

where

$$\boldsymbol{J}_{FC_i} = \begin{bmatrix} \boldsymbol{J}_{v_i} \\ \boldsymbol{J}_{\omega_i} \end{bmatrix} \in \mathbb{R}^{6\times 3}.$$

The angular velocity and translational velocity can be represented by

$$\dot{\boldsymbol{p}}_{s_i} = \boldsymbol{J}_{v_i}\dot{\boldsymbol{q}}_{F_i} - \boldsymbol{D}_{OC_i}^T\boldsymbol{v}_O, \tag{6.55}$$

$$\boldsymbol{\omega}_{s_i} = \boldsymbol{J}_{\omega_i}\dot{\boldsymbol{q}}_{F_i} - \boldsymbol{\omega}_O, \tag{6.56}$$

where $\boldsymbol{J}_{v_i} \in \mathbb{R}^{3\times 3}$ and $\boldsymbol{J}_{\omega_i} \in \mathbb{R}^{3\times 3}$. By obtaining $\dot{\boldsymbol{q}}_{F_i}$ from Eq. (6.55), and substituting it with $\boldsymbol{v}_O = [\dot{\boldsymbol{p}}_O^T, \boldsymbol{\omega}_O^T]^T$ into Eq. (6.56), it is expressed by

$$\begin{aligned} \boldsymbol{\omega}_{s_i} &= \boldsymbol{J}_{\omega_i}\boldsymbol{J}_{v_i}^{-1}(\dot{\boldsymbol{p}}_{s_i} + \boldsymbol{D}_{OC_i}^T\boldsymbol{v}_O) - \boldsymbol{\omega}_O \\ &= \boldsymbol{J}_{\omega_i}\boldsymbol{J}_{v_i}^{-1}\dot{\boldsymbol{p}}_{s_i} + (\boldsymbol{J}_{\omega_i}\boldsymbol{J}_{v_i}^{-1}\boldsymbol{D}_{OC_i}^T - [0, \boldsymbol{I}_{3\times 3}])\boldsymbol{v}_O. \end{aligned} \tag{6.57}$$

The constraint conditions of velocity at the contact using the curved surface coordinate are presented in Sec. 2.6.2 of Chapter 2. Given that the curved surface coordinate of the object at the contact point of the i-th finger is $\boldsymbol{\alpha}_{o_i} \in \mathbb{R}^2$, we get the following from Eqs. (2.52) and (2.53) of Chapter 2:

$$\dot{\boldsymbol{\alpha}}_{o_i} = \boldsymbol{M}_{o_i}^{-1}\boldsymbol{R}_{\varphi_i}(\boldsymbol{K}_{f_i} + \tilde{\boldsymbol{K}}_{o_i})^{-1}[\boldsymbol{K}_{f_i}\boldsymbol{D}_2, \boldsymbol{D}_1]\boldsymbol{v}_{s_i}, \tag{6.58}$$

where $\boldsymbol{M}_{o_i}$ is the metric tensor of the object surface, $\boldsymbol{K}_{f_i}$ is the curvature tensor of the finger surface, $\boldsymbol{R}_{\varphi_i}$ is the rotation matrix defined by Eq. (2.54) of Chapter 2, and $\boldsymbol{K}_{o_i}$ is the curvature tensor of the object,

$$\tilde{\boldsymbol{K}}_{o_i} = \boldsymbol{R}_{\varphi_i}\boldsymbol{K}_{o_i}\boldsymbol{R}_{\varphi_i}, \quad \boldsymbol{D}_1 = \begin{bmatrix} 0 & -1 & 0 \\ 1 & 0 & 0 \end{bmatrix}, \quad \text{and} \quad \boldsymbol{D}_2 = \begin{bmatrix} 1 & 0 & 0 \\ 0 & 1 & 0 \end{bmatrix}.$$

By substituting Eq. (6.57) into Eq. (6.58) and using the relation of Eq. (4.12) of Chapter 4, the following equation is obtained:

$$\dot{\boldsymbol{p}}_{s_i} = \boldsymbol{A}_i \dot{\boldsymbol{\alpha}}_{Oi} + \boldsymbol{B}_i \boldsymbol{T}_O \dot{\boldsymbol{r}}_O, \tag{6.59}$$

where $\boldsymbol{A}_i$ and $\boldsymbol{B}_i$ are given by

$$\boldsymbol{A}_i = (\boldsymbol{D}_1 \boldsymbol{J}_{\omega_i} \boldsymbol{J}_{v_i}^{-1} + \boldsymbol{K}_{f_i} \boldsymbol{D}_2)^{-1} (\boldsymbol{K}_{f_i} + \tilde{\boldsymbol{K}}_{o_i}) \boldsymbol{R}_{\varphi_i}^T \boldsymbol{M}_{o_i},$$

$$\boldsymbol{B}_i = (\boldsymbol{D}_1 \boldsymbol{J}_{\omega_i} \boldsymbol{J}_{v_i}^{-1} + \boldsymbol{K}_{f_i} \boldsymbol{D}_2)^{-1} \boldsymbol{D}_1 (\boldsymbol{J}_{\omega_i} \boldsymbol{J}_{v_i}^{-1} \boldsymbol{D}_{OC_1} - [0, \boldsymbol{I}_{3\times 3}]) \boldsymbol{T}_O \boldsymbol{v}_O.$$

Equation (6.59) represents the relationship between the sliding velocity, the velocity of the surface coordinate variables, and the object velocity.

Next, consider the friction force on the tangent plane at the contact point. Let it be given that the i-th finger is sliding on the object surface at the contact point C_i. At this time, the contact force $\boldsymbol{f}_{C_i} \in \mathbb{R}^3$ can be decomposed into a component of the normal direction $\boldsymbol{f}_{n_i}$ and a dynamic friction force on the tangent plane $\boldsymbol{f}_{t_i}$ as follows:

$$\boldsymbol{f}_{C_i} = \boldsymbol{f}_{n_i} + \boldsymbol{f}_{t_i}. \tag{6.60}$$

When the unit normal vector at the contact point C_i is expressed by $\boldsymbol{n}_i$, then $\boldsymbol{f}_{n_i} = \boldsymbol{n}_i \boldsymbol{n}_i^T \boldsymbol{f}_{C_i}$. It is assumed that the dynamic friction force is represented by

$$\boldsymbol{f}_{t_i} = \mu_{Ci} \boldsymbol{n}_i^T \boldsymbol{f}_{n_i} \frac{\dot{\boldsymbol{p}}_{s_i}}{\|\dot{\boldsymbol{p}}_{s_i}\|} + \mu_{d_i} \dot{\boldsymbol{p}}_{s_i}, \tag{6.61}$$

where $\dot{\boldsymbol{p}}_{s_i}$ is the sliding velocity at the contact point, μ_{C_i} is the Coulomb dynamic friction coefficient, and μ_{d_i} is a viscous friction coefficient. Eq. (6.61) is linear regarding the friction parameters μ_{Ci} and μ_{d_i}, therefore it can be represented as

$$\boldsymbol{f}_{t_i} = \boldsymbol{Y}_{ti}(\dot{\boldsymbol{p}}_{s_i}, \boldsymbol{n}_i, \boldsymbol{f}_{C_i}) \boldsymbol{\mu}_i, \tag{6.62}$$

where $\boldsymbol{\mu}_i = [\mu_{Ci}, \mu_{d_i}]^T$.

In order to compensate for the friction force at the sliding fingers, let us generate the desired friction force on the tangent plane based on an estimate of the friction parameter. First, we define a reference velocity of sliding motion by

$$\boldsymbol{v}_{s_i r} = \boldsymbol{A}_i(\dot{\boldsymbol{\alpha}}^d_{o_i} + \boldsymbol{\Lambda}_s(\boldsymbol{\alpha}^d_{o_i} - \boldsymbol{\alpha}_{o_i})) + \boldsymbol{B}_i \boldsymbol{T}_O \dot{\boldsymbol{r}}^d_O, \tag{6.63}$$

and a residual of the sliding velocity is given by

$$\boldsymbol{s}_{s_i} = \boldsymbol{v}_{s_i} - \boldsymbol{v}_{s_i r}, \tag{6.64}$$

where $\boldsymbol{\alpha}^d_{o_i}$ is a smooth desired position of the fingertip on the curved surface coordinates, and $\boldsymbol{\Lambda}_s > 0$ is a diagonal weight matrix. Then, a desired friction force and estimation law of the friction parameters can be given by

$$\boldsymbol{f}^d_{t_i} = \boldsymbol{Y}_{t_i}(\boldsymbol{v}_{s_i}, \boldsymbol{n}_i, \boldsymbol{f}_{C_i})\hat{\boldsymbol{\mu}}_i \tag{6.65}$$

$$\dot{\hat{\boldsymbol{\mu}}}_i = -\boldsymbol{\Gamma}_{t_i} \boldsymbol{Y}^T_{t_i}(\boldsymbol{v}_{s_i}, \boldsymbol{n}_i, \boldsymbol{f}_{C_i})\boldsymbol{s}_{s_i}, \tag{6.66}$$

where $\hat{\boldsymbol{\mu}}_i$ is an estimation of $\boldsymbol{\mu}_i$, and $\boldsymbol{\Gamma}_{t_i} > 0$ is a symmetrically adaptive gain matrix. It is noted that the desired friction force at the rolling contact is $\boldsymbol{f}^d_{t_i} = 0$.

Next, consider the desired contact force of the finger. The desired external force of the object is provided by Eq. (6.42) in the same way as adaptive control using a force sensor in Sec. 6.3.2; i.e.,

$$\boldsymbol{F}^d_O = \boldsymbol{Y}_O(\boldsymbol{r}_O, \dot{\boldsymbol{r}}_O, \boldsymbol{v}_{Or}, \dot{\boldsymbol{v}}_{Or})\,\hat{\boldsymbol{\sigma}}_O - \boldsymbol{K}_O \boldsymbol{s}_O. \tag{6.67}$$

The desired contact force to balance the desired external force should satisfy the following relation:

$$\boldsymbol{F}^d_O = \boldsymbol{D}^T_{OC}\, \boldsymbol{f}^d_C, \tag{6.68}$$

where $\boldsymbol{D}_{OC} = [\boldsymbol{D}^T_{OC_1}, \ldots, \boldsymbol{D}^T_{OC_k}]^T$ and $\boldsymbol{f}^d_C = [(\boldsymbol{f}^d_{C_1})^T, \ldots, (\boldsymbol{f}^d_{C_k})^T]^T$. The desired contact force $\boldsymbol{f}^d_{C_i}$ at the contact point C_i is then decomposed into the normal direction and force on the tangent plane, which is represented by

$$\boldsymbol{f}^d_{C_i} = f^d_{n_i}\boldsymbol{n}_i + \boldsymbol{f}^d_{t_i}, \tag{6.69}$$

where $\boldsymbol{f}^d_{n_i}$ is the size of the desired contact force in the normal direction and $\boldsymbol{f}^d_{t_i}$ is the desired force on the tangent plane. For the sliding-contact fingers, the desired external force is decomposed into the desired contact force in the normal direction and the desired friction force in the direction of the tangent. For the rolling-contact fingers, such decomposition is not performed. Then, the desired external force satisfies:

$$\boldsymbol{F}_O^d = \mathbf{G}\mathbf{f}_{Cn}^d + \boldsymbol{D}_{OC}\boldsymbol{f}_t^d, \tag{6.70}$$

where $\boldsymbol{f}_{Cn}^d \in \mathbb{R}^{3k-2s}$ is a vector that is the summation of the desired contact forces of the fingers with rolling contact, $\boldsymbol{G} \in \mathbb{R}^{6\times(3k-2s)}$ is a matrix with a list of $\boldsymbol{D}_{OC_i}$ corresponding to $\boldsymbol{f}_{Cn}^d$, and $\boldsymbol{f}_t^d = [(\boldsymbol{f}_{t_1}^d)^T, \ldots, (\boldsymbol{f}_{t_k}^d)^T]^T$. The contact force generates the internal forces that do not affect motion of the object. Therefore, the general solution for $\boldsymbol{f}_{Cn}^d$ is provided by

$$\boldsymbol{f}_{Cn}^d = \boldsymbol{G}^+(\boldsymbol{F}_O^d - \boldsymbol{D}_{OC}\boldsymbol{f}_t^d) + (\boldsymbol{I}_{m\times m} - \boldsymbol{G}^+\boldsymbol{G})\boldsymbol{f}_{int}^d, \tag{6.71}$$

where $\boldsymbol{G}^+(= \boldsymbol{G}^T(\boldsymbol{G}\boldsymbol{G}^T)^{-1})$ is a generalized inverse matrix of $\boldsymbol{G}$, and $\boldsymbol{f}_{int}^d$ is an arbitrary force vector that generates the internal force, which is continuous and bounded.

Finally, consider the control of the fingers. The reference velocity of the i-th finger is defined in the same way as Eq. (6.45):

$$\dot{\boldsymbol{q}}_{Fr_i} = \boldsymbol{J}_{C_i}^{-1}(\boldsymbol{G}_{C_i}^T\boldsymbol{v}_{Or} + \boldsymbol{\Omega}_i\boldsymbol{v}_{C_i} + \boldsymbol{\Psi}_i\boldsymbol{\eta}_{C_i}). \tag{6.72}$$

It is given that $\boldsymbol{v}_{C_i}$ is the output after filtering the contact force error $\Delta\boldsymbol{f}_{C_i} = \boldsymbol{f}_{C_i}^d - \boldsymbol{f}_{C_i}$, as in Eq. (6.24), and $\boldsymbol{\eta}_{C_i}$ is the integration of a normal component of contact force errors $\Delta\boldsymbol{f}_{C_i}$, which is represented by

$$\boldsymbol{\eta}_{C_i} = \eta_i\boldsymbol{n}_i \tag{6.73}$$

$$\eta_i = \int_0^t \boldsymbol{n}_i^T\Delta\boldsymbol{f}_{C_i}dt. \tag{6.74}$$

Using the residual of the i-th finger variable, $\boldsymbol{s}_i = \boldsymbol{q}_{F_i} - \boldsymbol{q}_{Fr_i}$, the control law and the estimation law of the kinetic parameters of the i-th finger are given by

$$\boldsymbol{\tau}_{F_i} = \boldsymbol{Y}_{F_i}(\boldsymbol{q}_{F_i}, \dot{\boldsymbol{q}}_{F_i}, \dot{\boldsymbol{q}}_{Fr_i}, \ddot{\boldsymbol{q}}_{Fr_i})\hat{\boldsymbol{\sigma}}_{F_i} + \boldsymbol{J}_{v_i}^T\boldsymbol{f}_{C_i}^d - \boldsymbol{K}_i\boldsymbol{s}_i \tag{6.75}$$

$$\dot{\hat{\boldsymbol{\sigma}}}_{F_i} = -\boldsymbol{\Gamma}_i\boldsymbol{Y}_{F_i}^T(\boldsymbol{q}_{F_i}, \dot{\boldsymbol{q}}_{F_i}, \dot{\boldsymbol{q}}_{Fr_i}, \ddot{\boldsymbol{q}}_{Fr_i})\boldsymbol{s}_i. \tag{6.76}$$

At this time, the control law and estimation law integrating the 1st to the k-th finger are expressed by

$$\boldsymbol{\tau}_F = \boldsymbol{Y}_F(\boldsymbol{q}_F, \dot{\boldsymbol{q}}_F, \dot{\boldsymbol{q}}_{Fr}, \ddot{\boldsymbol{q}}_{Fr})\hat{\boldsymbol{\sigma}}_F + \boldsymbol{J}_C^T \boldsymbol{f}_C^d - \boldsymbol{K}\boldsymbol{s}, \tag{6.77}$$

$$\dot{\hat{\boldsymbol{\sigma}}}_F = -\boldsymbol{\Gamma}\boldsymbol{Y}_F^T(\boldsymbol{g}_F, \dot{\boldsymbol{q}}_F, \dot{\boldsymbol{q}}_{Fr}, \ddot{\boldsymbol{q}}_{Fr})\boldsymbol{s}. \tag{6.78}$$

The difference of the control at the rolling contact is that the desired contact force is given based on the estimation of friction parameters. Other configurations are almost the same as the estimation law and control law in Sec. 6.3.2. The following theorem is then provided by this control (Ueki *et al.*, 2006).

Theorem 6.5. *Consider the control of a rigid body grasped by k fingers, each with three DOFs. For the hand system represented by Eq.* (6.12), *given that the parameter estimation law of the object is Eq.* (6.43), *the estimation law of friction parameters with sliding fingers at the contact point is Eq.* (6.61), *the desired external force of the object is Eq.* (6.63), *the desired force at the contact point is Eq.* (6.71), *the estimation law of the parameter of the finger is Eq.* (6.78), *and the control law is Eq.* (6.77). *Then, if β is sufficiently large, the closed-loop system is asymptotically stable in the following sense:*

(1) $\boldsymbol{r}_O \to \boldsymbol{r}_O^d$ *and* $\dot{\boldsymbol{r}}_O \to \dot{\boldsymbol{r}}_O^d$ *as* $t \to \infty$.
(2) $\boldsymbol{f}_C \to \boldsymbol{f}_C^d$ *as* $t \to \infty$.
(3) $\boldsymbol{\alpha}_{O_i} \to \boldsymbol{\alpha}_{O_i}^d$ *and* $\dot{\boldsymbol{\alpha}}_{O_i} \to \dot{\boldsymbol{\alpha}}_{O_i}^d$ *as* $t \to \infty$.

Outline of proof. To prove this theorem, a candidate Lyapunov function is given by

$$V = \frac{1}{2}\Big(\boldsymbol{s}_O^T \boldsymbol{M}_O \boldsymbol{s}_O + \Delta\boldsymbol{\sigma}_O^T \boldsymbol{\Gamma}_O^{-1} \Delta\boldsymbol{\sigma}_O + \frac{1}{\alpha}\boldsymbol{v}_C^T \boldsymbol{\Omega} v_C + \boldsymbol{\eta}_C^T \boldsymbol{\Psi} \eta_C + \boldsymbol{s}^T \boldsymbol{M}_F \boldsymbol{s} + \Delta\boldsymbol{\sigma}_F^T \boldsymbol{\Gamma}^{-1} \Delta\boldsymbol{\sigma}_F + \Delta\boldsymbol{\mu}^T \boldsymbol{\Gamma}_f^{-1} \Delta\boldsymbol{\mu}\Big), \tag{6.79}$$

where $\Delta\mu = \hat{\mu} - \mu$. Its time derivative is represented by

$$\dot{V} = -\boldsymbol{s}_O^T \boldsymbol{K}_O \boldsymbol{s}_O - \boldsymbol{s}^T \boldsymbol{K}\boldsymbol{s} - \boldsymbol{v}_C^T \boldsymbol{\Omega} v_C \leq 0. \tag{6.80}$$

Using a Lyapunov-like Lemma, the asymptotic stability can be demonstrated in a procedure similar to the case of Theorem 6.4. It should be noted that although this control is complicated, even more dexterous object manipulation is possible if the finger slides on the object surface.

Up to this point, adaptive control has been discussed in accordance with the contact conditions. However, in adaptive control, the shapes of the object and positions of the contact points are already known. Thus, study of adaptive controls with fewer restrictions is needed. Also, dynamic control by soft fingers will be a forthcoming challenge.

6.5. Grasp Imitating Human Grasping Reflex

This section describes a grasping strategy (Kawasaki *et al.*, 2002; Mouri *et al.*, 2007) for objects with unknown shapes that imitates the grasp reflex of infants. The **grasp reflex** occurs when a bar is placed on a newborn's open palm. Infants will close all five fingers around a bar even if the shape of the bar is unknown, and their grip on the bar will tighten when an attempt is made to remove the bar from their palm. This grasping reflex is considered an essential action for survival; it is a means of physically acquiring (a grasping action) and keeping (a withholding action) an object. It appears that the object is grasped with uniform contact force, even if the object's geometrical form is unknown. In a similar way, it is desirable that a robot hand with a grasping reflex should be able to easily grasp various unknown objects, including soft objects. In the following, it is supposed that the first joint motion at the base of the finger is adduction and abduction, and other joints motions are flexion and extension.

6.5.1. *Flexion and Extension*

When an object to be grasped comes into contact with the robotic hand, all joints except for the first joint of the thumb and fingers are controlled independently to match the contact force with the desired value. Figure 6.4 shows the contact state between the i-th finger and grasped object, where q_{ij} and F_{ik} represent the joint angle of j-th joint of i-th finger and the total contact force of Link k $(k = j - 1)$ of i-th finger, respectively. The finger and object are controlled by a

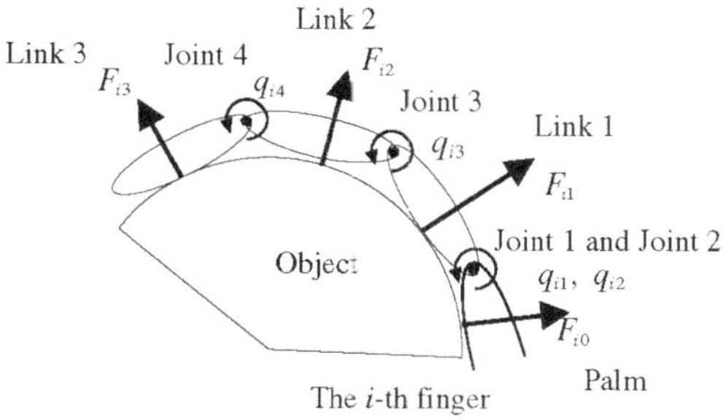

Fig. 6.4. Links of finger and contact force.

velocity control in a noncontact state, and a force control in contact state. Therefore, the joint torque of the j-th joint ($j \geq 2$) of the i-th finger is given by

$$\tau_{ij} = -K_{Vij}\left(\dot{q}_{ij} - \dot{q}_{dij}\right), \tag{6.81}$$

where K_{Vij} is a velocity feedback gain of the$j$-th joint of the i-th finger, and $\dot{q}_{dij}$ is a desired joint velocity of the j-th joint of the i-th finger. Based on the contact information obtained from the tactile sensor, $\dot{q}_{dij}$ is given by the following equation to be the velocity control in a noncontact state and the force control in a contact state:

$$\dot{q}_{dij} = \begin{cases} \dot{q}_{cij} & : \text{ Noncontact state} \\ -K_{fij}\boldsymbol{J}_{cik}^{T}\left(\boldsymbol{F}_{ik} - \boldsymbol{F}_{dik}\right) & \\ \qquad -K_{fIij}\int \boldsymbol{J}_{cik}^{T}\left(\boldsymbol{F}_{ik} - \boldsymbol{F}_{dik}\right)dt & : \text{ Contact state} \end{cases} \tag{6.82}$$

where $\dot{q}_{cij}$ is a constant desired joint velocity, K_{fik} is a force-feedback gain, K_{fIik} is a force integral gain, F_{dik} is a constant desired-contact force, and J_{cik} is the Jacobi matrix related with

$$\dot{\boldsymbol{r}}_{cik} = \boldsymbol{J}_{cik}\dot{\boldsymbol{q}}_{ij}, \tag{6.83}$$

where $\boldsymbol{r}_{cik}$ is a contact position. The desired velocity of each joint is set such that it is relatively larger closer to the base of the finger and smaller closer to the fingertip. This setting means that contact will start from the base link of the finger to the object. Furthermore, all desired contact forces should be the same value in order to grasp the object with a uniform contact force. Since the fourth joint of the Gifu Hand III is geared with the third joint of the finger, the bigger value in the contact force $\boldsymbol{F}_{i2}$ and $\boldsymbol{F}_{i3}$ is used for control of the third joint. This method controls fingers individually so as to envelop an object using the tactile information on the finger surface. If the desired velocity of the joint at the fingertip were set to be larger, a fingertip grasp would be feasible.

6.5.2. *Adduction and Abduction*

The first joint of the finger, used for the motion of adduction and abduction, is position-controlled to adapt to the size of the object, and the joint torque is given by

$$\tau_{i1} = -K_{Pi1}\left(q_{i1} - q_{di1}\right) - K_{Vi1}\dot{q}_{i1}, \tag{6.84}$$

where K_{Pi1} is the position feedback gain, and q_{di1} s the desired joint angle. As shown in Fig. 6.5, the desired joint angle, q_{di1}, is determined to match a least squares approximated straight line of distributed contact points with the center line of the finger. Note that this least squares approximated straight line is perpendicular to the rotation axis of the 1st joint. According to this control, a motion is generated that brings the contact points on the side of the finger closer to the centerline of the finger. As a consequence, a more stable grasp may be expected because the contact attaches along the centerline of the finger while avoiding a side contact with the finger.

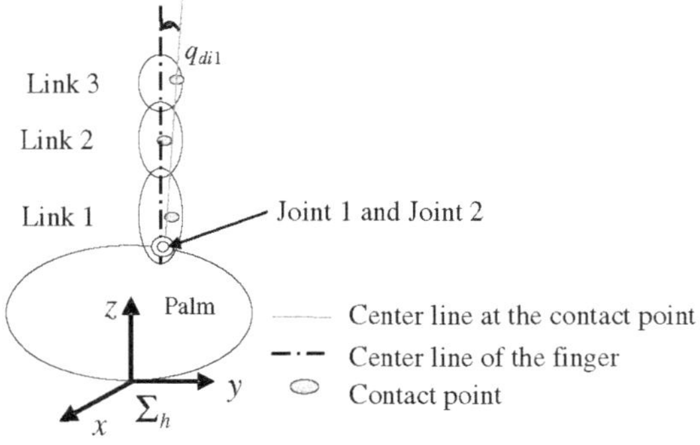

Fig. 6.5. Target angle of internal and external rotation.

6.5.3. *Holding Manipulation*

Consider the holding manipulation by the robot hand when the object is grasped and then removed. When external forces are applied to the object, the robotic hand will not be able to continually grasp it if the desired contact force $\boldsymbol{F}_{dik}$ is given by a constant $\boldsymbol{F}_{dik\_cons}$, as shown in Eq. (6.82). In order to keep the grasp and resist the external force, the desired contact force $\boldsymbol{F}_{dik}$ must be set to be proportional to the Euclidean norm of velocity of the contact point $\|\boldsymbol{v}_c\|$, as given by

$$\boldsymbol{F}_{dik} = (1 + K_{VCik}\|\boldsymbol{v}_c\|)\boldsymbol{F}_{dik\_const}, \tag{6.85}$$

where K_{VCik} is a constant number. Thus, the forces of the contact points are controlled in accordance with external forces.

Such grasping manipulations increase the possibility for the number of contact points to be maximized — an object is to be wrapped with the thumb and fingers, and an unknown-shaped object is to be grasped. Moreover, the wrapping grasp enables a powerful grasp. This is a very simple method; however, grasping an unknown three-dimensional (3D) object is feasible because it does not require geometric information about the object. Figure 6.6 presents examples of a five-fingered hand grasping objects of unknown shape using the grasp approach that imitates the human grasping reflex. In all cases, the objects are grasped properly in accordance with the object shape.

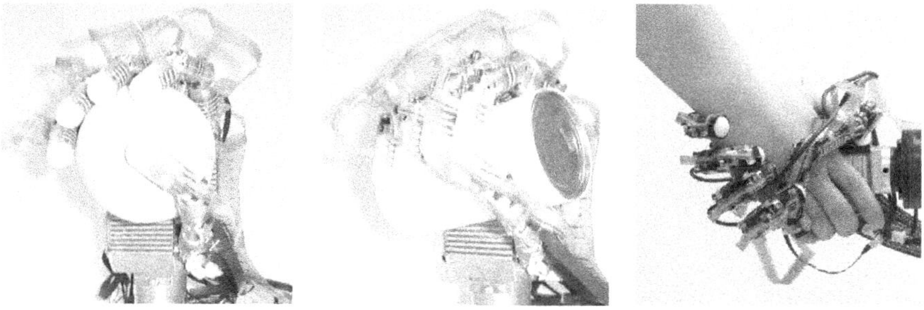

(a) Spherical object (b) Cylindrical object (c) Hand of a person

Fig. 6.6. Examples of grasping unknown shaped object.

Exercises

6.1. Prove Theorem 6.2.

6.2. Prove Theorem 6.3.

6.3. When the dynamic parameter vector of the finger, $\boldsymbol{\sigma}_F$, is known and the dynamic parameter vector of the object, $\boldsymbol{\sigma}_O$, is unknown, the control law that is modified for adaptive control without a force sensor as in Sec. 6.3.1 is given as

$$\boldsymbol{\tau}_F = \boldsymbol{Y}_F(\boldsymbol{q}_F, \dot{\boldsymbol{q}}_F, \dot{\boldsymbol{q}}_{Fr}, \ddot{\boldsymbol{q}}_{Fr})\boldsymbol{\sigma}_F + \boldsymbol{J}_C^T \boldsymbol{F}_C^d - \boldsymbol{K}\boldsymbol{s}. \tag{1}$$

Then, using the candidate Lyapunov function given by

$$V = \frac{1}{2}(\boldsymbol{s}_O^T \boldsymbol{M}_O \boldsymbol{s}_O + \Delta\sigma_O^T \Gamma_O \Delta\sigma_O + \boldsymbol{s}^T \boldsymbol{M}_F \boldsymbol{s}), \tag{2}$$

prove the following theorem, where it is supposed that the grasping conditions A1–A4 shown in Sec. 4.8 of Chapter 4 are satisfied.

Theorem 6.6. *Consider a rigid body that is grasped by k fingers. For the system represented by Eq.* (6.12), *given that the parameter estimation law of the object is Eq.* (6.32), *the desired external force of the object is Eq.* (6.31), *the desired force at the contact point is Eq.* (6.34), *the residual is Eq.* (6.37), *and the control law is Eq.* (1). *The closed-loop system is then asymptotically stable in the following sense:*

(1) $\boldsymbol{r}_O \to \boldsymbol{r}_O^d$ *and* $\dot{\boldsymbol{r}}_O \to \dot{\boldsymbol{r}}_O^d$ *as* $t \to \infty$.

(2) *The force error at the contact point* $\boldsymbol{F}_O^d - \boldsymbol{F}_O$ *is bounded.*

6.4. Prove Theorem 6.4.

Chapter 7

Multi-Fingered Haptic Interface

Haptic interfaces are used to present force and tactile feeling to humans in cases of telecontrol for robots, manipulating objects in a virtual reality (VR) environment, playing games with force feeling, and so on. Some of these interfaces provide three-dimensional (3D) force feelings to the human fingertips, and a few haptic interfaces cover the workspace of the human arm. This chapter focuses on the control of a multi-fingered haptic interface in a VR environment, and on applications using the multi-fingered haptic interface.

7.1. Indirect Control and Direct Control

Haptic interfaces are classified into two types: wearable-type interfaces (Burdea, 1996), which are mounted firmly on the human hand, and grounded-type interfaces (Kawasaki, 2007b), which are located in positions that match and are opposed to the human hand. The dynamics of the later type are almost the same as those of hand robots; i.e., as shown in Sec. 4.3, the dynamics of the haptic interface is represented by,

$$\boldsymbol{M}\ddot{\boldsymbol{q}} + \boldsymbol{C}(\boldsymbol{q}, \dot{\boldsymbol{q}})\dot{\boldsymbol{q}} + \boldsymbol{q}(\boldsymbol{q}) = \boldsymbol{\tau} - \boldsymbol{J}^T(\boldsymbol{q})\boldsymbol{F}_H, \tag{7.1}$$

where $\boldsymbol{F}_H$ is the force that the operator exerts on the haptic interface. When the operator manipulates virtual objects, the force feeling experienced by the operator is called the haptic rendering. Control methods for haptic rendering are classified broadly into indirect control and direct control. Indirect control is also called impedance control (Kawai and Yoshikawa, 2004), while direct control is also called admittance control.

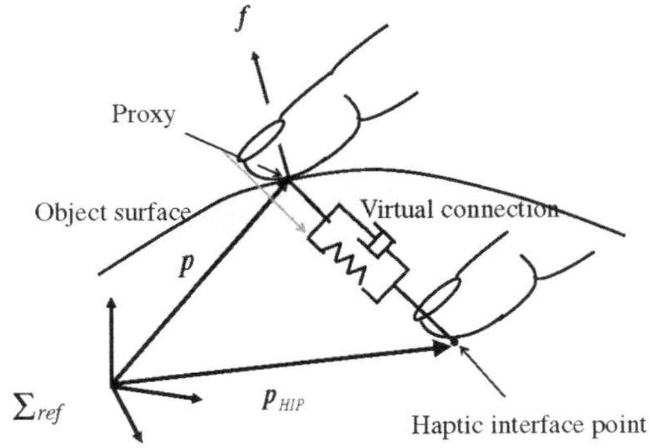

Fig. 7.1. Proxy and haptic rendering.

In indirect control (Colgate *et al.*, 1993), the force generated by the interference between the operator's finger and the object in the VR environment is relayed to the operator by position and velocity feedback control. This method is suitable for haptic interfaces that have less friction in mechanisms and are without force sensors. It can communicate the hardness and softness of virtual objects by adjusting the impedance characteristics. Now, consider a contact point P between the operator's finger and virtual object as shown in Fig. 7.1. The point lies on the object surface and is called a proxy. The finger of the operator, meanwhile, penetrates the virtual object. The finger point below the surface of the virtual object at the penetration depth is called the **haptic interface point** (HIP). We assume that there is a spring and damper mechanism to connect the proxy and the HIP, whose impedance characteristic is represented by,

$$\boldsymbol{f} = m_v(\ddot{\boldsymbol{p}}_{HIP} - \ddot{\boldsymbol{p}}) + d_v(\dot{\boldsymbol{p}}_{HIP} - \dot{\boldsymbol{p}}) + k_v(\boldsymbol{p}_{HIP} - \boldsymbol{p}), \tag{7.2}$$

where $\boldsymbol{p}$ is a position vector of the proxy, $\boldsymbol{p}_{HIP}$ is a position vector of HIP, and m_v, d_v, and k_v are mass, damping, and stiffness parameters of the virtual object, respectively. It is possible to display hard or soft objects by adjusting these parameters. To generate the contact force, the input of the haptic interface is given by,

$$\boldsymbol{\tau} = \boldsymbol{J}^T \boldsymbol{f} + \hat{\mathbf{g}}, \tag{7.3}$$

where $\boldsymbol{\tau}$ is the input, $\boldsymbol{f}$ is the contact force, and $\hat{\boldsymbol{g}}$ is a gravitational compensation term. By substituting (7.3) into (7.1) and supposing

$\hat{\boldsymbol{g}} = \boldsymbol{g}$, then the dynamics of haptic interface are represented by,

$$\boldsymbol{M}\ddot{\boldsymbol{q}} + \boldsymbol{C}(\boldsymbol{q}, \dot{\boldsymbol{q}})\dot{\boldsymbol{q}} = \boldsymbol{J}^T(\boldsymbol{q})(\boldsymbol{f} - \boldsymbol{F}_H),$$

where $\boldsymbol{F}_H$ is an external force that the operator exerts on the haptic interface. When the mass, joint angle, and joint acceleration of the haptic interface are negligibly small, it can be represented by,

$$m_v(\ddot{\boldsymbol{p}}_{HIP} - \ddot{\boldsymbol{p}}) + d_v(\dot{\boldsymbol{p}}_{HIP} - \dot{\boldsymbol{p}}) + k_v(\boldsymbol{p}_{HIP} - \boldsymbol{p}) = \boldsymbol{F}_H. \tag{7.4}$$

The $\boldsymbol{F}_H$ is the force that the operator exerts on the haptic interface, and at the same time, receives back through the haptic interface. Most commercial haptic interfaces use indirect control, adopting a rope-pulley reduction mechanism. These experience less backlash and friction and do not need a force sensor; however, the mechanism tends to be large.

With direct control, the force is transmitted to the operator's finger by the force feedback control, in which the desired force is the interference force between the operator's finger and object in the VR environment. This type of control can realize small haptic interfaces, because it can adopt the high reduction mechanism of the motor, but it requires a force sensor. The force error is given by,

$$\Delta\boldsymbol{F} = \boldsymbol{F}_d - \boldsymbol{F}_H, \tag{7.5}$$

where $\boldsymbol{F}_d$ is the desired force, and the external force $\boldsymbol{F}_H$ is measured by the force sensor. The control input is given by,

$$\boldsymbol{\tau} = \boldsymbol{J}^T\left(k_1\Delta\boldsymbol{F} + k_2\int_0^t \Delta\boldsymbol{F}dt + \boldsymbol{F}_d\right) + \hat{\boldsymbol{g}}, \tag{7.6}$$

where $k_1, k_2 > 0$ are the force error and integral of force error feedback gains, respectively. By substituting (7.3) into (7.1) and supposing $\hat{\boldsymbol{g}} = \boldsymbol{g}$, the dynamics of haptic interface are represented by,

$$\boldsymbol{M}\ddot{\boldsymbol{q}} + \boldsymbol{C}(\boldsymbol{q}, \dot{\boldsymbol{q}})\dot{\boldsymbol{q}} = \boldsymbol{J}^T(\boldsymbol{q})((k_1 + 1)\Delta\boldsymbol{F} + k_2\int_0^t \Delta\boldsymbol{F}dt).$$

As in the case of indirect control, when the mass, joint angle, and joint acceleration of the haptic interface are negligibly small, it can be represented by

$$(k_1 + 1)\Delta \boldsymbol{F} + k_2 \int_0^t \Delta \boldsymbol{F} dt = 0. \tag{7.7}$$

By setting k_1 and k_2 properly, force error converges to zero. This means that the desired force is displayed to the operator's finger.

It is noted that stability of haptic interfaces which are coupled with human operators and virtual environments are affected by the dynamics of human (Adams and Hannaford, 2002).

7.2. Control of Multi-Fingered Haptic Interface

Haptic interfaces that present force and tactile feeling to the fingertips of humans have been investigated extensively over the last two decades. Some of these interfaces provide 3D force feelings to the user's fingertip, and a few cover the workspace of the human arm. Such interfaces can be classified into two types: Wearable-type interfaces, which are mounted firmly on the human hand, and grounded-type interfaces, which are located in positions that match and are opposed to the human hand. The wearable-type haptic interface has a large workspace, but with this interface, it is difficult to present three-directional forces to the user, or to simulate the weight of virtual objects on the fingertips, because the mechanism is mounted on the back of the hand, and the exerted force is one-directional. In contrast, the grounded-type haptic interface generally has a comparatively small workspace, but such an interface consisting of an arm and fingertips could be used in a large workspace. A haptic interface must be safe, function in a wide operational space, and represent the weight of virtual objects as well as the force at the contact points. In addition, it should not produce an oppressive feeling when attached to a person, nor should it represent its own weight.

A multi-fingered haptic interface called HIRO III (Endo *et al.*, 2011) was developed to present three-directional force at each of the five human fingertips, reduce oppressive feelings, and increase safety. It is placed opposite the human hand, and the haptic fingertips are connected to the human fingertips through finger holders and passive spherical magnet joints (see Sec. 1.5). The HIRO III is a redundant interface robot with 21 degrees of freedom (DOFs). When five human fingertips are connected to the HIRO III through the passive spherical joints, the haptic hand posture is non-unique because of the redundancy of HIRO III. To take this redundancy into account, redundant force control and hand manipulability-based control have been developed.

7.2.1. *Redundant Force Control*

This section presents redundant force control in which the haptic hand and haptic arm are force controlled separately (Mouri *et al.*, 2006). The haptic hand is force controlled to follow the desired 3D force at each fingertip, and the haptic arm is force controlled to follow the desired force and moment at the wrist of the haptic interface, which are equivalent to the sum of the desired fingertip forces.

We first present the force control of the fingers. It is assumed that the number of fingers is k, the DOF of each finger is three, and the desired force of the i-th haptic finger, $\boldsymbol{F}_i^d \in \mathbb{R}^3$, is computed by the interference between the human finger and virtual object. Let us define the i-th fingertip force as $\boldsymbol{F}_i \in \mathbb{R}^3$, the i-th finger joint torque as $\boldsymbol{\tau}_{F_i} \in \mathbb{R}^3$, and the kinematic Jacobian of the i-th finger as $\boldsymbol{J}_{F_i} \in \mathbb{R}^{3\times 3}$. The force error at i-th fingertip is then given by,

$$\Delta \boldsymbol{F}_i = \boldsymbol{F}_i^d - \boldsymbol{F}_i, \tag{7.8}$$

and its equivalent joint torque can be represented by $\boldsymbol{J}_{F_i}^T \Delta \boldsymbol{F}_i$. Now, let us define $\boldsymbol{F} = [\boldsymbol{F}_1^T, \cdots, \boldsymbol{F}_k^T]^T, \boldsymbol{F}^d = [(\boldsymbol{F}_1^d)^T, \cdots, (\boldsymbol{F}_k^d)]^T, \boldsymbol{\tau}_F = [\boldsymbol{\tau}_{F_1}^T, \cdots, \boldsymbol{\tau}_{F_k}^T]^T$, and $\boldsymbol{J}_F = block\ diag[\boldsymbol{J}_{F_1}, \cdots, \boldsymbol{J}_{F_k}]$. The control law of the finger is then given as a PI (Proportional and Integral) force control in the joint space presented by,

$$\tau_F = \boldsymbol{K}_{F1} \boldsymbol{J}_F^T \Delta \boldsymbol{F} + \boldsymbol{K}_{F2} \int \boldsymbol{J}_F^T \Delta \boldsymbol{F} \mathbf{dt} - \boldsymbol{K}_{F3} \dot{\boldsymbol{q}}_F + \boldsymbol{q}_F + \boldsymbol{J}_F^T \boldsymbol{F}^d, \tag{7.9}$$

where $\boldsymbol{K}_{F1}, \boldsymbol{K}_{F2}, \boldsymbol{K}_{F3} > 0$ are feedback diagonal gain matrices of force error, integral of force error, and velocity, respectively, and $\boldsymbol{g}_F$ is a gravity compensate term. The first and second terms on the right side of Eq. (7.9) are PI force control terms; the third term is a damping compensation term to give the control system stability;

the fourth term is a gravity compensation term; and the fifth term is a feed-forward term of the desired force. A point of difference from Eq. (7.6) is that the feedback gain matrices are given in the joint space. This is because the feedback gain adjustment becomes easy when the dynamic characteristics of each finger joint are taken into account. When the mass of the haptic finger mechanism is negligibly small and the joint velocity and joint acceleration are not large, the force error has the following characteristics:

$$(\boldsymbol{K}_{F1} + \boldsymbol{I}_{3\times 3})\boldsymbol{J}_F^T \Delta \boldsymbol{F} + \boldsymbol{K}_{F2} \int \boldsymbol{J}_F^T \Delta \boldsymbol{F} dt = 0. \tag{7.10}$$

This shows that each fingertip force converges to the desired force independently.

Next, the force control of the haptic arm is presented. It is supposed that the DOF of the arm is 6. The force and moment errors at the wrist of the haptic arm, equivalent to those at the i-th fingertip, are given by $\Delta \boldsymbol{F}_i$ and $\boldsymbol{p}_{Fi} \times \Delta \boldsymbol{F}_i$, where $\boldsymbol{p}_{Fi} \in \mathbb{R}^3$ is a position vector from the wrist to the i-th fingertip. Hence, the arm joint torque that is equivalent to the sum of fingertip force errors can be given by,

$$\boldsymbol{\tau}_e = \boldsymbol{J}_A^T \begin{bmatrix} \sum_{i=1}^{k} \Delta \boldsymbol{F}_i \\ \sum_{i=1}^{k} \boldsymbol{p}_{Fi} \times \Delta \boldsymbol{F}_i \end{bmatrix}, \tag{7.11}$$

where $\boldsymbol{J}_A \in \mathbb{R}^{6\times 6}$ is a kinematic Jacobian matrix of the arm. The desired force and moment at the wrist that are equivalent to the sum of the desired fingertip force are $\sum_{i=1}^{k} \boldsymbol{F}_i^d$ and $\sum_{i=1}^{k} \boldsymbol{p}_{Fi} \times \boldsymbol{F}_i^d$, respectively. Hence, the force control law of arm is given by,

$$\tau_A = \boldsymbol{K}_{A1}\tau_e + \boldsymbol{K}_{A2} \int \tau_e dt - K_{A3}\dot{\boldsymbol{q}}_A + \boldsymbol{q}_A + \boldsymbol{J}_A^T \begin{bmatrix} \sum_{i=1}^{k} \boldsymbol{F}_i^d \\ \sum_{i=1}^{k} \boldsymbol{p}_{Ai} \times \boldsymbol{F}_i^d \end{bmatrix}, \tag{7.12}$$

where $\boldsymbol{K}_{A1}, \boldsymbol{K}_{A2}, \boldsymbol{K}_{A3} > 0$ are feedback diagonal gain matrices of force error, integral of force error, and velocity, respectively, and $\boldsymbol{g}_A$ is an arm gravity compensate term. Equation (7.12) shows that the arm is controlled such that the sum of force error converges to zero. Therefore, the arm will not move when the fingertip force error is controlled to be zero by the force control of the haptic finger.

The force control laws given by Eqs. (7.9) and (7.12) are a redundant force control for the multi-fingered haptic interface, receiving forces at multiple points, and thus different from conventional redundant force control for manipulators that receive force at only one point. It is possible to prove the stability of the system when the inertia matrix of the haptic interface and velocity square term are negligibly small, but when the haptic interface moves very quickly, dynamic compensators of inertia force and centrifugal force are required for stability. Moreover, when the force responsibility of the haptic finger differs significantly from that of the haptic arm, the haptic interface joint with high force responsibility tends to move to the limit of the joint quickly. Hence, an adjustment of force responsibilities is required, because the arm control will not recover from reaching to the movable limit of the fingers.

7.2.2. *Hand Manipulability-Based Control*

The haptic hand posture is non-unique when the five fingertip positions are fixed, because of the redundancy of HIRO. To take the redundancy into account, hand manipulability-based control is adopted. This control method combines finger force control and arm position control. Each finger is controlled independently by a PI force control using a force error at the fingertips. In arm position control, a desired hand posture is determined to maximize the hand manipulability measure in order to respond to an operator's various hand poses. The hand manipulability measure is defined as follows:

$$PI = \sum_{i=1}^{k} (\alpha_i W_i + \beta_i P_i) + Q, \qquad (7.13)$$

where α_i and β_i are weighting coefficients; W_i is a manipulability measure of the i-th haptic finger given by,

$$W_i = \sqrt{\det(\boldsymbol{J}_{F_i}^T \boldsymbol{J}_{F_i})}, \tag{7.14}$$

where $\boldsymbol{J}_{Fi}$ is a kinematic Jacobian of the i-th haptic finger. P_i is a penalty function of the i-th haptic finger to keep the joint angle within the range of movement given by,

$$P_i = -\sum_{i}^{3} \gamma_i (\boldsymbol{e}^{-\mu(q_{ij}-a_{ij})} + \boldsymbol{e}^{\mu(q_{ij}-b_{ij})}), \tag{7.15}$$

where γ_i is the weighting coefficient, μ is the parameter to adjust an exponential function, and a_{ij} and b_{ij} are the lower and upper limits of the j-th joint angle of the i-th finger, respectively. The form of the penalty function will be non-limiting. Q is a penalty function to prevent a large arm motion, which sometimes confuses the operator because it creates the illusion that the device is malfunctioning. It is given by,

$$Q = -\frac{1}{2}(\boldsymbol{q}_{Ad} - \boldsymbol{q}_A)^T \boldsymbol{\Gamma}\, (\boldsymbol{q}_A^d - \boldsymbol{q}_A), \tag{7.16}$$

where $\boldsymbol{q}_A \in \mathbb{R}^6$ is the current arm joint angle vector, $\boldsymbol{q}_A^d \in \mathbb{R}^6$ is a desired arm joint angle vector, and $\boldsymbol{\Gamma} > 0$ is a weighting matrix.

The desired arm joint angle is defined so as to maximize Eq. (7.13) under a constraint condition in which the five haptic fingertip positions are fixed to the operator's fingertip positions. The optimum arm joint angle is computed by the conjugate gradient method in line with the convergence criterion $PI[k] - PI[k-1] < \delta$, where k is the iteration number and δ is the scalar constant. The haptic arm is position-controlled by,

$$\boldsymbol{\tau}_A = \boldsymbol{K}_{A1}(\boldsymbol{q}_A^d - \boldsymbol{q}_A) + \boldsymbol{K}_{A2}(\dot{\boldsymbol{q}}_A^d - \dot{\boldsymbol{q}}_A) + \boldsymbol{g}_A(\boldsymbol{q}_A) + \boldsymbol{J}_A^T \boldsymbol{F}_{Ae}, \tag{7.17}$$

where $\boldsymbol{\tau}_A$ is the arm control law, $\boldsymbol{K}_{A1}, \boldsymbol{K}_{A2} > 0$ are the position and velocity feedback gains, respectively, and $\boldsymbol{F}_{Ae}$ is a feed forward compensate term of external forces acting on fingertips given by,

$$\boldsymbol{F}_{Ae} = \begin{bmatrix} \sum_{i=1}^{k} \boldsymbol{F}_i \\ \sum_{i=1}^{k} \boldsymbol{p}_{Fi} \times \boldsymbol{F}_i \end{bmatrix}. \quad (7.18)$$

When the arm moves with high speed, the controller will be required to add the dynamic compensate term for the inertial and centrifugal force.

The finger is the PI force controlled in the movable joint angle and the PD (Proportional and Derivative) position control at the limit of the movable joint angle. After reaching the limit of the movable range, the finger that has been switched to position control is switched back to force control when the direction of the joint torque input is in the same direction as that of the joint angle, apart from the limit of the movable angle. In this force control, the feedback gain is constant. When users manipulate the haptic interface in free space, they feel the inertia and friction of the interface; therefore, high-gain feedback is desirable. On the other hand, when a user uses the haptic interface within a constrained space — i.e., when a user contacts a virtual object — the high-gain feedback will create a vibrating response, and so low-gain feedback is preferable. Therefore, the haptic finger is force controlled by using a non-constant gain. The controller uses high-gain feedback when the desired force is zero, and low-gain feedback when the desired force is not zero.

7.3. Medical Application Systems

The use of multi-fingered haptic interfaces contributes to the development of novel medical application systems, such as a hand rehabilitation support system (Hioki *et al.*, 2011), a measurement system for breast engorgement (Mouri *et al.*, 2012), a surgery training system using plural devices (Kawasaki *et al.*, 2007a; Endo *et al.*, 2013a), a breast palpation training system (Daniulaitis *et al.*, 2004; Alharabi *et al.*, 2005; Kawai *et al.*, 2013), and a skill transfer system (Endo *et al.*, 2013b). These systems are outlined below.

7.3.1. *Hand Rehabilitation Support System*

Finger function is very important in daily life, and finger injuries can be a very serious problem. Patients with finger injuries need timely and persistent rehabilitation to recover their lost abilities and to resume their normal lives. Long rehabilitation training sessions with therapists, who are in relatively short supply, are not always available. One solution to this problem could be a robotic finger rehabilitation system for use by individual patients. Recently, many robotic systems have been studied for rehabilitation purposes. Many aspects of robotic arm rehabilitation therapy, including clinical tests, have been reported.

Hand rehabilitation, however, is particularly difficult, because the hand possesses many DOFs of motion, and the device must be relatively small. Many types of patients experience some kind of finger paralysis: for example, carpal-tunnel syndrome, peripheral nerve disease, and hemiplegia by cerebrovascular accident (CVA). Such patients need rehabilitation to resume daily life activities. The system structure of hand rehabilitation will differ depending on the type of patient. In general, hand rehabilitation systems are expected to provide three-directional motions to the fingertips. For hemiplegia caused by a CVA, an exoskeleton-type device with 18 DOFs has been developed as a support system. In this system, finger joint motion of the affected side is generated by that of unaffected side (Ueki *et al.*, 2012).

For patients with peripheral nerve injury of the arm, a hand rehabilitation system has been developed using the HIRO (Fig. 7.2) (Hioki *et al.*, 2011). To achieve effective rehabilitation for patients, this system utilizes surface electromyography (sEMG) control. A finger joint angle estimation system that utilizes sEMG has been developed (see Sec. 8.5), and is helpful for complex exercises. This research could result in a challenging clinical trial, and the integration of sEMG into rehabilitation robotics systems for the fingers is of great significance.

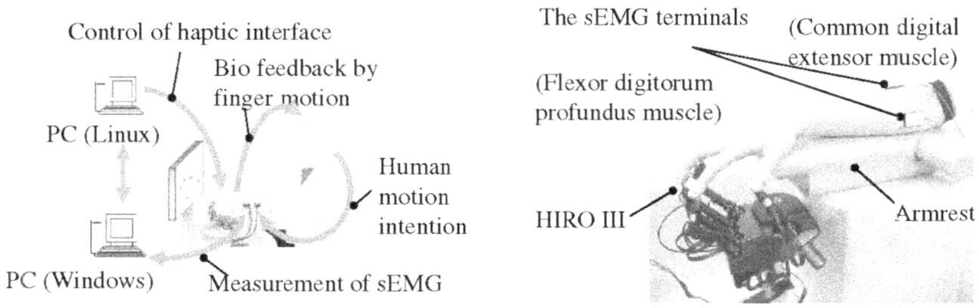

(a) Outline of the hand rehabilitation system (b) System setup

Fig. 7.2. Hand rehabilitation system.

7.3.1.1. *System configuration*

The system uses the HIRO hand for the fingertips, and sEMG information from the patient's forearm or hand. Since this haptic device is quite small, it can be used on a desk while the patient is seated. The system first decides on the proper fingertip exercises based on the sEMG results; the device then provides exercises at the fingertips. For example, if the sEMG of the muscles for finger extension (extensor digitorum muscle) activates and crosses over a certain threshold value, the haptic device will exert 3-DOF forces at each fingertip to open the fingers.

7.3.1.2. *sEMG control for hand rehabilitation*

The sEMG measures bioelectrical signals from muscles in response to voluntary contractions. These signals provide various types of information about a person's intent. It is therefore hoped that the sEMG will function as a new type of human–machine interface. In the study, the haptic device was controlled by the sEMG. First, the motion intent of the user was extracted from the sEMG; the fingertip exercises and the intent were then provided to the user by the HIRO. This biofeedback is very easy to image for the patient compared with waveforms or sounds.

For signal processing of the sEMG, the system adopts an extracting feature value method, determining the mean absolute value (MAV) or root mean square (RMS) with segmentation and

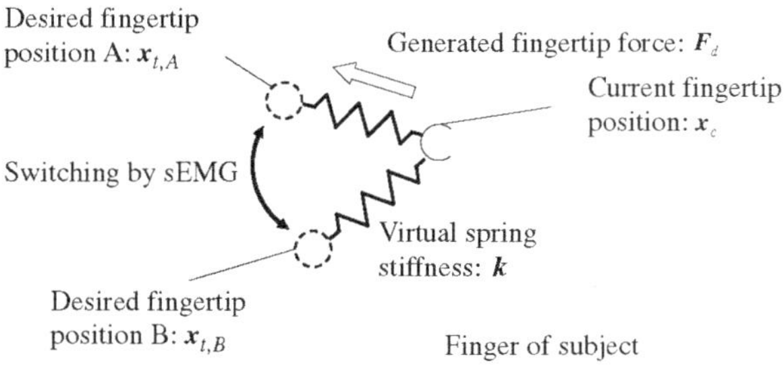

Fig. 7.3. Virtual spring model.

normalizing. After obtaining the desired fingertip exercise by the above method, the system must transform from the exercise into a continuous fingertip trajectory. The haptic device HIRO provides fingertip forces depending on the desired fingertip trajectory. For rehabilitation, force control is more appropriate than position control in that the force can be adjusted. As a simple example, one switches target positions based on the intent. It is necessary to measure these positions beforehand. When the joints of a patient do not have contraction, the target positions can be measured with the help of a therapist or by the patient. When the joints have contraction, the length of the patient's finger link and range of motion are measured, and solving the kinematics with these parameters will identify the target positions. The simplest method for generating the desired fingertip forces is to use the virtual spring model, as shown in Fig. 7.3.

This method requires only the target position and the current position. The desired force vector, $\boldsymbol{F}_d$, of one finger is calculated by,

$$\boldsymbol{F}_d = \boldsymbol{k}(\mathbf{x}_d - \mathbf{x}_c), \tag{7.19}$$

where $\boldsymbol{k}$ is the spring coefficient diagonal matrix, $\mathbf{x}_c$ is the current fingertip position of a certain finger, and $\mathbf{x}_d$ shows the target or desired fingertip position. Current fingertip positions can be calculated by forward kinematics. Furthermore, in order to adjust the forces to the fingertip, the desired forces are applied with a low pass filter (LPF). The haptic finger of HIRO III will be controlled by the PI control law for fingertip forces. First, the error of torque vector $\boldsymbol{\tau}_e$ is defined as follows:

$$\boldsymbol{\tau}_e = \boldsymbol{J}^T (\boldsymbol{F}_d - \boldsymbol{F}) \in \mathbb{R}^3, \quad (7.20)$$

where $\boldsymbol{J}$ is the kinematical Jacobian matrix of a haptic finger and $\boldsymbol{F}$ is the measured fingertip force vector from a three-axes force sensor located at the top of the haptic finger. The control input of HIRO III is given by,

$$\boldsymbol{\tau} = \boldsymbol{K}_P \boldsymbol{\tau}_e + \boldsymbol{K}_I \int \boldsymbol{\tau}_e dt, \quad (7.21)$$

where $\boldsymbol{\tau}$ is the applied torque vector onto a haptic finger, and $\boldsymbol{K}_P$ and $\boldsymbol{K}_I$ are the PI gain matrices, respectively. The gains or parameters are calibrated in advance.

As a pretest for the clinical trial, the system has been tested on able-bodied subjects. The experimental results show that the system provided three-directional forces at each fingertip corresponding to the subject's sEMG, and that the subjects felt natural movement support. The next issue will be to carry out a clinical trial with the proposed system.

7.3.2. *Robotic Measurement System for Breast Engorgement*

Manual breast palpation is a commonly used diagnostic technique to diagnose a new mother's degree of breast engorgement. Breast engorgement occurs in the mammary glands due to expansion and pressure exerted by the synthesis and storage of breast milk, and can cause mammary neuralgia, blood stasis, indurations, and edema of the breasts. Such engorgement is frequently painful and long lasting, and it may discourage new mothers from breastfeeding their infants. Problems such as neuralgia and blood stasis can be prevented or treated early with proper diagnosis, but learning materials and methods for acquiring knowledge of breast engorgement are inadequate. Moreover, the subjective nature of manual breast palpation results in wide variation among healthcare professionals' diagnostic skills with regard to breast engorgement. To solve this problem, a measurement system for achieving quantitative evaluations of breast engorgement

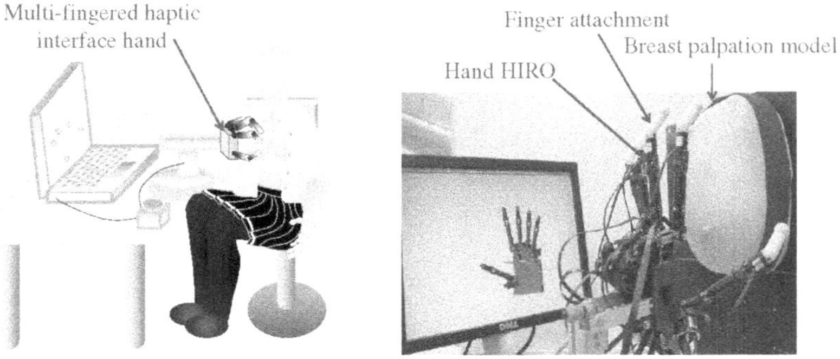

(a) Concept of measurement system (b) Environment of the evaluation test

Fig. 7.4. Developed measurement system for breast engorgements.

has been proposed using a multi-fingered haptic interface robot HIRO III (Mouri *et al.*, 2012).

7.3.2.1. *System configuration*

Figure 7.4 shows the developed measurement system for breast engorgement. A patient is seated next to the table on which the system is located. The system was designed by taking into account the following aspects and requirements: (a) Quantification of stiffness — this is a key factor of the measurement system for breast engorgement; (b) simultaneous multipoint measurement — this enhances the system's accuracy and decreases measurement time; (c) easy operability — necessary for nursing students and new users; and (d) sense of safety and comfort — the system should not make patients anxious and must not cause any pain.

The developed system measures the stiffness of a breast by active sensing of positions and forces at five contact points. A finger attachment with a spherical tip is fixed to the haptic fingertip to make contact comfortable.

7.3.2.2. *Measurement*

Measurement is executed automatically. First, the haptic fingers move to a surface of the breast with a grasping motion. After contacting the surface, each haptic finger is position controlled and moves step by step in the penetration direction until the finger force meets the threshold level. At this stage, fingertip forces are measured. Finally, the haptic fingers return to their initial position. A control input of the haptic fingers at the measurement is given by,

$$\boldsymbol{\tau} = \boldsymbol{K}_1 \boldsymbol{q}_e - \boldsymbol{K}_2 \dot{\boldsymbol{q}} + \boldsymbol{K}_3 \int \boldsymbol{q}_e dt, \tag{7.22}$$

where $\boldsymbol{q} \in \mathbb{R}^{15}$ is a finger joint angle vector, $\boldsymbol{q}_e = \boldsymbol{q}_d - \boldsymbol{q} \in \mathbb{R}^{15}$ is a deviation angle vector from a desired angle $\boldsymbol{q}_d = \Delta \boldsymbol{q} t \in \mathbb{R}^{15}$, $\Delta \boldsymbol{q} \in \mathbb{R}^{15}$ is a given constant vector, and $\boldsymbol{K}_1, \boldsymbol{K}_2, \boldsymbol{K}_3 \in \mathbb{R}^{15 \times 15}$ are feedback gain matrices of position, velocity, and position integral, respectively. Each finger joint is controlled to follow the desired joint angle independently. From the measurement data, stiffness is evaluated by a second-order least squares approximation polynomial using penetration depth from the contact surface of the breast. Figure 7.5 shows an example of a stiffness measurement. In this experiment, a breast palpation model to detect breast cancers (Kyoto Kagaku Co.), as shown in Fig. 7.4(b), is used as a test subject. The results showed that the measurement system was able to obtain quantitative evaluations of breast engorgement. This measurement system could be extended to a stiffness measurement system for biologic bodies.

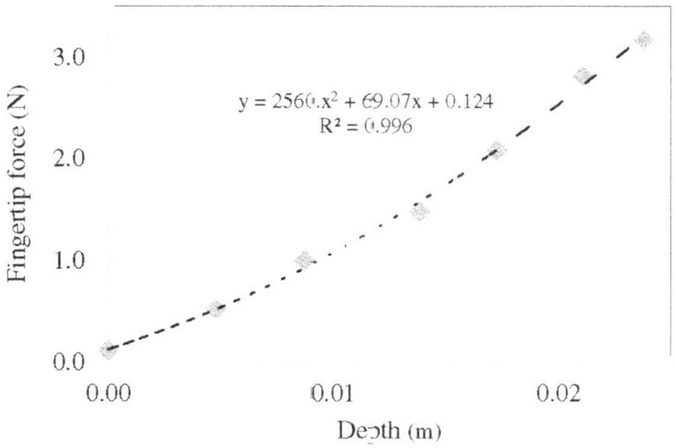

Fig. 7.5. Example of measurement and curve fitting.

7.3.3. *Medical Training System Using Plural Devices*

Haptic interfaces that present force and tactile feeling have also been utilized in the area of medical training and evaluation. In the case of application for surgical training, medical doctors need many surgical tools. Many researchers have developed single-tool-type haptic interfaces with scissors, surgical knives, injectors, and so on, for use in limited objective tasks. However, presentation of force feelings of multiple surgical tools requires many single-tool-type haptic interfaces. Providing so many interfaces would require an appropriate installation location and would be very costly. Moreover, most single-tool-type haptic devices are fixed to a base, and thus cannot be displaced. The multi-fingered haptic interface has the advantage of force feeling for plural surgical devices (Kawasaki *et al.*, 2007a).

7.3.3.1. *Conceptual idea*

Plural tool-type devices require that each tool-type device be easy to put on and take off. Moreover, the overall system should be small and available for use in a large workspace. These requirements are fulfilled when the multi-fingered haptic interface is equipped with tool-type devices. Figure 7.6 shows the concept for presenting force feelings with a variety of tools. The multi-fingered haptic interface is equipped

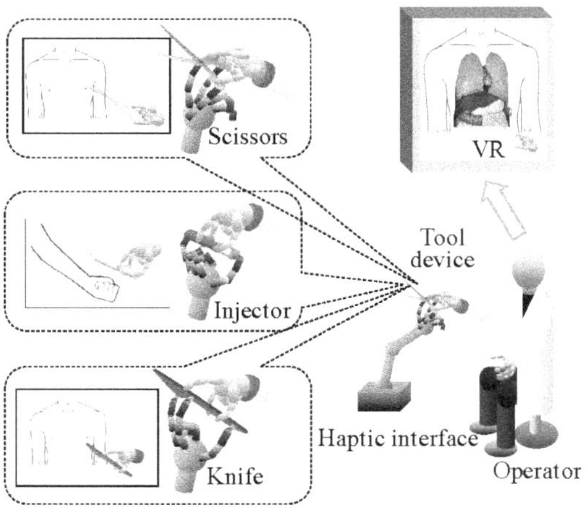

Fig. 7.6. Concept of plural surgical tool devices.

with exchangeable surgical tool-type devices. The proposed haptic system has the following advantages: (a) each tool-type device is very simple and, in some cases, has features that are superior to those of the equivalent conventional devices; (b) permanent magnets at the fingertips make it easy to exchange the plural tool-type devices; (c) it occupies less space than previous single tool-type haptic systems; (d) the tool-type devices can be used over a wide area; and (f) force and position can be measured and controlled by the multi-fingered haptic interface. With these many advantages, the proposed system has greater potential for the force feelings of plural tool devices than does a single-tool-type interface.

7.3.3.2. *Cutting simulation*

Many surgical tools are scissors-type devices. Using scissors tool-type devices with HIRO, it is easy to express various surgical scissors by changing the knife blade of the scissors in VR. The developed scissors tool-type device is shown in Fig. 7.7(a). It weighs 70 g. Three iron balls are attached, one at a supporting point and two at working points, to connect to the HIRO fingers. As a result of analyzing the workspace of the devices, the system is attached at the thumb, index finger, and ring finger, as shown in Fig. 7.7(b).

A computational method for cutting sheets has previously been proposed (Honma *et al.*, 2004). The present system uses the same method. Haptic fingers follow the desired forces through PI force feedback control, which are computed in the VR environment. The

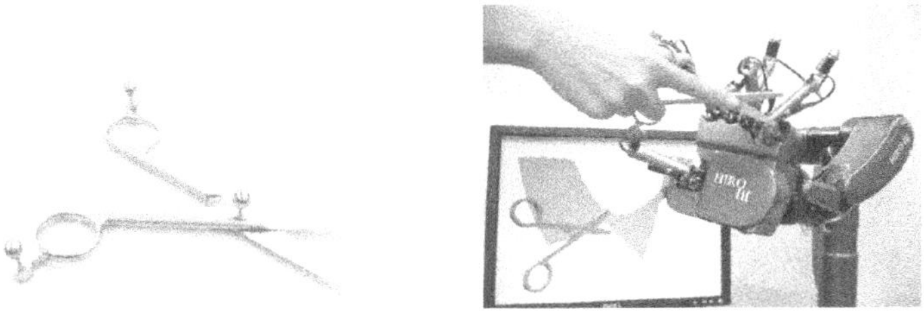

(a) Medical scissors for training in VR

(b) HIRO connected to a medical scissors-type device

Fig. 7.7. VR training system of medical scissors for surgical operations.

operator cuts a virtual sheet using virtual scissors with reactive feeling. The device displays the force feeling for the cutting sheet. Experimental results showed the successful presentation of force feeling, depending on the sheet characteristic. However, the error of the force was not small enough for use in surgical training, and thus a new design approach for tool-type devices was proposed (Endo *et al.*, 2013a).

7.3.3.3. *Optimum connection*

In previous research, the connection between the HIRO finger and the tool-type device was determined by trial and error, and its optimal connection was not clarified. Technically, clarifying the optimal connection is important with regard to presenting the operator with the force feeling through the tool-type device. To solve this problem, an optimal connection method was developed from the point of view of mobility and singularity. Real-world tools can be divided into two main classes: tools with no joints and tools with joints. For example, a knife has no joints, and thus its DOF is 6. On the other hand, tweezers and scissors have 1 joint, allowing 1 DOF, and thus their DOF is 7.

The HIRO has five haptic fingers. When HIRO is connected to the tool-type device via several haptic fingers, the system acts as a parallel mechanism as shown in Fig. 7.7(b). Having the features of a parallel mechanism, the system is highly accurate, with high output force, a high level of stiffness, and other advantageous features that are important elements for a haptic interface. In the parallel mechanism, the number of joints does not correspond to the DOF of the overall system. In this case, the mobility index M implies the DOF of the overall system that is given by,

$$M = 6(N - m - 1) + \sum_{i=1}^{m} l_i, \tag{7.23}$$

where N is the number of links, m is the number of joints, and l_i is the DOF of the i-th joint. To realize the tool-type device's DOFF_{tool}, the mobility must satisfy

$$M \geq F_{tool}. \tag{7.24}$$

To control all DOFs of the overall system, the number of active joints D must satisfy

$$D \geq M. \tag{7.25}$$

From these two conditions, the required number of haptic fingers is determined. This leads to the number of haptic fingers for the connection, but does not clarify which fingers should be used. It is well known that a parallel mechanism has the drawback of singularity, where the system becomes uncontrollable. Thus, avoidance of singularity is an important factor in the selection of fingers. When the input–output relationship of the parallel mechanism can be written as

$$\boldsymbol{A}\dot{\boldsymbol{q}} = \boldsymbol{B}\boldsymbol{v}, \tag{7.26}$$

where $\dot{\boldsymbol{q}} \in \mathbb{R}^n$ is the actuated joint velocity vector, $\mathbf{v} \in \mathbb{R}^m$ is the platform Cartesian coordinate velocity vector, and $\boldsymbol{A} \in \mathbb{R}^{n\times n}$ and $\boldsymbol{B} \in \mathbb{R}^{n\times m}(n \geq m)$ are appropriate matrixes defined by the kinematics. The singularity of the parallel mechanism consists of three kinds of singularities. The first kind of singularity is defined when $\boldsymbol{A}$ is singular, the second kind of singularity is defined when $\boldsymbol{B}$ is singular, and the third kind of singularity is defined when $\boldsymbol{A}$ and $\boldsymbol{B}$ are singular at the same time. To avoid the three kinds of singularity, the following performance indexes are introduced:

$$PI = \alpha W_A W_B + \beta W_q, \tag{7.27}$$

and

$$\begin{aligned} W_A &= |\det \boldsymbol{A}|, \quad W_B = \det(\boldsymbol{B}^T\boldsymbol{B}), \\ W_q &= -\sum_{i=1}^{p} \gamma_i (e^{-\mu(q_i - a_i)} + e^{\mu(q_i - b_i)}), \end{aligned} \tag{7.28}$$

where α, β, and γ_i are the weighting coefficients of W_A, W_B, W_q and the i-th actuated joint, respectively. p is the number of the actuated joint, q_i is the angle of the i-th actuated joint, μ is the parameter to adjust an exponential function, and a_i and b_i are the lower and upper limits of the angle of the i-th actuated joint, respectively.

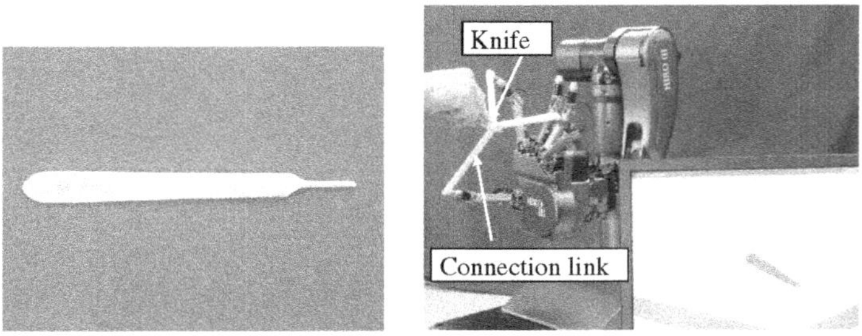

(a) Original surgical knife (b) Optimum connection to HIRO and VR environment

Fig. 7.8. Knife-type device.

In Eq. (7.27), W_A is a well-known manipulability index, and the configuration of the system is away from the first kind of singularity when the value of W_A becomes large. W_B evaluates the second kind of singularity, and if the value of W_B becomes large, the system is away from the second kind of singularity. If the configuration of the system is away from the first and second kinds of singularity, i.e., if $W_A W_B$ becomes large, then this corresponds to the third kind of singularity. Note here that the robot joint usually has a limit to its movable range. Therefore, the penalty function W_q is added to the performance index, which corresponds to the limits of the joint angles. From these points of view, by maximizing *PI*, one can decide which haptic fingers should be used.

Figure 7.8 shows a knife-type device optimally connected to HIRO. The knife-type device is connected to the thumb, index finger, and little finger of HIRO via a connection link. A detailed analysis of this setup and a description of an associated experiment are presented in a previous study (Endo *et al.*, 2013a).

7.3.4. *Breast Palpation Training System*

Despite the potential utility of clinical breast examination (CBE), doctors' palpation skills are often inadequate, and it is difficult to train doctors in this skill. The sense of touch is extremely important when training physicians to perform CBE. Breast palpation is a practiced tactile skill that does not necessarily correlate with cognitive knowledge about breast cancer detection. Training that emphasizes tactile skills is known to improve examination sensitivity. With a combination of technologies, such as VR and haptic interfaces, it is possible to greatly extend the capabilities and effectiveness of training simulators. A simulation can record kinematics, touch, and force feedback for later display to a trainee. This technology could be used to train medical students before they palpate a real patient, and could also be used by trained physicians to improve their skill.

This section focuses on the development of a realistic palpation tool for lesions, where the user can touch soft tissue looking for a tumor beneath the surface (Kawai *et al.*, 2013). In this work, a realistic model was built using the finite element method (FEM), which was fast enough for a relatively complex model in real time. To simulate a medical palpation, the simulation has to handle the deformation caused by multiple fingers contacting the surface at the same time. The system allows the user to specify the stiffness for each finite element in order to situate tumors with arbitrary shapes and positions up to the mesh resolution.

7.3.4.1. *System overview*

Figure 7.9 shows the overview of the developed training system. It is comprised of a HIRO installed below a 3D display, based on

Fig. 7.9. Breast cancer palpation training system.

a liquid crystal display (LCD), with a half mirror placed in-between. A user wearing a pair of 3D shutter glasses sits in front of the half mirror to view the 3D images of a female chest, including breasts projected on the half mirror as a stereoscopic image. At the same time, all five of the user's fingers are connected to the HIRO placed behind the half mirror, so that he/she can receive haptic sensations for each finger individually while seeing his/her own hand in the mirror. The 3D images of the PC-generated deformed virtual breast are overlaid on top of the user's hand, thereby creating a mixed-reality effect. This effect is expected to enhance the user's training experience.

7.3.4.2. *Deformable body simulation*

Many parts of a human body, including organs and/or tissues, are modeled as a deformable body. Deformable body simulation is generally a difficult task, since deformable bodies have infinite DOFs. Therefore, we usually have to discretize the deformable body using numerical methods such as FEM. Our past breast palpation simulator (Daniulaitis *et al.*, 2004; Alharabi *et al.*, 2005) was based on an FEM. It realized a real-time simulation of a deformable body based on precomputation of the body's displacement and reaction forces; the deformations of the entire body and the reaction forces were precomputed for unit displacements for each vertex of the FE model. At runtime, the deformation of the entire body and the reaction forces for each finger were computed as a linear combination of these precomputed displacements, and given a set of contact points. The drawback of this method is the need for lengthy precomputation and lack of dynamics.

Our new method simulates the dynamics of a deformable body in real time using general purpose computing on graphics processing units (GPGPU), and at the same time interacts with HIRO for haptic rendering so that more realistic results can be obtained on the fly. The simulation algorithm is comprised of (a) collision detection and response and (b) FEM computation. The first step for the algorithm is to receive the current position of the fingertips from HIRO and perform collision detection between the fingertip spheres and the deformable body surface. If a collision is detected, its contact forces are computed and used as boundary conditions for the FEM. At the same time, the contact forces will be sent to HIRO for haptic rendering. Finally, the time step in the simulator is proceeded and the next time step's physical states are calculated.

In the simulator, the deformation of the breast is simulated using an FEM based on a corotational formulation on the GPU (Allart *et al.*, 2011). The corotational formulation accounts for geometric non-linearities of the deformations. It can avoid so-called "inflation" artifacts when subject to large deformations. The idea of corotational FEM is to rotate the stiffness matrix by a rotational matrix that brings it to the original coordinate space before deformation. This eliminates the negative influence of the rotational component of the deformation to the linear strain that leads to an inflation effect. The system of ordinary differential equations is solved by implicit time integration based on the backward Euler method for stability. The backward Euler method updates the velocity using the acceleration at the end of the time step. This requires that a system of linear equations be solved at every time step, a task accomplished by a conjugate gradient solver.

HIRO has five fingers, and each fingertip is modeled as a rigid sphere. The spheres and the deformable body must avoid interpenetration. A penalty method is used for simplicity. Proximity is checked for each sphere center against all the vertices within the deformable mesh. The penalty force is computed by a spring and damping model. For each sphere, all the penalty forces exerted on the sphere are applied to the deformable body as external forces at each contact point. The penalty forces for each sphere are summed and converted into one 3D force vector for haptic rendering.

The entire FEM and collision detection and response are implemented based on GPGPU. The use of GPUs allows us to simulate large-scale models in real time that are intractable to conventional CPU implementations. These features enable the user to quickly change the virtual models without lengthy offline precomputations. This means that users can train using a wide variety of virtual patients with different medical conditions.

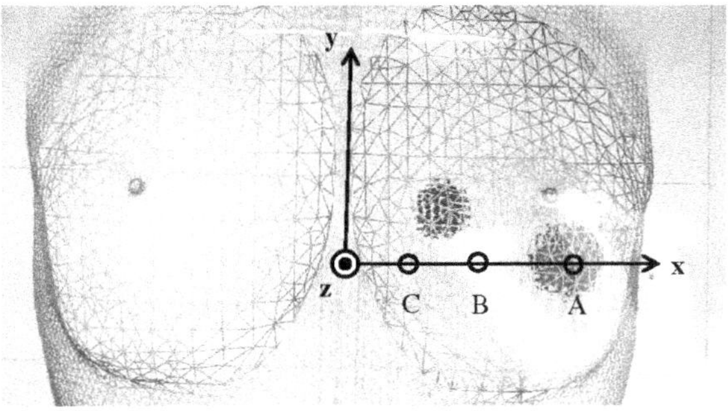

Fig. 7.10. Tetrahedral mesh and tumor.

7.3.4.3. *Experimental results*

The frequency of the haptic control loop was kept to 1 kHz and the graphics frame rate was set to 30 Hz. A virtual breast model was constructed as follows. A real silicone palpation-training model was scanned. After the data was cleaned up, it was processed with mesh-generation software to create a triangular surface mesh and a tetrahedral volumetric mesh. The breast model was represented with the surface mesh (6,004 triangles), as shown in Fig. 7.10. A volumetric mesh (44,208 tetrahedra, 9,097 nodes) was then constructed from the surface mesh. The Young's modulus of the entire breast — except for the tumor — was set to $E = 19{,}000$ Pa, and the tumor was set to 107,000 Pa, so that the tumor would feel stiffer than the other parts of the breast. Figure 7.10 also shows the size and location of the tumor used in this experiment. The tumor's radius was set to 15 mm.

Although the user can generally use up to five fingers simultaneously, the results of only one finger case are shown here for simplicity. The palpation was done with his/her index finger along the z direction translating his/her hand at three points A, B, and C, where A is a pint on the tumor and B and C are normal points. Figure 7.11 shows the relations between the push-in displacement and the corresponding z component of the reaction forces measured by the haptic interface. As shown in the figure, the reaction force became noticeably high at the position A. This meant that around this location,

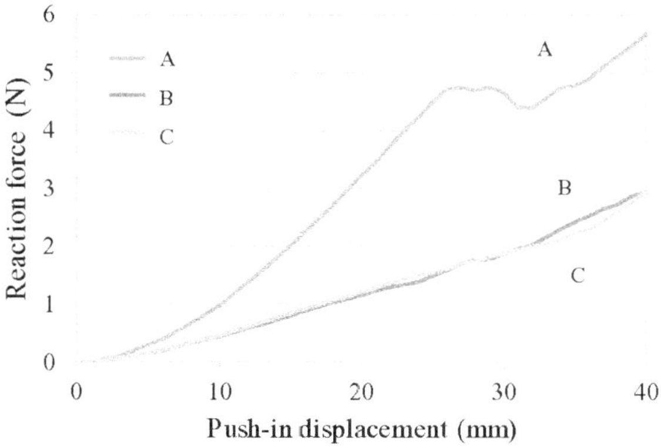

Fig. 7.11. Push-in displacement at index fingertip position vs. Z component of reaction force of index fingertip measured in a virtual palpation session.

the user should have been receiving significant reaction forces in the +z direction, and feeling the existence of the tumor. This verified that the location of the tumor was reasonably well represented, thus showing the effectiveness of the system.

7.3.5. *Skill Transfer System*

In the medical fields, expert skills such as surgical techniques, palpation techniques and the like are obtained through long training, and skill is normally acquired by the experience of working with actual patients. However, it is difficult for residents and medical students to train with actual patients because of a decrease in volunteers willing to cooperate in the training and the risk of medical error. To transfer expert skills from a trainer (senior doctor) to a trainee (medical student), the trainer has to teach the trainee: (1) how to move the hands and (2) how to exert the exact amount of force with the fingertips. It is difficult to teach accurate data regarding position and force using only words. Because of these challenges, a skill transfer system that uses VR and a haptic interface has been researched aggressively. The benefits of such a system include: (1) the movement of the trainer's hand and the operation of the trainer's force can be recorded, so that accurate information can be transmitted to the trainee using a screen and the haptic interface; (2) training can be selected according to the trainee's skill level, and the effect of the training can be presented to the user; and (3) training can occur at a remote site via a network terminal, and several trainees can receive training at the same time.

In performing activities in our daily lives, we usually use multiple fingers, so it is important to exert force at multiple fingertips to make a sensation highly realistic. To design and develop a skill transfer system in which the operation of the trainer's force is presented to the trainee, it is important to consider the transfer method based on knowledge of human perception. The force that a human being exerts using his/her hands can be expressed by direction vector and magnitude. Perception ability is likely to alter based on the spatial change of the presented force, and on the time used to present the force. This section considers the skill transfer method for the multi-finger system based on human perception (Endo *et al.*, 2010; 2013b).

7.3.5.1. *Transfer method of force and position information*

In the skill transfer system, the trainer's work is recorded and reproduced in VR space. The trainee is then taught to imitate the trainer's work. The trainee's goal is to make his/her fingertip positions and forces track the trainer's positions and forces. A configuration of the expert skill transfer system is the same as the breast palpation training system shown in Fig. 7.9. For fingertip-position tracking, two approaches were experimentally evaluated: The first involved visual cues like those shown in Fig. 7.12; the second included virtual

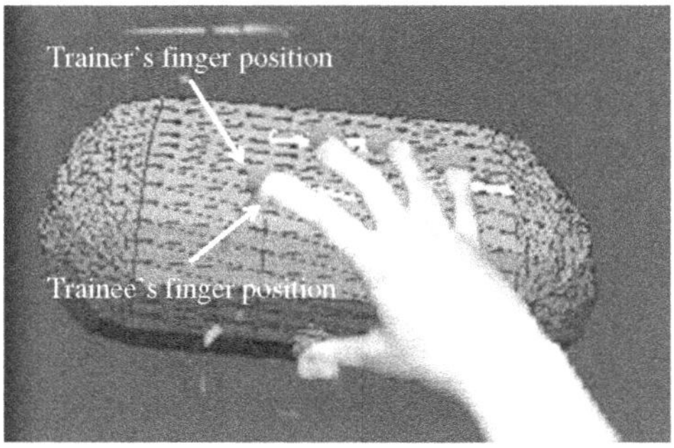

Fig. 7.12. Skill transfer system using visual cues.

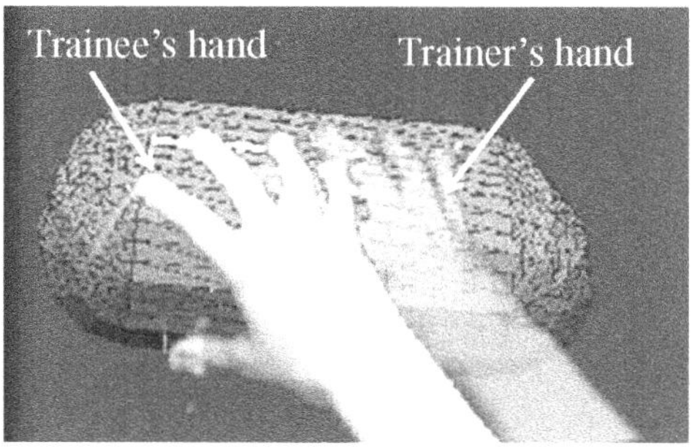

Fig. 7.13. Skill transfer system using virtual trainer's hand.

hands like those shown in Fig. 7.13. In the former, the five fingertip positions of the trainer and trainee are shown as small circles in VR space. In the latter, the trainee's real hand and the trainer's virtual hand are shown in VR space. The trainee controlled his/her fingertip positions to track the trainer's hand positions.

In fingertip forces tracking, two kinds of forces are transferred to the user:

(a) The reaction force $\boldsymbol{F}_r$ from the virtual object (reaction force).
(b) The force $\boldsymbol{F}_t$ that a trainer exerts on an object (trainer's force).

In the force transfer method, the reaction force, $\boldsymbol{F}_r$, and the trainer's force, $\boldsymbol{F}_t$, are presented to the user and switched over time, as shown in Fig. 7.14. If the time taken to show the trainer's force is long, the user will not be able to feel the reaction force, so it is necessary to shorten the time for presenting $\boldsymbol{F}_t$ as much as possible. The time duration should be decided based on the responsibility of the HIRO and force perception ability of the human. Based on an analysis of these, $\boldsymbol{F}_t$ was set to be displayed to the user for 0.5 s, and $\boldsymbol{F}_r$ for 0.2 s. These forces were then displayed repeatedly. In practice, if the force such as that shown in Fig. 7.14 was presented to the user, the user would feel the pulse force that is the difference between the trainer's force and the trainee's force. Thus, if the user regulates his/her fingertip forces so that the pulse forces become small, the force transfer would be achieved. That is, if the user grasps the virtual object using the same force as the trainer, the pulse forces would

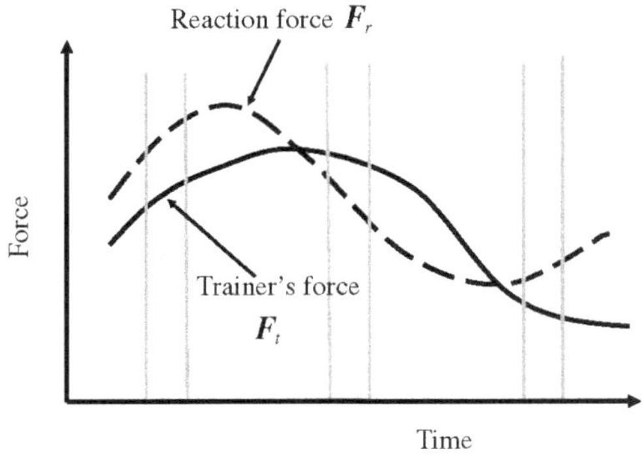

Fig. 7.14. Presentation of reaction force and trainer's force.

disappear. It is an advantage of this method that both force $\boldsymbol{F}_t$ and $\boldsymbol{F}_r$ can be recognized even if the force changes periodically.

7.3.5.2. *Experimental results*

For the comparison, the following three kinds of skill transfer method were evaluated:

Method 1: The fingertip positions of the trainer and the trainee were shown as small circles in Fig. 7.12, and the trainer's fingertip forces are presented to the trainee using the proposed method, as shown in Fig. 7.14.

Method 2: The hands of the trainer and the trainee were shown as in Fig. 7.13, and the trainer's fingertip forces were presented to the trainee using the proposed method.

Method 3: The hands of the trainer and the trainee were shown as in Fig. 7.13, and the trainee's fingertip force only was presented to the trainee.

The experimental results of breast palpation by six subjects are shown in Fig. 7.15. In the figure, FEN is the average integral value of the fingertip force error norm, PEN is the average integral value of the fingertip position error norm, pre-test is the first trial, and post-test is the last of 10 time trials. A two-tailed paired t-test was carried out between the pre-test and the post-test for each transfer method. For Method 1 and Method 3, there was no significant difference between

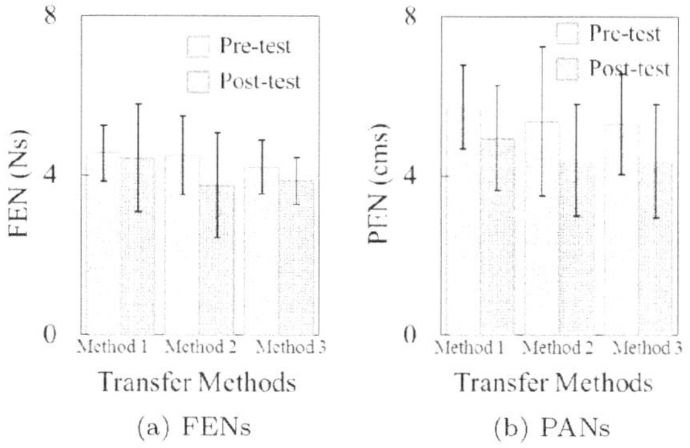

Fig. 7.15. Experimental results of breast palpation. Error bars show ±SD (From Endo *et al.*, 2013b).

the pre-test and post-test. On the other hand, there was a significant difference between the pre-test and post-test in Method 2. These results demonstrated that Method 2 had a training effect and the force magnitude error of the fingertip decreased. In contrast, Methods 1 and 3 had no training effect.

Exercise

7.1. Survey the history of haptic interface.

Chapter 8

Teleoperation of Robot Hands

Teleoperation of robots is often used in hazardous environments, such as outer space, the ocean floor, and nuclear power settings. A **master–slave system** is the type of teleoperation technology generally used, where the master is a haptic interface operated by a human, and the slave is a remotely located robot. The slave then moves according to motions generated through the master system. Human biological signals have also been used as a type of master system for the command of robots. One typical application is the myoelectric controlled hand. The electroencephalogram (EEG) and surface electromyogram (sEMG) have been used in many studies as a brain–machine Interface (BMI) because of their direct relationship to human motion. This chapter presents the fundamentals of the master–slave system, teleoperation of robot hands through a multi-fingered haptic interface, and prophetic hand control by sEMG.

8.1. Fundamentals of Bilateral Control

Teleoperation methods in the master–slave system may involve unilateral or bilateral control, as shown in Fig. 8.1. Basically, the master generates a motion command for the slave based on human motion or human motion intention. In the case of **unilateral control**, the motion command is sent from the master to the slave; no information regarding the slave motion state is sent from slave to master. Because of this scheme, it is not easy to execute complex tasks using unilateral control. On the other hand, in **bilateral control**, the slave moves according to motion commands from the master, and then sends information involving the slave motion state back to the master,

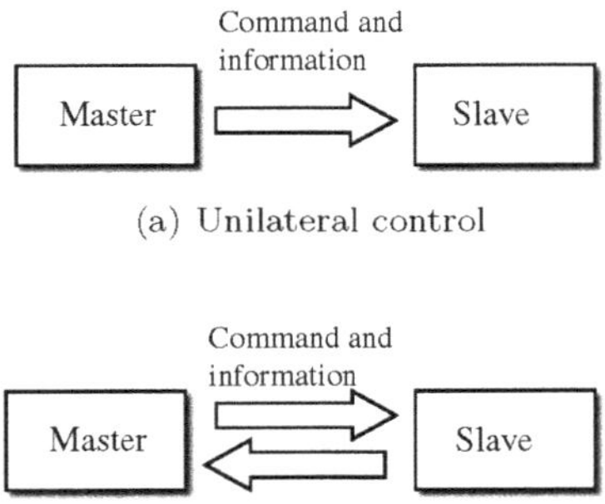

(a) Unilateral control

(b) Bilateral control

Fig. 8.1. Control scheme for telecontrols.

thereby having a significant impact on the motion of the master. Thus, bilateral control has a high potential for executing complex tasks. For robots designed to operate in hazardous environments, bilateral control in which forces acting on the robot hand are sent to the human hand as a force feeling is expected to enhance operability and efficiency.

In bilateral control, there are four targets — the master position, $\boldsymbol{x}_m$, a force acting on the master, $\boldsymbol{F}_m$, a slave position, $\boldsymbol{x}_s$, and a force acting on the slave, $\boldsymbol{F}_s$. The position and force in the slave are controlled to follow the position and force in the master. Let us suppose that the gravitation term and velocity square term of the master robot are compensated to cancel their effects and the dynamics model of the master is expressed by a linear secondary differential equation given by,

$$\boldsymbol{M}_m \ddot{\boldsymbol{x}}_m + \boldsymbol{D}_m \dot{\boldsymbol{x}}_m = \boldsymbol{u}_m + \boldsymbol{F}_m, \tag{8.1}$$

where the suffix m means the master, $\boldsymbol{M}_m$ is an inertia matrix of the master, $\boldsymbol{D}_m$ is a damping matrix of the master, and $\boldsymbol{u}_m$ is a control input of the master. Similarly, suppose that a dynamics model of the slave is expressed by,

$$\boldsymbol{M}_s \ddot{\boldsymbol{x}}_s + \boldsymbol{D}_s \dot{\boldsymbol{x}}_s = \boldsymbol{u}_s - \boldsymbol{F}_s, \tag{8.2}$$

where the suffix s means the slave, $\boldsymbol{M}_s$ is an inertia matrix of the slave, $\boldsymbol{D}_s$ is a damping matrix of the slave, and $\boldsymbol{u}_s$ is a control input

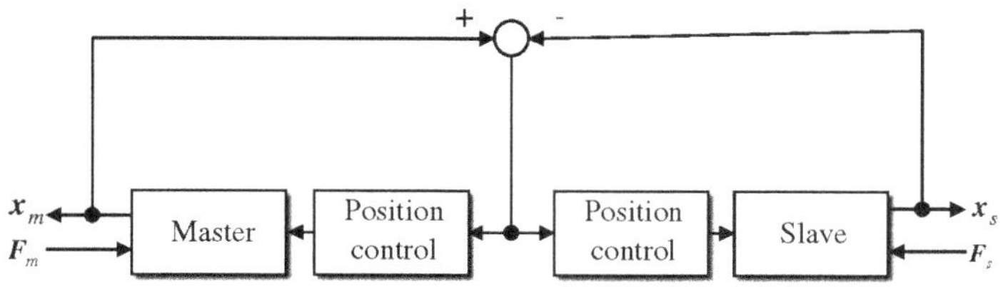

Fig. 8.2. Symmetric-type bilateral control.

of the slave. Let us define a position error $\boldsymbol{e}_p$ and a force error $\boldsymbol{e}_f$ of the master–slave system as follows:

$$\boldsymbol{e}_p = \boldsymbol{x}_m - \boldsymbol{x}_s \tag{8.3}$$

$$\boldsymbol{e}_f = \boldsymbol{F}_m - \boldsymbol{F}_s. \tag{8.4}$$

There are then three methods of typical bilateral control: symmetric-type bilateral control, force-reflection bilateral control, and force-feedback-type bilateral control.

8.1.1. *Symmetric-Type Bilateral Control*

A block diagram of the **symmetric-type bilateral control** system is presented in Fig. 8.2. Both the master and slave are position controlled. A relative displacement between the master and slave is derived by the master motion, and is controlled to be zero. As a result, the drive force is generated in the master and the constraint force is generated in the slave. The control input of the master and slave can then be represented by,

$$\boldsymbol{u}_m = -\boldsymbol{K}_p\boldsymbol{e}_p - \boldsymbol{K}_d\dot{\boldsymbol{e}}_p \tag{8.5}$$

$$\boldsymbol{u}_s = \boldsymbol{K}_p\boldsymbol{e}_p + \boldsymbol{K}_d\dot{\boldsymbol{e}}_p = -\boldsymbol{u}_m, \tag{8.6}$$

where $\boldsymbol{K}_p > 0$ and $\boldsymbol{K}_d > 0$ are a diagonal position and velocity feedback gain matrices, respectively. By adding Eq. (8.1) and Eq. (8.2), the force error is represented by,

$$\boldsymbol{e}_f = \boldsymbol{M}_m\ddot{\boldsymbol{x}}_m + \boldsymbol{D}_m\dot{\boldsymbol{x}}_m + \boldsymbol{M}_s\ddot{\boldsymbol{x}}_s + \boldsymbol{D}_s\dot{\boldsymbol{x}}_s. \tag{8.7}$$

The force error is affected by the dynamics of the master and slave, but it becomes zero at the steady state; that is, the force of the

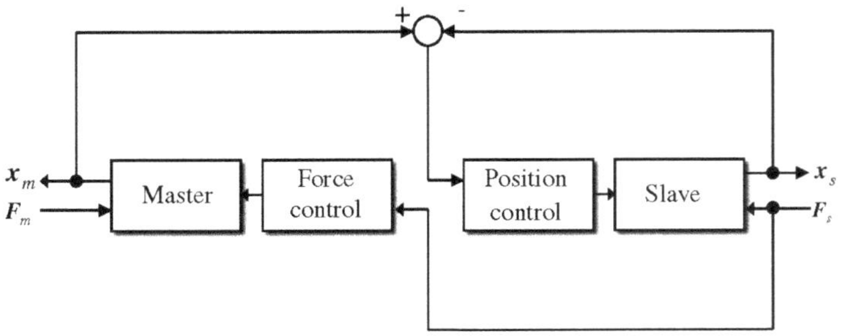

Fig. 8.3. Force-reflection-type bilateral control.

master coincides with the force of the slave. Symmetric-type bilateral control has a simple control scheme because it does not require a force sensor, but it is easily affected by both inertial force and friction force. Moreover, relative displacement is inevitably generated when an external force acts on the master or the slave.

8.1.2. *Force-Reflection Bilateral Control*

A block diagram of the **force-reflection-type bilateral control** is shown in Fig. 8.3. The slave is position controlled to follow master's position and the force of the slave is sent to the master. The master, meanwhile, is open controlled by the reaction force from the slave, which is measured by a force sensor in the slave. The control inputs of the master and slave are represented by,

$$\boldsymbol{u}_m = -\boldsymbol{K}_f \boldsymbol{F}_s \tag{8.8}$$

$$\boldsymbol{u}_s = \boldsymbol{K}_p \boldsymbol{e}_p + \boldsymbol{K}_d \dot{\boldsymbol{e}}_p, \tag{8.9}$$

where $\boldsymbol{K}_f > 0$ is a diagonal force-feedback gain matrix. Substituting Eq. (8.8) into Eq. (8.1), the force of the slave is represented by,

$$\boldsymbol{F}_m = \boldsymbol{K}_f \boldsymbol{F}_s + \boldsymbol{M}_m \ddot{\boldsymbol{x}}_m + \boldsymbol{D}_m \dot{\boldsymbol{x}}_m. \tag{8.10}$$

The reaction force from the slave is sent to the master with $\boldsymbol{K}_f$ amplification, but the operational feeling on the master side will be heavy, because the force of the master is added to the dynamical forces of the master. The operator will feel the reaction force surely when the slave connects with something. Furthermore, if $\boldsymbol{K}_f$ is an

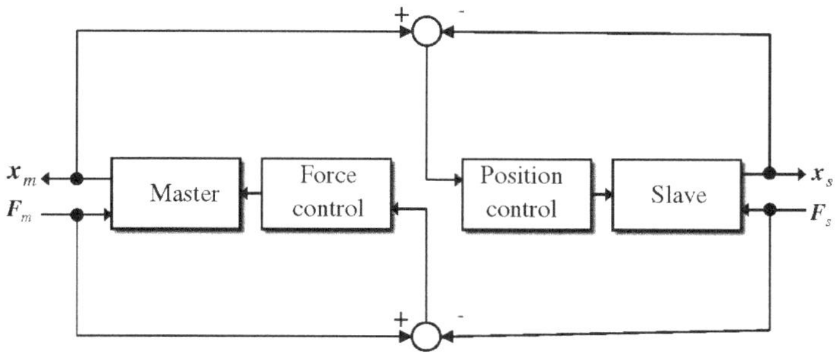

Fig. 8.4. Force-feedback-type bilateral control.

unit matrix, then the force error is represented by,

$$\boldsymbol{e}_f = \boldsymbol{M}_m \ddot{\boldsymbol{x}}_m + \boldsymbol{D}_m \dot{\boldsymbol{x}}_m. \tag{8.11}$$

This shows that the force of the slave converges to the force of the master at the steady state; that is, $\boldsymbol{e}_f = 0$.

8.1.3. *Force-Feedback-Type Bilateral Control*

A block diagram of the **force-feedback-type bilateral control** is shown in Fig. 8.4. Its control scheme is similar to that of the force-reflection-type bilateral control, but here, the master is a force-feedback control. The control inputs of the master and slave are represented by,

$$\boldsymbol{u}_m = \boldsymbol{K}_f \boldsymbol{e}_f \tag{8.12}$$

$$\boldsymbol{u}_s = \boldsymbol{K}_p \boldsymbol{e}_p + \boldsymbol{K}_d \dot{\boldsymbol{e}}_p. \tag{8.13}$$

By substituting Eq. (8.12) into Eq. (8.1), the force of the master is represented by,

$$\boldsymbol{F}_m = (\boldsymbol{K}_f + \mathbf{I})^{-1}(\boldsymbol{K}_f \boldsymbol{F}_s + \boldsymbol{M}_m \ddot{\boldsymbol{x}}_m + \boldsymbol{D}_m \dot{\boldsymbol{x}}_m). \tag{8.14}$$

When the force-feedback gain is sufficiently large and the master finger is at rest, the force of the master converges to the force of the slave; that is, $\boldsymbol{F}_s \to \boldsymbol{F}_m$ as $\boldsymbol{K}_f \to \infty$. Practically, a vibration generated by the master will be large when the magnitude of the force-feedback gain is large. Hence, the gain is limited in the range where no vibration is evoked. When the slave is not constrained by the environment — that is, when $\boldsymbol{F}_s = \boldsymbol{0}$ — then the control input of the master has the effect of a power assist, because the force of the slave is operated by the small force in the master side given by Eq. (8.14).

8.1.4. *Uniform Expression*

In the previous sections, transfer ratios of displacement and force between the master side and the slave side have been 1:1. Changing the transfer ratios of displacement and force contributes enhanced operability when the slave is a micro robot or heavy work robot. Let us define a transfer ratio of displacement as s_p and a transfer ratio of force as s_f; then, $\boldsymbol{x}_m$ and $\boldsymbol{F}_s$ in the input of the master can be replaced by $s_p\boldsymbol{x}_m$ and $s_p\boldsymbol{F}_s$, respectively. As a result, control inputs of the above three bilateral controls can be expressed uniformly as follows:

$$\boldsymbol{u}_m = \begin{bmatrix} \boldsymbol{K}_{mp1} + \boldsymbol{K}_{md1}\dfrac{d}{dt} & \boldsymbol{K}_{mf1} \end{bmatrix} \begin{bmatrix} s_p\boldsymbol{x}_m \\ \boldsymbol{F}_m \end{bmatrix} - \begin{bmatrix} \boldsymbol{K}_{mp2} + \boldsymbol{K}_{md2}\dfrac{d}{dt} & \boldsymbol{K}_{mf2} \end{bmatrix} \begin{bmatrix} \boldsymbol{x}_s \\ s_f\boldsymbol{F}_s \end{bmatrix} \quad (8.15)$$

$$\boldsymbol{u}_s = \begin{bmatrix} \boldsymbol{K}_{sp1} + \boldsymbol{K}_{sd1}\dfrac{d}{dt} & \boldsymbol{K}_{sf1} \end{bmatrix} \begin{bmatrix} s_p\boldsymbol{x}_m \\ \boldsymbol{F}_m \end{bmatrix} - \begin{bmatrix} \boldsymbol{K}_{sp2} + \boldsymbol{K}_{sd2}\dfrac{d}{dt} & \boldsymbol{K}_{sf2} \end{bmatrix} \begin{bmatrix} \boldsymbol{x}_s \\ s_f\boldsymbol{F}_s \end{bmatrix}, \quad (8.16)$$

where $\boldsymbol{K}_{mpi}, \boldsymbol{K}_{mdi}, \boldsymbol{K}_{mfi}$ $(i = 1, 2)$ are feedback gain matrices of position, velocity, and force, respectively, of the master side, and $\boldsymbol{K}_{spi}, \boldsymbol{K}_{sdi}, \boldsymbol{K}_{sfi}$ $(i = 1, 2)$ are feedback gain matrices of position, velocity, and force, respectively, of the slave side. These feedback gains in the above three bilateral controls are set as follows:

(1) Symmetric type

$$\boldsymbol{K}_{mpi} = -\boldsymbol{K}_{spi} = -\boldsymbol{K}_p, \quad \boldsymbol{K}_{mdi} = -\boldsymbol{K}_{sdi} = -\boldsymbol{K}_d, \quad \boldsymbol{K}_{mfi} = \boldsymbol{K}_{sfi} = 0. \quad (i = 1, 2)$$

(2) Force-reflection type

$$\boldsymbol{K}_{mf2} = \boldsymbol{K}_f, \quad \boldsymbol{K}_{spi} = \boldsymbol{K}_p, \quad \boldsymbol{K}_{sdi} = \boldsymbol{K}_d,$$
$$\boldsymbol{K}_{mpi} = \boldsymbol{K}_{mdi} = \boldsymbol{K}_{mf1} = \boldsymbol{K}_{sfi} = 0. \quad (i = 1, 2)$$

(3) Force-feedback type

$$\boldsymbol{K}_{mfi} = \boldsymbol{K}_f, \quad \boldsymbol{K}_{spi} = \boldsymbol{K}_p, \quad \boldsymbol{K}_{sdi} = \boldsymbol{K}_d,$$
$$\boldsymbol{K}_{mpi} = \boldsymbol{K}_{mdi} = \boldsymbol{K}_{sfi} = 0. \quad (i = 1, 2)$$

8.2. Improved Master–Slave Control

Force-feedback-type bilateral control has traditionally been used in mater–slave systems. It can, however, sometimes become unstable, because the master and slave are not controlled individually, but closed-loop controlled as a unit. To address this issue, a parallel control method and impedance control method have been proposed.

8.2.1. *Parallel Control*

A block diagram of the **parallel control** method is shown in Fig. 8.5. The master and slave are parallel closed-controlled individually. This control scheme enhances the stability of the master–slave system because the phase delay of the control corresponds to that of one robot system. The master and slave are position controlled to follow a desired position command, $\boldsymbol{x}_d$, that is generated in the position

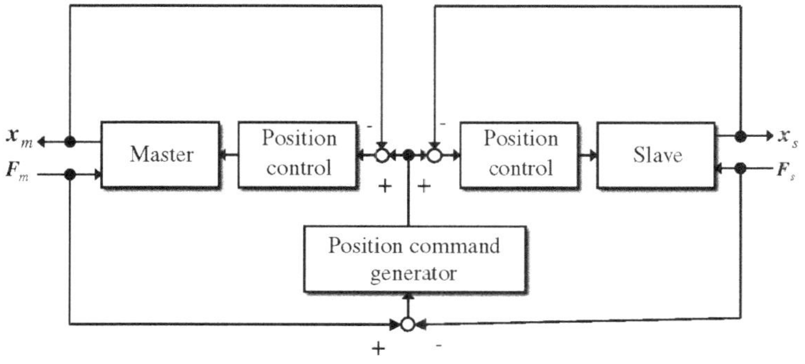

Fig. 8.5. Parallel control.

command generator. Let us define position errors of the master and slave as follows:

$$\boldsymbol{e}_{mp} = \boldsymbol{x}_d - \boldsymbol{x}_m, \tag{8.17}$$

$$\boldsymbol{e}_{sp} = \boldsymbol{x}_d - \boldsymbol{x}_s. \tag{8.18}$$

The control inputs of the master and slave are given by,

$$\boldsymbol{u}_m = \boldsymbol{K}_p \boldsymbol{e}_{mp}, \tag{8.19}$$

$$\boldsymbol{u}_s = \boldsymbol{K}_p \boldsymbol{e}_{sp}. \tag{8.20}$$

The desired position command is calculated from a difference between the forces of the master and slave, and is given by,

$$\boldsymbol{x}_d = \boldsymbol{K}_f \boldsymbol{e}_f. \tag{8.21}$$

Then, the force of the master can be given by,

$$\boldsymbol{F}_m = (\boldsymbol{K}_p \boldsymbol{K}_f + \mathbf{I})^{-1}(\boldsymbol{K}_p \boldsymbol{K}_f \boldsymbol{F}_s + \boldsymbol{M}_m \ddot{\boldsymbol{x}}_m + \boldsymbol{D}_m \dot{\boldsymbol{x}}_m + \boldsymbol{K}_p \boldsymbol{x}_m). \tag{8.22}$$

This shows that the force of the master is similar to that of the force-feedback-type bilateral control. When $\boldsymbol{K}_p \boldsymbol{K}_f$ is sufficiently large, then the force of the master approaches the force of the slave, i.e., $\boldsymbol{F}_s = \boldsymbol{F}_m$. Furthermore, this control easily enhances stability because the position control given by Eqs. (8.19) and (8.20) makes up individual closed loops.

8.2.2. *Impedance Control*

A block diagram of **impedance control** in which dynamic characteristics of the master and slave are taken into account is shown in Fig. 8.6. In this control, the master and slave are controlled such that

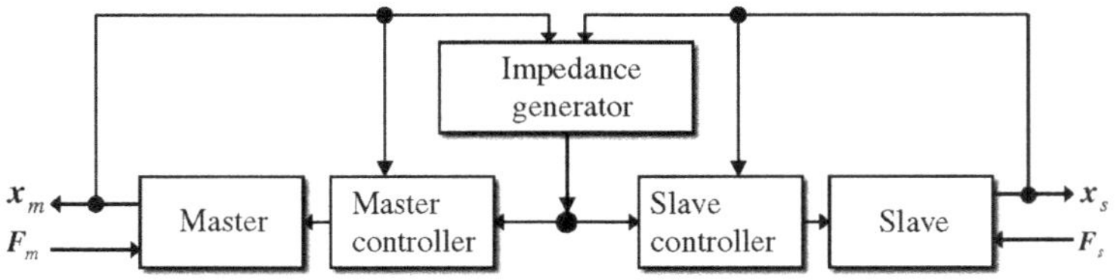

Fig. 8.6. Impedance control.

the impedance characteristics of both the master and slave coincide. The control inputs of the master and slave are then given by,

$$\boldsymbol{u}_m = \boldsymbol{M}_m\ddot{\boldsymbol{x}}_m + \boldsymbol{D}_m\dot{\boldsymbol{x}}_m - (\boldsymbol{M}_d\ddot{\boldsymbol{e}}_p + \boldsymbol{D}_d\dot{\boldsymbol{e}}_p + \boldsymbol{K}_d\boldsymbol{e}_p) \quad (8.23)$$

$$\boldsymbol{u}_s = \boldsymbol{M}_s\ddot{\boldsymbol{x}}_s + \boldsymbol{D}_s\dot{\boldsymbol{x}}_s + (\boldsymbol{M}_d\ddot{\boldsymbol{e}}_p + \boldsymbol{D}_d\boldsymbol{e}_p + \boldsymbol{K}_d\boldsymbol{e}_p), \quad (8.24)$$

where $\boldsymbol{M}_d, \boldsymbol{D}_d, \boldsymbol{K}_d$ are an inertial matrix, viscous matrix, and stiffness matrix, respectively, of a desired impedance characteristic. By substituting these into Eqs. (8.1) and (8.2), the impedance of the master and slave can be given by,

$$\boldsymbol{M}_d\ddot{\boldsymbol{e}}_p + \boldsymbol{D}_d\dot{\boldsymbol{e}}_p + \boldsymbol{K}_d\boldsymbol{e}_p = \boldsymbol{F}_m \quad (8.25)$$

$$\boldsymbol{M}_d\ddot{\boldsymbol{e}}_p + \boldsymbol{D}_d\dot{\boldsymbol{e}}_p + \boldsymbol{K}_d\boldsymbol{e}_p = \boldsymbol{F}_s. \quad (8.26)$$

These equations show that the master and slave have the same impedance characteristics. If the master–slave system is stabilized by the appropriate selection of $\boldsymbol{M}_d, \boldsymbol{D}_d, \boldsymbol{K}_d$, it leads to $\boldsymbol{F}_m = \boldsymbol{F}_s$. When the reaction force from the slave is zero, that is, $\boldsymbol{F}_s = 0$, then $\boldsymbol{e}_p$ converges to zero, and $\boldsymbol{x}_m \to \boldsymbol{x}_s$ as $t \to \infty$. This control scheme does not need a force sensor, but acceleration signals of the master and slave are necessary. It should be noted that such acceleration signals have a high tendency to include signal noise.

8.3. Bilateral Control with Time Delay

In a master–slave system in which the master is connected to the slave through the network, there exists a time delay at the transporting position and force data. It is well known that such time delays cause instability in a master–slave system. In such systems, proof of strict stability is difficult to achieve because all the dynamics of the master and slave, and characteristics of the network, should be handled as a unit. In previous research, the stability of such systems has been presented using the network theory, passivity of robot systems, scattering theory, and Lyapunov stability theory. In this section, an outline of a parallel position control (Kawai and Fujita, 2008) is presented.

Figure 8.7 shows a block diagram of bilateral control with **time delay**. It is supposed that the master states, $\boldsymbol{x}_m, \dot{\boldsymbol{x}}_m$, are transmitted

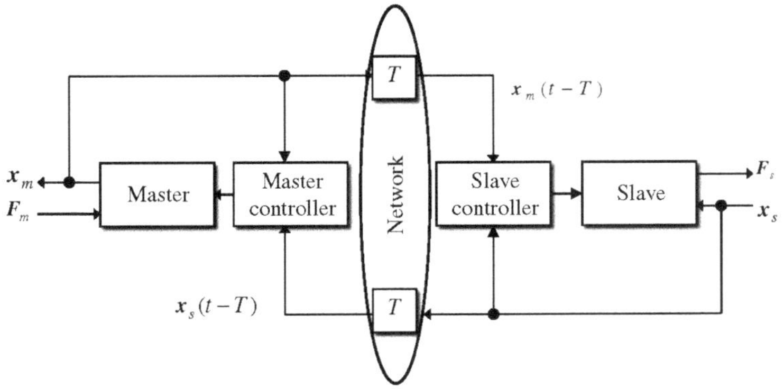

Fig. 8.7. Bilateral control with time delay of telecommunication.

to the slave with time delay T, and that the slave states, $\boldsymbol{x}_s, \dot{\boldsymbol{x}}_s$, are similarly transmitted to the master with time delay T. Position errors of the master and slave are then defined by,

$$\boldsymbol{e}_{mp}(t) = \boldsymbol{x}_s(t - T) - \boldsymbol{x}_m \tag{8.27}$$

$$\boldsymbol{e}_{sp}(t) = \boldsymbol{x}_m(t - T) - \boldsymbol{x}_s, \tag{8.28}$$

and control inputs of the master and slave are given by,

$$\boldsymbol{u}_m = \boldsymbol{K}_p \boldsymbol{e}_{mp} + \boldsymbol{K}_d \dot{\boldsymbol{e}}_{mp} \tag{8.29}$$

$$\boldsymbol{u}_s = \boldsymbol{K}_p \boldsymbol{e}_{sp} + \boldsymbol{K}_d \dot{\boldsymbol{e}}_{sp}. \tag{8.30}$$

It is assumed that the master is passive, and the following is comprised:

$$-\int_0^t \boldsymbol{F}_m \dot{\boldsymbol{x}}_m dt \geq 0. \tag{8.31}$$

It is also assumed that the slave is passive and the following is comprised:

$$\int_0^t \boldsymbol{F}_s \dot{\boldsymbol{x}}_s dt \geq 0. \tag{8.32}$$

Then, using a candidate of the Lyapunov function given by,

$$V = \frac{1}{2}\left(\boldsymbol{M}_m \dot{\boldsymbol{x}}_m^2 + \boldsymbol{M}_s \dot{\boldsymbol{x}}_s^2 + \boldsymbol{K}_d \int_{t-T}^t (\dot{\boldsymbol{x}}_m^2 + \dot{\boldsymbol{x}}_s^2) d\tau + \boldsymbol{K}_p (\boldsymbol{x}_m - \boldsymbol{x}_s)^2\right)$$

$$+ \int_0^t (\boldsymbol{F}_s \dot{\boldsymbol{x}}_s - \boldsymbol{F}_m \dot{\boldsymbol{x}}_m) dt \geq 0, \tag{8.33}$$

it has been proven that the time derivative of the Lyapunov function is negative and semi-definite, and that $\boldsymbol{e}_{sp}$ is bounded.

In this theory, the boundedness of the position error was proven, but the asymptotic convergence to zero of the position error was not. In an actual network, the time delays of the master side and the slave side vary, and information will go missing due to packet loss. A stability theorem including these is expected.

8.4. Teleoperation of Multi-Fingered Robot Hand

Teleoperation for robots has been actively researched over the last two decades. Many successful control schemes have been proposed, mainly for linear robotic teleoperations (Lawrence, 1993). Several control techniques have been presented to cope with uncertainty, time delay in telecommunication (Chopra *et al.*, 2006), parameter estimation of the environment on the slave side (Cortesao *et al.*, 2006), and non-linearity of robot dynamics (Lee and Li, 2005). Experimental results have been offered mainly for robot arms but also for multi-fingered robot hands (Mouri and Kawasaki, 2007). Teleoperation of a multi-fingered robot hand requires a multi-fingered haptic interface. The ability to convey three-dimensional (3D) contact force on the slave side through a multi-fingered haptic interface would better enable a human operator to manipulate objects with telepresence.

8.4.1. *Teleoperation Using HIRO*

This section presents the system configuration, bilateral control method, and experimental results for the haptic teleoperation of a humanoid robot hand by a multi-fingered haptic interface robot using 3D force feedback (Mishima *et al.*, 2009).

8.4.1.1. *System configuration*

In the experiment, the master was a multi-fingered haptic interface robot named HIRO III (see Chapter 7), with five haptic fingers, each with three DOFs, connected to the operator's fingers with finger holders. The slave was a humanoid robot hand named KH Hand (Mouri and Kawasaki, 2007), with a thumb and four fingers, each with three

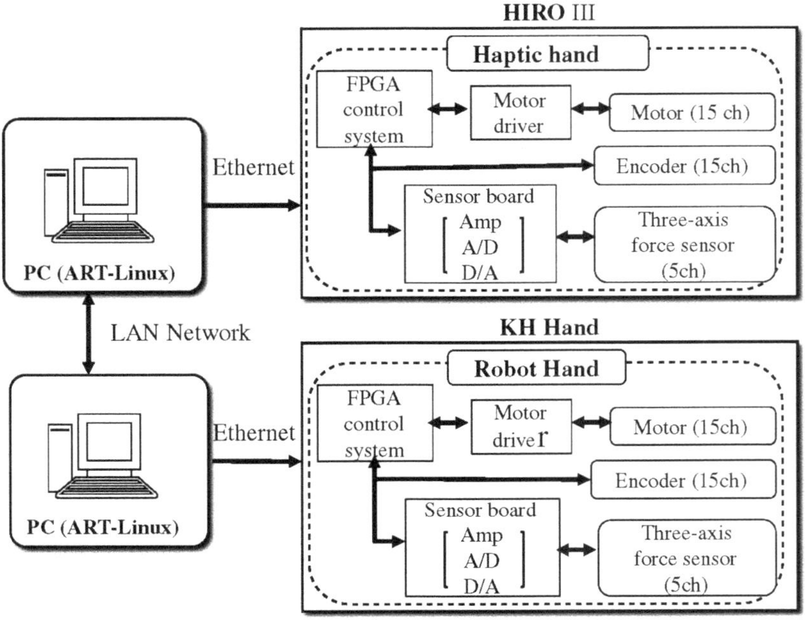

Fig. 8.8. Teleoperation system of a humanoid robot hand.

DOFs. Both robot hands were equipped with a 3D force sensor at each fingertip link. This combination enabled us to teleoperate the humanoid robot hand with 3D force feedback. The teleoperation system of the KH hand through HIRO III is shown in Fig. 8.8. The two devices were connected by a LAN. The sampling cycle of the LAN was about 1 ms.

8.4.1.2. *Control formation*

The dynamics of the master and slave are represented by,

$$\boldsymbol{M}_m(\ddot{\boldsymbol{q}}_m)\dot{\boldsymbol{q}}_m + \boldsymbol{C}_m(\boldsymbol{q}_m, \dot{\boldsymbol{q}}_m)\dot{\boldsymbol{q}}_m + \boldsymbol{g}_m(\boldsymbol{q}_m) = \boldsymbol{\tau}_m - \boldsymbol{J}_m^T\boldsymbol{f}_m \quad (8.34)$$

$$\boldsymbol{M}_s(\ddot{\boldsymbol{q}}_s)\dot{\boldsymbol{q}}_s + \boldsymbol{C}_s(\boldsymbol{q}_s, \dot{\boldsymbol{q}}_s)\dot{\boldsymbol{q}}_s + \boldsymbol{g}_s(\boldsymbol{q}_s) = \boldsymbol{\tau}_s - \boldsymbol{J}_s^T\boldsymbol{f}_s, \quad (8.35)$$

where the lower suffix m indicates HIRO III on the master side, the lower suffix s indicates the KH hand on the slave side, $\boldsymbol{\tau}$ is the joint torque, $\boldsymbol{q}$ is the joint angle, $\boldsymbol{f}$ is the contact force, $\boldsymbol{J}$ is the kinematic Jacobian, $\boldsymbol{M}$ is the inertia matrix, $\boldsymbol{C}\dot{\boldsymbol{q}}$ is the velocity square term, and $\boldsymbol{g}$ is the gravitational term. The master receives the operation force $\boldsymbol{f}_m$ from the human operator, and the slave receives the contact force $\boldsymbol{f}_s$ from the environment. It is supposed that communication delay is negligible; as a result, we adopted the Cartesian coordinate master–slave control given by,

$$\boldsymbol{\tau}_m = \boldsymbol{J}_m^T \left(\boldsymbol{K}_1 \boldsymbol{e}_f + \boldsymbol{K}_2 \int_0^t \boldsymbol{e}_f dt + \boldsymbol{f}_m \right) + \boldsymbol{g}_m(\boldsymbol{q}_m) \tag{8.36}$$

$$\boldsymbol{\tau}_s = \boldsymbol{J}_s^T \left(\boldsymbol{K}_p \boldsymbol{e}_x + \boldsymbol{K}_i \int_0^t \boldsymbol{e}_x dt + \boldsymbol{f}_s \right) - \boldsymbol{K}_v \dot{\boldsymbol{q}}_s + \boldsymbol{g}_s(\boldsymbol{q}_s), \tag{8.37}$$

where $\boldsymbol{K}_1$ and $\boldsymbol{K}_2$ are the feedback gain matrices of force error and integral of force error, respectively, while $\boldsymbol{K}_p, \boldsymbol{K}_v$, and $\boldsymbol{K}_i$ are the feedback gain matrices of position error, velocity, and integral of position error, respectively. The force error $\boldsymbol{e}_f$ in the master and the position error $\boldsymbol{e}_x$ in the slave are given by

$$\boldsymbol{e}_f = \boldsymbol{S}_f^m \boldsymbol{f}_s - \boldsymbol{f}_m \tag{8.38}$$

$$\boldsymbol{e}_x = \boldsymbol{S}_x^s \boldsymbol{x}_m(\boldsymbol{q}_m) - \boldsymbol{x}_s(\boldsymbol{q}_s), \tag{8.39}$$

where the upper suffix indicates the coordinate system, $\boldsymbol{x}$ is the fingertip position, $\boldsymbol{S}_f$ is the force-scaling matrix, and $\boldsymbol{S}_x$ is the position-scaling matrix. The control input of the master is a force PI control with a gravitational compensator in which the desired force is given by the contact force of the slave. The control input of the slave is a position proportional-integral-derivative (PID) control with a gravitational compensator in which the desired position is given by the position on the slave side. This control scheme is a force-feedback-type of bilateral control. A block diagram of the teleoperation system is presented in Fig. 8.9.

8.4.1.3. *Experiment*

The human operator rotated a stiff spherical object on the slave side using four fingers. The diameter and mass of the object were 0.1 m and 0.23 N, respectively. This teleoperation was relatively difficult, because representation of the 3D force at the fingertip was required. Figure 8.10 shows the sequences of the teleoperation. The operator grasped the spherical object located on a base, then rotated it using

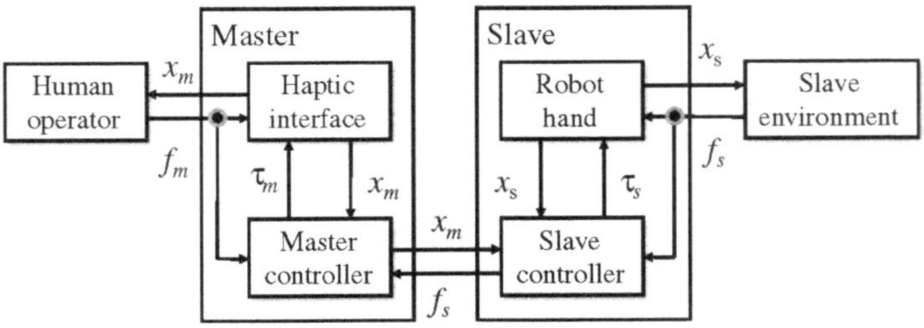

Fig. 8.9. Block diagram of the teleoperation system.

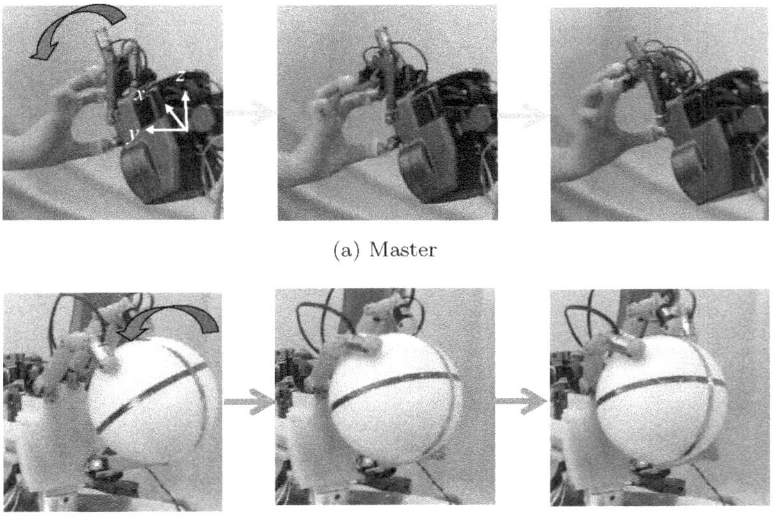

(a) Master

(b) Slave

Fig. 8.10. Rotating teleoperation of a spherical object.

four fingers, and finally released it. Figure 8.11 shows the experimental force responses of the thumb and ring finger. This demonstrated that the direction of the contact force at the operator fingertip varied according to the rotational angle of the object. When the robot fingers contacted the object, the root-mean-square (RMS) force error at the thumb was 13.5 N, while that at the ring finger was 0.46 N. The human operator was able to successfully rotate the spherical object feeling not only normal contact force, but also the frictional force in the plane tangential to the object surface.

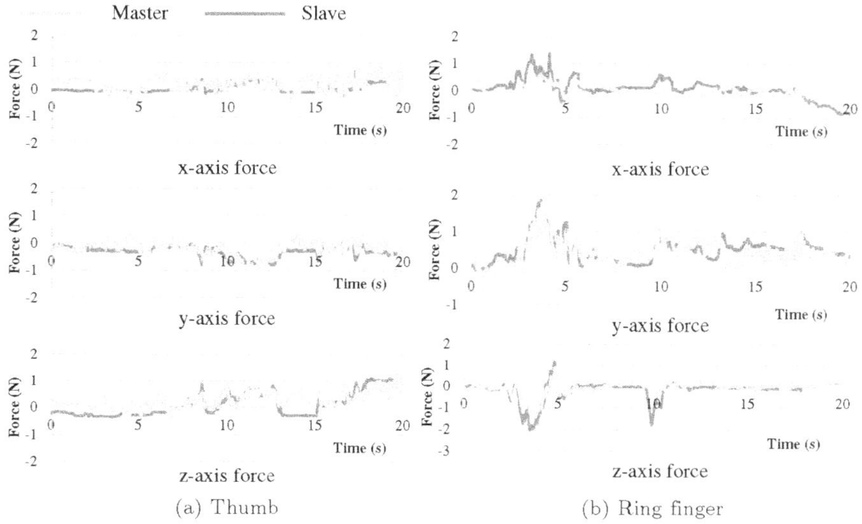

Fig. 8.11. Force responses of the thumb and ring at teleoperating a spherical object.

8.4.2. *Toward Whole-Hand Manipulation*

The arms of the master and slave should be controlled when the robot hand is attached to a robot arm. When contact points between the slave robot hand and the environment are limited to the fingertips, the PID position control is one of the stable controls where the desired arm positions of the multi-fingered haptic interface and the slave robot are generated based on the hand manipulability optimization presented in Sec. 7.2. This control does not affect the stability of the master–slave system because it makes a local closed-loop. Furthermore, the abilities of the haptic interface hand and the robot hand are utilized to the maximum degree because they are placed in positions where the performance index based on hand manipulability is optimum.

When the contact points between the slave robot hand and the environment are not limited to the fingertips, as in a power grasp, a hand haptic interface that presents force feeling not only to the human fingertips but also to the human finger pad and/or palm is necessary (Fig. 8.12) (Kawasaki *et al.*, 2012). The hand haptic

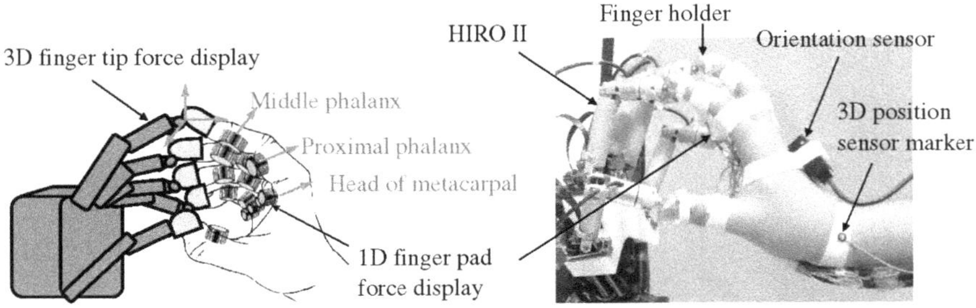

(a) Conceptual figure (b) Developed system

Fig. 8.12. Hand haptic interface.

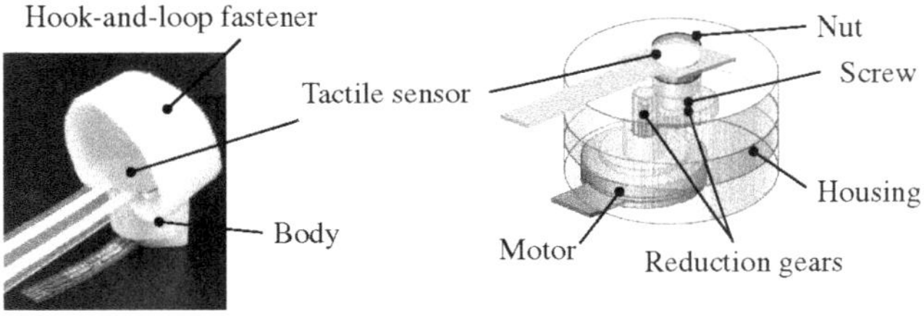

(a) Photograph (b) Mechanical structure of body

Fig. 8.13. Finger pad force display.

interface consisted of a multi-fingered haptic interface called HIRO and a finger-pad force display; it was capable of displaying 3D force to human fingertips and 1D force to human finger pads. The finger-pad force display consisted of a body, a hook-and-loop fastener, and a tactile sensor. The mechanical structure of the device is shown in Fig. 8.13. It consisted of a flat-type brushless motor equipped with a Hall sensor, reduction gears, and a screw mechanism. The nut of the screw mechanism moved up and down with the rotation of the motor. The range of movement was 4 mm. The mass and output forces were 4.8 g and 2.6 N, respectively. This interface will contribute to the enhanced reality of hand manipulation. Moreover, it will be necessary to establish a hand haptic interface control that takes into account the differences of mechanical size and joint configuration between the human hand and robot hand.

8.5. Design and Control of Prosthetic Hand

Electromyogram (EMG) **prosthetic hands**, with actuators to realize multi-degree operationality, controlled by a user's **surface electromyogram sEMG**, are necessary to maintain quality of life for forearm/hand amputees. This is one of the applications for robot hands. It is important that the prosthetic hand be light and unitized, because amputees usually feel uncomfortable if the hand feels massive, and a unitized prosthetic hand is easy to apply and maintain. Moreover, the hand should have a grasping force that will allow an amputee to manipulate objects encountered in daily life, such as plastic bottles, cups, and plates. It is also desirable for the hand to have many DOFs for dexterous manipulation. In particular, both wrist motion and finger motion are required, because coordinated motion between fingers and wrist is necessary for tasks such as opening doors or writing letters. Wrist motion can facilitate various types of object handling.

To satisfy these needs, an EEG prosthetic hand with a grasping strategy consisting of thumb opposition with high grasping force has been developed (Hioki *et al.*, 2011; Ebisawa *et al.*, 2013).

8.5.1. *Mechanical Design*

The prosthetic hands shown in Fig. 8.14 were designed to satisfy the following specifications: a thumb with three joints and two DOFs for extension/flexion and opposition motion; four fingers with three joints and one DOF for flexion/extension; a wrist with two DOFs for pronation/supination and dorsal/palmar motion; a total mass of 300 g or less; a gripping force of 20 N or more; all motors and field programmable gate array (FPGA) control circuit mounted in the hand frame; a size almost the same as the hand of a Japanese adult female.

Powerful motors are necessary for realizing strong grasp forces; however, generally, this type of motor is too heavy to employ for each finger. Therefore, a novel grasping strategy was invented in which the four fingers kept a resting state without back motion while the thumb pushed the object to the virtual wall constructed by the four fingers

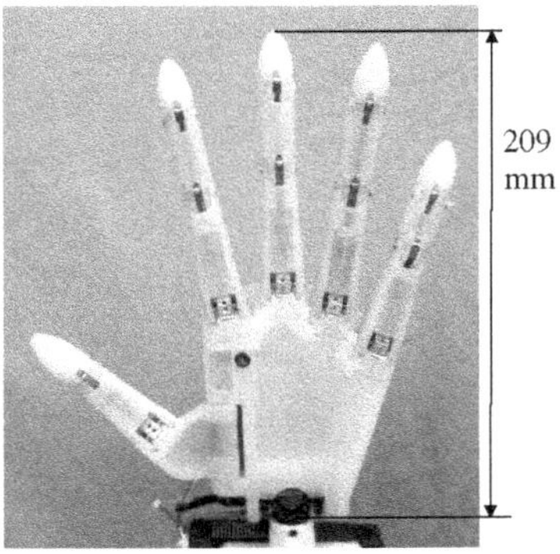

(a) Hand body

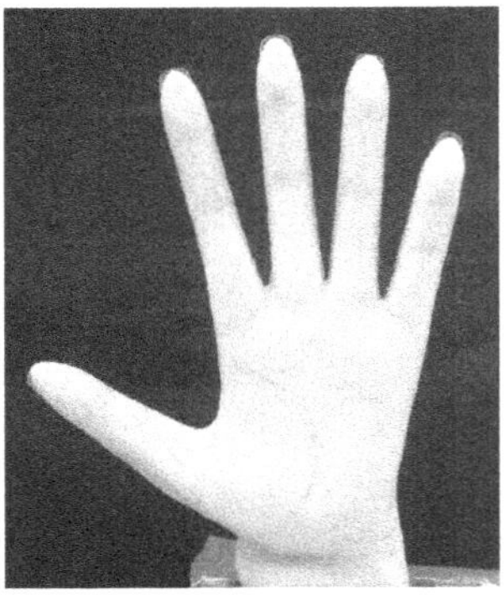

(b) Hand covered with artificial skin

Fig. 8.14. Developed prosthetic hand.

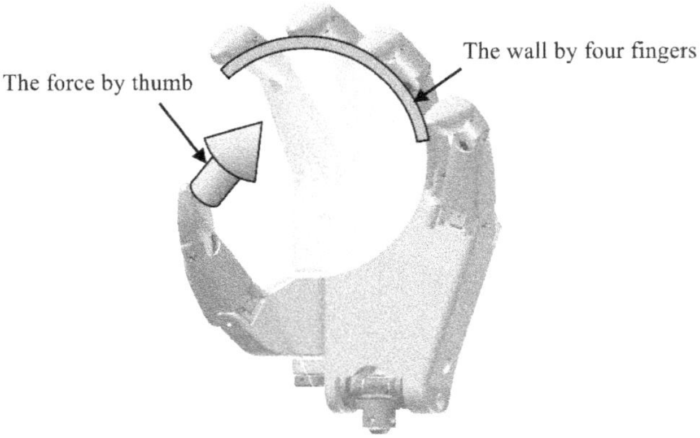

Fig. 8.15. Grasping strategy.

(see Fig. 8.15). To realize the grasping strategy, a worm gear system without back-drivability was used as a reduction gear mechanism. By adopting this mechanism, the fingers could hold their resting state even if external forces were applied, and therefore the hand could use small motors for finger flexion/extension. Only the thumb motor was high power. This was effective for high grasping forces and light mass. In addition, the mechanism without back-drivability was effective in saving electrical energy, important in a prosthetic hand system.

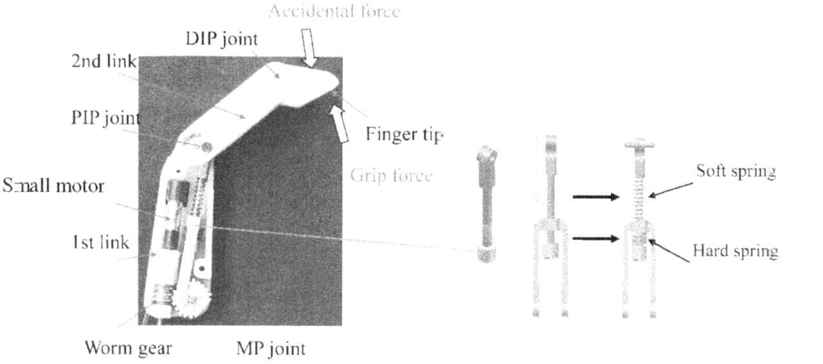

(a) Finger mechanism (b) Link rod mechanism with two type springs

Fig. 8.16. Passive flexion/extension mechanism.

It was possible, however, that this mechanism would result in a stiff or hard grasp. To deal with this grasp issue, a passive flexion/extension mechanism, shown in Fig. 8.16, was added to the link mechanism. The passive mechanism had a link rod with two types of spring (soft and hard). The hard spring worked when external forces were applied at the fingertip along the 'extension direction'. The finger could closely fit the frame of the object by this hard spring. This supported the hand in grasping a target object with all fingers. The soft spring worked when external forces were applied along the 'flexion direction'. If the finger hit an obstacle, the finger could absorb the shock and work as a safety mechanism.

The palm of the prosthetic hand was equipped with an FPGA control circuit system, as shown in Fig. 8.14(b). The circuit implemented pulse width modulation (PWM) control for eight motors and communicated with a PC by TCP/IP, which are abbreviations of the Transmission Control Protocol (TCP) and the Internet Protocol (IP). With this circuit system, the only cables coming out of the hand were the LAN and power cables.

To make the proposed prosthetic hand look as much like a human hand as possible, it was covered with an artificial skin (mass 64 g) as shown in Fig. 8.17. The skin was composed of thermoplastic styrene elastomer (TPE). Artificial skin is also useful for increasing friction forces or soft touching for the stable holding of objects.

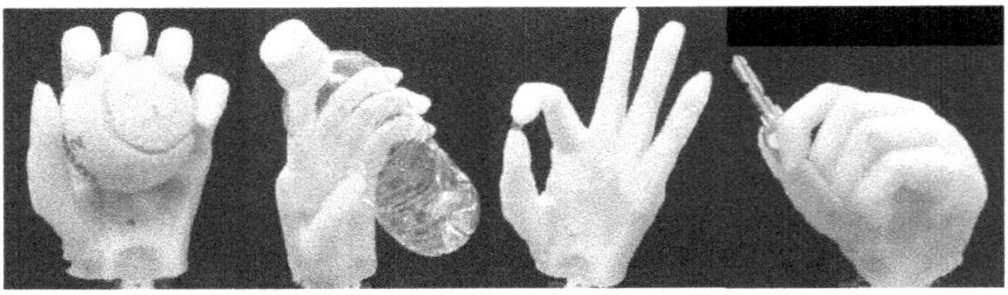

Fig. 8.17. Grasping various type of objects with artificial skin. (From left, tennis ball, plastic bottle, screw nut, and key.)

8.5.2. *Control by EMG*

8.5.2.1. *Control based on finger motion pattern recognition*

In daily life, humans make combination motions with their fingers and wrist joints. For example, the use of a key to lock or unlock a door requires a side grasp with the thumb and index finger and wrist pronation or supination motion. Therefore, the possibility of key operation by EMG was examined as an experimental target task using a **neural network** (NN) system. In general, an NN system has the following useful characteristics for estimating input–output relations:

(a) The NN does not need a strict system model because it can approximate the input–output non-linear function using training data (Fig. 8.18).
(b) It is possible to apply the NN to a multi-input/multi-output system because modifying the dimensions of the input and/or output vectors is quite easy.
(c) By using the sigmoid function as a unit output function in the output layer of the NN, it is possible to limit the output values from the network to (0, 1).

(a) sEMG signal processing

The sEMG was measured at the subject's forearm. The measurement conditions were 1 ms sampling, 12-bit data and four channels. After sEMG acquisition, it was segmented into short time signals of time

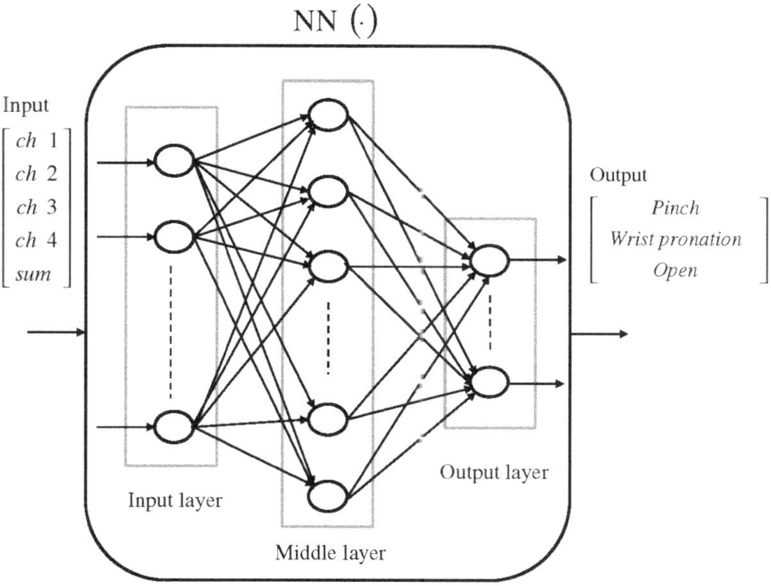

Fig. 8.18. NN for recognition of finger motion patterns.

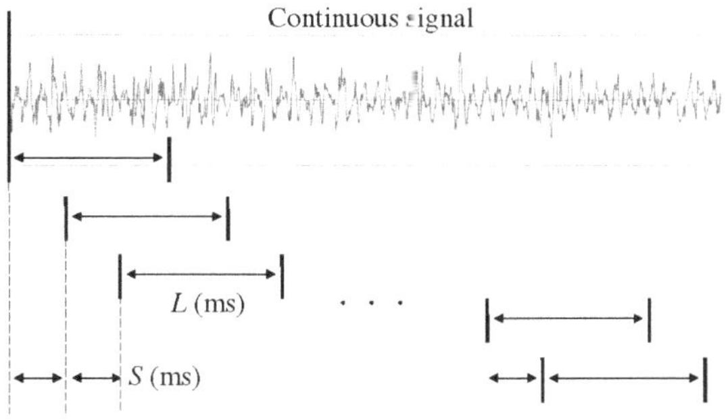

Fig. 8.19. Segmentation of signal.

length L ms and shift length S ms, as shown in Fig. 8.19. This segmentation permitted us to get feature values in both the time and frequency domains. The time length showed how long in time previous signals were used, and related frequency resolution. The shift length related a down sampling that contributed to reducing the amount of information.

The feature values of sEMG are obtained as a wavelength of segmented signal by,

$$WL = \sum_{i=1}^{N-1} |EMG[i] - EMG[i-1]|, \tag{8.40}$$

where N is a number for the data in the segmented signal, and $EMG[i]$ is the i-th signal of the raw sEMG. The WL expresses a wavelength of the segmented signal and reflects both the absolute of signal and the frequency of signal (Arvetti *et al.*, 2007).

Attention should be paid to magnitudes of input, output, and learning signal in NN systems because these can cause bias in learning. It is recommended that input and output signals be normalized using the minimum and maximum of the signal. Doing so contributes to removing the differences among individuals. A maximal voluntary contraction method is adopted for normalization of sEMG. That is, sEMG is measured with the person in a resting state without strain, and then in a state of maximal voluntary contraction. These then constitute the minimum and maximum of the signal.

A normalized WL of the i-th channel (window width is 500 points) was calculated and described as the EMG_i. The ratio *chi* was determined by the following equations:

$$\text{sum} = \sum_{i=1}^{4} EMG_i, \tag{8.41}$$

$$\text{chi} = \frac{EMG_i}{\text{sum}} (i = 1, \ldots, 4). \tag{8.42}$$

To control the prosthetic hand, it was necessary to generate the desired joint angles from the sEMG feature values. To estimate the finger motion patterns from sEMG information, an NN system like that shown in Fig. 8.18 was used. Input of the NN system must be above five signals; in this experiment, the number of middle layer unit was 50. The output of the NN indicates finger motion patterns. This experiment targeted three motions: pinch with thumb and index finger (wrist maintains current state), wrist pronation (fingers maintain current state), and open the hand (all joints move to initial state shown in Fig. 8.14).

(b) NN learning

The experimental process was divided into two sections — the training phase and the real-time estimation phase. During the training phase, the user held the three hand states (pinch, wrist pronation, and open fingers) while the sEMG signals were measured in 4 s. Feature values were extracted from raw sEMG, and teacher vectors were set as [0.9 0 0] for pinch, [0 0.9 0] for wrist pronation, and [0 0 0.9] for the move to the initial position. The system then received 12,000 data sets (4,000 for each hand state). When measurements were completed, the NN trained with these data sets. The training was finished when training reached 200 epochs.

(c) Estimation of finger motion patterns

After training with the teaching input–output data set, the NN was capable of estimating finger motion patterns from the sEMG. Using the estimated finger motion patterns, the desired joint angles were generated, and the prosthetic hand was controlled to produce the desired angles. The results are shown in Fig. 8.20. Figure 8.20(a) presents the feature values extracted from sEMG, while 8.20(b) shows the joint angles of the prosthetic hand, where the desired joint angles are generated based on motion-pattern recognition. These results indicate the accomplishment of the target task.

The finger motion pattern in this system was limited in the key manipulation in which the thumb, index finger, and wrist were all used. For dexterous manipulation by the prosthetic hand, the estimation system for various finger motion patterns would need to be extended.

8.5.2.2. *Control based on finger joint angle estimation*

Pattern recognition of the hand and finger state from sEMG may be able to function as a computer interface. The possible number of patterns to be recognized will be limited, however. To operate a multi-fingered robot hand according to the operator's intent, it is more important to estimate finger joint angles than to determine finger motion patterns. The system that estimates finger joint angles

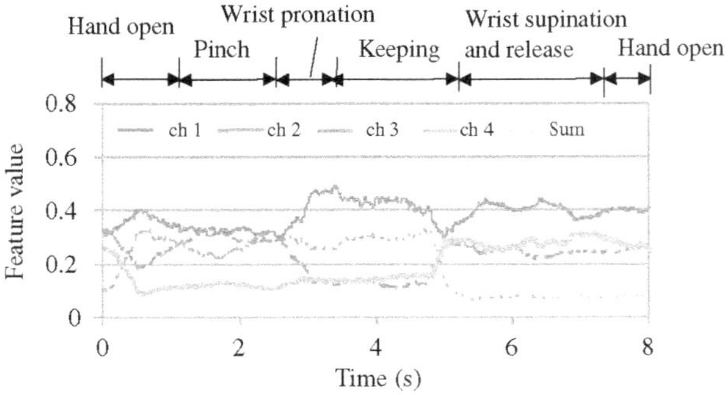

(a) Feature values extracted from sEMG

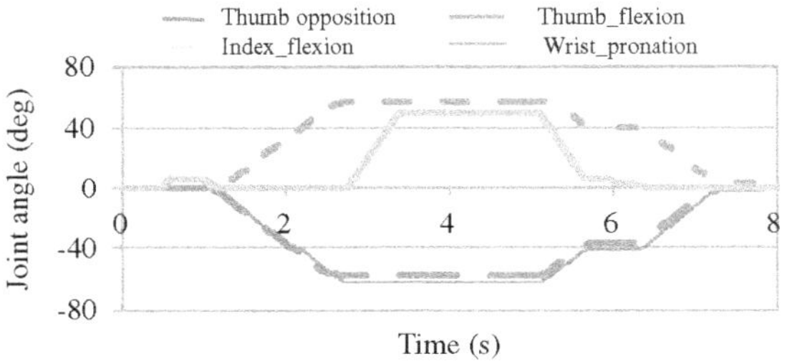

(b) Joint angles of prosthetic hand

Fig. 8.20. Experimental result of control of the prosthetic hand by sEMG.

is able to recognize every hand state, and therefore has high general versatility. Instead of using a pattern recognition approach, finger joint angles from sEMG were estimated using the NN (Hioki and Kawasaki, 2012). In this research, an NN was used to estimate a total of five joint angles (the metacarpophalangeal (MP) joint angles of five fingers) from sEMG during the individual movement of each joint. The estimating precision was ~11%. However, certain estimated joint angles may vary even if the joint angles are not moving. This may be because the dynamics of the estimation system cannot be trained, as the system uses a simple feed-forward NN.

(a) Dynamic NN system

In order to learn the dynamics of finger motion, an estimation system has been proposed consisting of a three-layered NN, a feedback stream of past output, and a time-delay factor of input, as shown

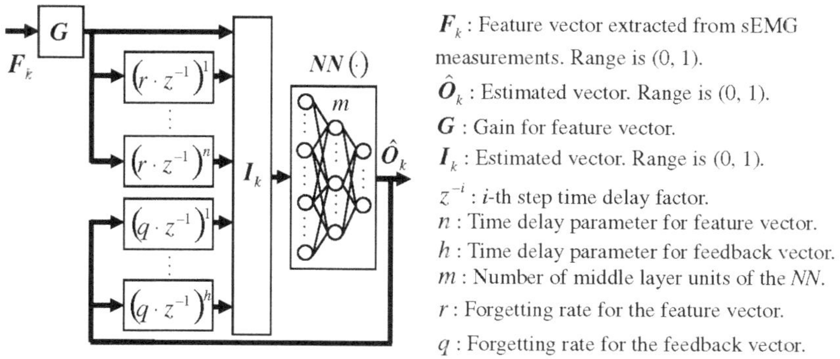

Fig. 8.21. Structure of the estimating system.

in Fig. 8.21. In this system, time delay parameters for feature vector n and feedback vector h indicate the extent of previous feature vectors and feedback vectors, respectively. In Fig. 8.21, $\boldsymbol{NN}(\cdot)$ is a three-layered neural network, and the output of the system can be given by,

$$\begin{aligned}\hat{\boldsymbol{O}}_k &= \boldsymbol{NN}(\boldsymbol{I}_k)\\ &= \boldsymbol{NN}(\boldsymbol{F}_k, \boldsymbol{F}_{k-1}, \ldots, \boldsymbol{F}_{k-n}, \boldsymbol{O}_{k-1}, \ldots, \mathrm{O}_{k-h},). \end{aligned} \tag{8.43}$$

The input vector into the NN is composed of the following factors: newest feature vector $\boldsymbol{F}_k$, earlier feature vectors $\boldsymbol{F}_{k-1}, \ldots, \boldsymbol{F}_{k-n}$, and earlier output vectors $\boldsymbol{O}_{k-1}, \ldots, \boldsymbol{O}_{k-h}$. If parameters n and h are both zero, the estimation system becomes ordinary, as shown in Fig. 8.18.

This type of NN can train any input–output relation if the number of the middle layer unit m is adequate. The system adopts the standard sigmoid function as the activating function of each unit in NN, and the steepest-descent and back-propagation methods as the training methods. This approach would be useful both for humans operating robot hands and for those with a prosthetic hand. The target finger motion groups were individual finger flexion and extension.

(b) sEMG signal processing

Measured raw sEMG data was segmented after processing with a band-pass filter (pass band: 10–350 Hz). This signal processing is useful and powerful for extracting feature vectors from raw sEMG data. The system adopted the following segmentation parameters: segment length, 1,024, and shift length, 32; in other words, the sEMG was performed with down sampling into 32 ms samples from 1 ms samples. This feature value included both time domain and frequency domain. The raw sEMG vector had four elements. After extracting the feature vector, the vector was normalized in a range of 0–1. The parameters necessary for normalizing (maximum and minimum values) were obtained and calculated from advanced measuring data, which took the relaxed state of the muscle as the minimum value and the state of extreme muscle contraction as the maximum value.

(c) NN system learning

Teacher signal vectors are required to train the NN. Because the system aims to estimate finger joint angles, the teacher vectors have been extracted from finger joint angles measured with a 32-ms sampling time by a data glove measurement system. The measured finger joints were the MP joint, the proximal interphalangeal (PIP) joint, the distal interphalangeal (DIP) joint, and the abduction–adduction (Abd) joint. Therefore, the total number of measured finger joints was 20. This training method changes the coefficients of the internal unit, and it can be shown simply as follows:

$$\Delta \boldsymbol{w} = \alpha \cdot \delta \cdot \boldsymbol{x}, \tag{8.44}$$

where Δw is the amount of change of a certain coefficient of the unit, α is the training rate and adequate small value, δ is a back-propagation term, and $\boldsymbol{x}$ is the input into this unit. During network training, modifying the value of the input vector into the network is also important for highly precise functioning. The gain G is a constant value matrix whose value is 1 or more than 1. G makes the value of the feature vector components large rather than those of the feedback vector components because NN training depends on sEMG information rather than on feedback information. The forgetting rates — r for feature vectors and q for feedback vectors — have a value of 1 or less than 1. These parameters compel the estimation

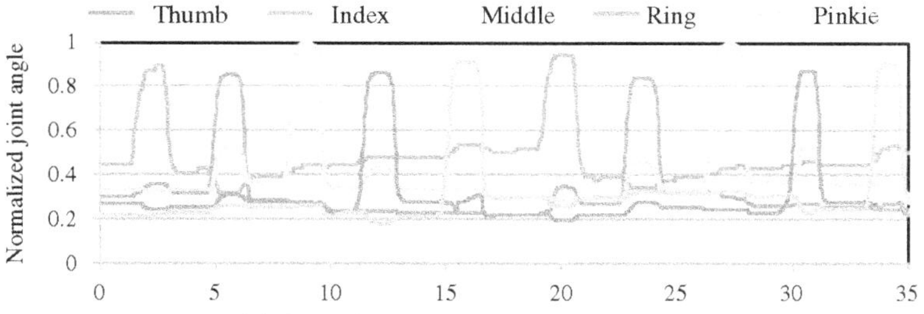

(a) Actual finger joint angle from data glove

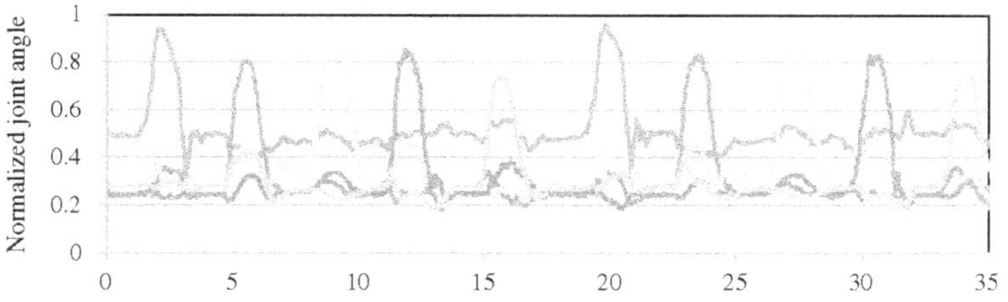

(b) Estimated finger joint angle

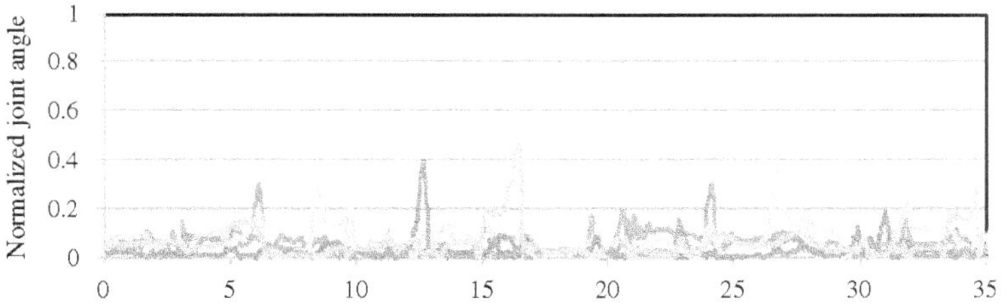

(c) Absolute finger joint angle error

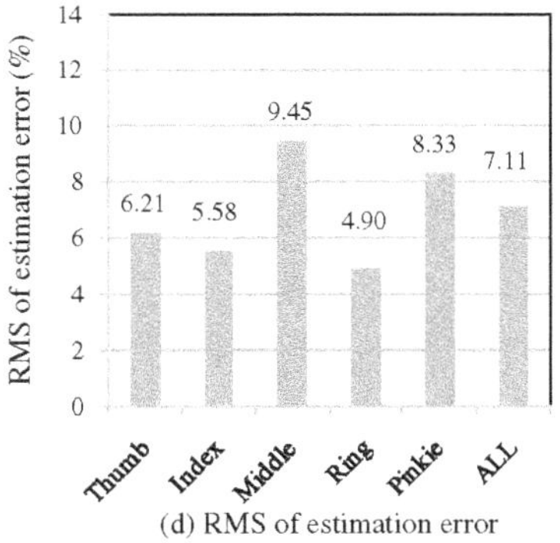

(d) RMS of estimation error

n = 2: Time delay parameter for feature vector.
h = 6: Time delay parameter for feedback vector.
m = 40: Number of middle layer unit of NN.
$\boldsymbol{G}$ = 12.5: Gain for feature vector.
r = 0.75: Forgetting rate for feature vector.
q = 0.01: Forgetting rate for feedback vector.

Fig. 8.22. Estimation results of best performance results.

system to weigh newer information as more important than older information.

(d) Experiment

The system has six parameters that influence the results of joint angle estimation. These parameters are set such that estimated joint angle error by root mean square approximation (RMS) is minimum by random search. In Fig. 8.22, estimation results are shown for a motion set of finger flexions. This shows that the system was able to recognize the finger movements and the resulting RMS of error was 7.1%. We conclude that the proposed system is able to integrate human finger dynamics because the estimation error was much below that of the feed-forward NN.

Exercise

8.1. Survey the teleoperation technology for robots.

Appendix

Linear Algebra

Linear algebra is useful in the study of mechanical control systems in general, and in the state–space approach in particular. The subject of linear algebra is complex and is based on a rigorous mathematical foundation. In this appendix, the basics of vectors and matrices are reviewed.

A.1. Vectors and Matrices

The symbol $\mathbb{R}$ will denote the set of real numbers, and $\mathbb{R}^n$ will denote the usual vector space of n-tuples over $\mathbb{R}$. We use bold lowercase letters $\boldsymbol{a}, \boldsymbol{b}, \boldsymbol{x}, \boldsymbol{y}$, etc., to denote vectors in $\mathbb{R}^n$. Bold uppercase letters $\boldsymbol{A}, \boldsymbol{B}, \boldsymbol{X}, \boldsymbol{Y}$, etc., denote matrices in $\mathbb{R}^{m\times n}$. The scalar component of a vector or a matrix will be denoted by lowercase letters a, b, x, y, etc. Unless otherwise stated, vectors will be defined as column vectors. Thus, the statement $\boldsymbol{a} \in \mathbb{R}^n$ means that

$$\boldsymbol{a} = \begin{bmatrix} a_1 \\ \vdots \\ a_n \end{bmatrix}, \quad \text{with } a_i \in \mathbb{R} \quad \text{for } i = 1, \ldots, n. \tag{A.1}$$

We will frequently denote this as $\boldsymbol{a} = [a_1, \ldots, a_n]^T$, where the superscript T denotes transpose.

A system of vectors X is a **vector space** on the field of real numbers if the operation of the sum of two vectors of X and product of a scalar α by a vector of X have values in X and the following properties hold for $\forall\, \boldsymbol{a}, \boldsymbol{b}, \boldsymbol{c} \in X$ and $\forall\, \alpha, \beta \in \mathbb{R}$:

Commutativity of addition	$\boldsymbol{a} + \boldsymbol{b} = \boldsymbol{b} + \boldsymbol{a}$
Associativity of addition	$\boldsymbol{a} + (\boldsymbol{b} + \boldsymbol{c}) = (\boldsymbol{a} + \boldsymbol{b}) + \boldsymbol{c}$

Identity element of addition	$\exists 0 \in X; \boldsymbol{a} + 0 = \boldsymbol{a}$
Inverse elements of addition	$\exists (-\boldsymbol{a}) \in X; \boldsymbol{a} + (-\boldsymbol{a}) = 0$
Compatibility of scalar multiplication	$\alpha(\beta \boldsymbol{a}) = (\alpha\beta)\boldsymbol{a}$
Distributivity of scalar sum	$(\alpha + \beta)\boldsymbol{a}) = \alpha\boldsymbol{a} + \beta\boldsymbol{a}$
Distributivity of vector sums	$\alpha(\boldsymbol{a} + \boldsymbol{b}) = \alpha\boldsymbol{a} + \alpha\boldsymbol{b}$
Identity element of scalar multiplication	$1\boldsymbol{a} = \boldsymbol{a}$

where $\forall$ means 'for any' and $\exists$ means 'there exists'.

A matrix of dimension $m \times n$ is an array of a_{ij} arranged into m rows and n columns:

$$\boldsymbol{A} = (a_{ij}) = \begin{bmatrix} a_{11} & \cdots & a_{1n} \\ \vdots & \vdots & \vdots \\ a_{m1} & \cdots & a_{mn} \end{bmatrix} \in \mathbb{R}^{m \times n} \tag{A.2}$$

If $m = n$, the matrix is said to be a **square matrix**.

The **transpose** of a matrix $\boldsymbol{A} \in \mathbb{R}^{m \times n}$ is defined as

$$\boldsymbol{A}^T = \begin{bmatrix} a_{11} & \cdots & a_{m1} \\ \vdots & \vdots & \vdots \\ a_{1n} & \cdots & a_{mn} \end{bmatrix} \in \mathbb{R}^{n \times m}, \tag{A.3}$$

which is obtained from the original matrix by interchanging its rows and columns.

A square matrix $\boldsymbol{A} \in \mathbb{R}^{n \times n}$ is said to be **symmetric** if $\mathbf{A}^T = \boldsymbol{A}$.

A square matrix $\boldsymbol{A} \in \mathbb{R}^{n \times n}$ is said to be **skew-symmetric** if $\boldsymbol{A}^T = -\boldsymbol{A}$, which implies $a_{ij} = -a_{ji}$ for $i \neq j$ and $a_{ii} = 0$, leading to

$$\boldsymbol{A} = \begin{bmatrix} 0 & a_{12} & \cdots & a_{1n} \\ -a_{12} & 0 & \cdots & a_{2n} \\ \vdots & \vdots & \ddots & \vdots \\ -a_{1n} & -a_{2n} & \cdots & 0 \end{bmatrix}. \tag{A.4}$$

A square matrix $\boldsymbol{A} \in \mathbb{R}^{n \times n}$ is said to be **orthogonal** if $\boldsymbol{A}^T\boldsymbol{A} = \boldsymbol{A}\boldsymbol{A} = \boldsymbol{E}_{n \times n}$, where $\boldsymbol{E}_{n \times n}$ is the $n \times n$ identity matrix. This means that each column (row) vector is orthogonal to other column (row) vectors.

A **partitioned** matrix is a matrix whose elements are matrices (blocks) of proper dimensions:

$$\boldsymbol{A} = \begin{bmatrix} \boldsymbol{A}_{11} & \cdots & \boldsymbol{A}_{1m} \\ \vdots & \vdots & \vdots \\ \boldsymbol{A}_{n1} & \cdots & \boldsymbol{A}_{nm} \end{bmatrix} \tag{A.5}$$

If $\boldsymbol{A}_{ij} = 0$ for $i \neq j$, the partitioned matrix is called **block diagonal,** and is expressed by:

$$blok\ diag(\boldsymbol{A}_1, \boldsymbol{A}_2, \ldots, \boldsymbol{A}_n) = \begin{bmatrix} \boldsymbol{A}_1 & 0 & \cdots & 0 \\ 0 & \boldsymbol{A}_2 & \cdots & 0 \\ \vdots & \vdots & \ddots & \vdots \\ 0 & 0 & \cdots & \boldsymbol{A}_n \end{bmatrix}. \tag{A.6}$$

A.2. Vector Operations

A.2.1. *Scalar Product*

The **scalar product** of vectors $\boldsymbol{a} \in \mathbb{R}^n$ and $\boldsymbol{b} \in \mathbb{R}^n$, denoted $\boldsymbol{a}^T\boldsymbol{b}$, is a real number defined by

$$\boldsymbol{a}^T\boldsymbol{b} = \sum_{i=1}^{n} a_i b_i. \tag{A.7}$$

The scalar product of vectors is commutative; i.e.,

$$\boldsymbol{a}^T\boldsymbol{b} = \boldsymbol{b}^T\boldsymbol{a}. \tag{A.8}$$

A.2.2. *Norm*

A norm on vector space is a scalar function with the following properties:

(i) $\|\boldsymbol{a}\| \geq 0$ and $\|\boldsymbol{a}\| = 0$ if and only if $\boldsymbol{a} = 0$.
(ii) $\|\alpha\boldsymbol{a}\| = |\alpha|\|\boldsymbol{a}\|$ for any scalar α.
(iii) $\|\boldsymbol{a} + \boldsymbol{b}\| \leq \|\boldsymbol{a}\| + \|\boldsymbol{b}\|$ for any vectors $\boldsymbol{a}, \boldsymbol{b}$.

The Euclidian **norm** of a vector $\boldsymbol{a} \in \mathbb{R}^n$ is defined as

$$\|\boldsymbol{a}\| = (\boldsymbol{a}^T\boldsymbol{a})^{\frac{1}{2}} = \left(\sum_{i=1}^{n} a_i^2\right)^{\frac{1}{2}}, \tag{A.9}$$

which is called the l^2-norm.

For a vector in $\mathbb{R}^2$ or $\mathbb{R}^3$, the scalar product can be represented as,

$$\boldsymbol{a}^T\boldsymbol{b} = \|\boldsymbol{a}\|\|\boldsymbol{b}\|\cos\theta, \tag{A.10}$$

where θ is the angle between the vectors $\boldsymbol{a}$ and $\boldsymbol{b}$.

A.2.3. *Unit Vector*

A **unit vector** $\tilde{\boldsymbol{a}}$ is a vector whose norm is unity, i.e., $\tilde{\boldsymbol{a}}^T\tilde{\boldsymbol{a}} = 1$. Given a vector $\boldsymbol{a}$, its unit vector is obtained by

$$\tilde{\boldsymbol{a}} = \frac{\boldsymbol{a}}{\|\boldsymbol{a}\|}. \tag{A.11}$$

The vector $\tilde{\boldsymbol{a}}$ is said to be the normalized vector of $\boldsymbol{a}$.

A.2.4. *Outer Product*

The **outer product** of vectors $\boldsymbol{a} \in \mathbb{R}^n$ and $\boldsymbol{b} \in \mathbb{R}^n$ is a matrix defined by

$$\boldsymbol{a}\boldsymbol{b}^T = \begin{bmatrix} a_1b_1 & \cdots & a_1b_n \\ \vdots & \vdots & \vdots \\ a_nb_1 & \cdots & a_nb_n \end{bmatrix} \in \mathbb{R}^{n\times n}. \tag{A.12}$$

From Eq. (A.7) we can see that the scalar product and the outer product are related by

$$\boldsymbol{a}^T\boldsymbol{b} = tr(\boldsymbol{a}\boldsymbol{b}^T), \tag{A.13}$$

where the function $tr(\cdot)$ denotes the **trace** of a matrix, which is the sum of the diagonal element of the matrix.

A.2.5. *Cross Product*

A vector $\boldsymbol{a} \in \mathbb{R}^3$ with respect to a right-hand coordinate system $\Sigma = \{\boldsymbol{x}, \boldsymbol{y}, \boldsymbol{z}\}$, where $\boldsymbol{x}, \boldsymbol{y}, \boldsymbol{z}$ are unit vectors in $\mathbb{R}^3$, is described as

$$\begin{aligned}\boldsymbol{a} &= \begin{bmatrix} a_x \\ a_y \\ a_z \end{bmatrix} \\ &= a_x\boldsymbol{x} + a_y\boldsymbol{y} + a_z\boldsymbol{z}.\end{aligned} \tag{A.14}$$

The **vector product** or **cross product** $\boldsymbol{a} \times \boldsymbol{b}$ of two vectors $\boldsymbol{a}$ and $\boldsymbol{b}$ belonging to $\mathbb{R}^3$ is a vector defined by

$$\begin{aligned}\boldsymbol{a} \times \boldsymbol{b} &= \det \begin{bmatrix} \boldsymbol{x} & \boldsymbol{y} & \boldsymbol{z} \\ a_x & a_y & a_z \\ b_x & b_y & b_z \end{bmatrix} \\ &= (a_y b_z - a_z b_y)\boldsymbol{x} + (a_z b_x - a_x b_z)\mathbf{y} + (a_x b_y - a_x b_y)\boldsymbol{z}.\end{aligned} \tag{A.15}$$

This cross product can be expressed by

$$\boldsymbol{a} \times \boldsymbol{b} = \|\boldsymbol{a}\|\|\boldsymbol{b}\| \sin\theta \boldsymbol{e}, \tag{A.16}$$

where θ is the angle between $\boldsymbol{a}$ and $\boldsymbol{b}$, and $\boldsymbol{e}$ is the unit vector that is orthogonal to $\boldsymbol{a}$ and $\boldsymbol{b}$. These are defined by the right-hand rule, as shown in Fig. A.1. The cross product can also be expressed by a multiplication of matrix and vector as

$$\boldsymbol{a} \times \boldsymbol{b} = \boldsymbol{S}(\boldsymbol{a})\boldsymbol{b}, \tag{A.17}$$

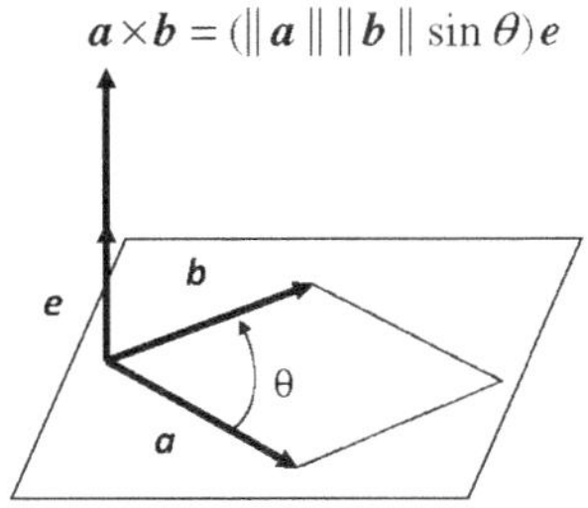

Fig. A.1. Cross product of vectors.

where $\boldsymbol{S}(\boldsymbol{a})$ is a skew-symmetric matrix expressed as

$$\boldsymbol{S}(\boldsymbol{a}) = \begin{bmatrix} 0 & -a_z & a_y \\ a_z & 0 & -a_x \\ -a_y & a_x & 0 \end{bmatrix}. \tag{A.18}$$

The following properties hold:

$$\begin{aligned} &\boldsymbol{a} \times \boldsymbol{b} = -\boldsymbol{b} \times \boldsymbol{a} \\ &\boldsymbol{S}^T(\boldsymbol{a}) = -\boldsymbol{S}(\boldsymbol{a}) \\ &\boldsymbol{S}(\boldsymbol{a})\boldsymbol{a} = \boldsymbol{S}^T(\boldsymbol{a})\boldsymbol{a} = 0 \\ &\boldsymbol{S}(\alpha\boldsymbol{a} + \beta\boldsymbol{b}) = \alpha\boldsymbol{S}(\boldsymbol{a}) + \beta\mathbf{S}(\boldsymbol{b}) \end{aligned} \tag{A.19}$$

We often describe $\boldsymbol{S}(\boldsymbol{a})$ by $[\boldsymbol{a}\times]$.

A.2.6. *Triple Products*

Given three vectors $\boldsymbol{a}, \boldsymbol{b}, \boldsymbol{c} \in \mathbb{R}^3$, the following expressions hold for **triple products**:

(i) $$\boldsymbol{a}^T(\boldsymbol{b} \times \boldsymbol{c}) = \boldsymbol{b}(\boldsymbol{c} \times \boldsymbol{a}) = \boldsymbol{c}^T(\boldsymbol{a} \times \boldsymbol{b}) = \begin{vmatrix} a_x & a_y & a_z \\ b_x & b_y & b_z \\ c_x & c_y & c_z \end{vmatrix} \tag{A.20}$$

(ii) $$\boldsymbol{a} \times (\boldsymbol{b} \times \boldsymbol{c}) = (\boldsymbol{c}^T\boldsymbol{a})\boldsymbol{b} - (\boldsymbol{a}^T\boldsymbol{b})\boldsymbol{c} = (\boldsymbol{c}^T\boldsymbol{a}\boldsymbol{E}_{3\times 3} - \boldsymbol{c}\boldsymbol{a}^T)\boldsymbol{b} \tag{A.21}$$

(iii) $$\boldsymbol{a} \times (\boldsymbol{b} \times \boldsymbol{c}) + \boldsymbol{b} \times (\boldsymbol{c} \times \boldsymbol{a}) + \boldsymbol{c} \times (\boldsymbol{a} \times \boldsymbol{b}) \tag{A.22}$$

A.2.7. *Differentiation of Vectors*

Suppose that the vector $\boldsymbol{a}(t) = [a_1(t), a_2(t), \ldots, a_n(t)]^T \in \mathbb{R}^n$ is a function of time. Then, the time derivative of $\boldsymbol{a}(t)$ is the vector

$$\dot{\boldsymbol{a}}(t) = [\dot{a}_1(t), \dot{a}_2(t), \ldots, \dot{a}_n(t)]^T. \tag{A.23}$$

Similar statements hold for the integration of vectors and matrices. The scalar and vector products satisfy the following product rules for differentiation similar to the product rules for differentiation of ordinary function.

(i) $$\frac{d}{dt}(\boldsymbol{a}^T\boldsymbol{b}) = \dot{\boldsymbol{a}}^T\boldsymbol{b} + \boldsymbol{a}^T\dot{\boldsymbol{b}} \tag{A.24}$$

(ii) $$\frac{d}{dt}(\boldsymbol{a}\times\boldsymbol{b}) = \dot{\boldsymbol{a}}\times\boldsymbol{b} + \boldsymbol{a}\times\dot{\boldsymbol{b}} \tag{A.25}$$

A.2.8. *Linear Independence*

A set of vectors $\{\boldsymbol{a}_1, \boldsymbol{a}_2, \ldots, \boldsymbol{a}_n\}$ is said to be **linearly independent** if and only if

$$\sum_{i=1}^{n} \alpha_i \boldsymbol{a}_i = 0 \quad \text{implies } \alpha_i = 0 \quad \text{for all } i. \tag{A.26}$$

The **basis** of a vector space X is a linearly independent set of vectors $\{\boldsymbol{e}_1, \boldsymbol{e}_2, \ldots, \boldsymbol{e}_n\}$ such that every vector $\boldsymbol{a} \in X$ can be written as a linear combination:

$$\boldsymbol{a} = a_1\boldsymbol{e}_1 + \cdots a_n\boldsymbol{e}_n. \tag{A.27}$$

The dimension of the vector space X is the number of basis vectors; i.e., $\dim(X) = n$.

A.2.9. *Gram–Schmidt Orthogonalization*

Given a set of vectors $\{\boldsymbol{x}_1, \boldsymbol{x}_2, \ldots, \boldsymbol{x}_n\}$ that are linearly independent in X, we consider a set of orthogonal and normalized vectors $\{\tilde{\boldsymbol{y}}_1, \tilde{\boldsymbol{y}}_2, \ldots, \tilde{\boldsymbol{y}}_n\}$, which are a linear combination of $\tilde{\mathbf{x}}_i$.

Start $\tilde{\boldsymbol{y}}_1 = \tilde{\mathbf{x}}_1 = \frac{x_1}{\|x_1\|}$.

Then, $\boldsymbol{y}_i = \boldsymbol{x}_i - \sum_{j=1}^{i-1}(\boldsymbol{x}_i^T\tilde{\boldsymbol{y}}_j)\tilde{\boldsymbol{y}}_j$, for $i = 1, 2, \ldots, n$.

Normalize $\boldsymbol{y}_i$ to produce $\tilde{\boldsymbol{y}}_i$.

A.3. Matrix Operations

A.3.1. *Rank of a Matrix*

The **rank** of a matrix $\boldsymbol{A} \in \mathbb{R}^{m\times n}$ is the largest number of linearly independent rows (or columns) of $\boldsymbol{A}$, and is denoted by $rank(\boldsymbol{A})$. It can be no greater than the minimum of n and m.

A.3.2. *Sum and Product*

Two matrices $\boldsymbol{A} = (a_{ij})$ and $\boldsymbol{B} = (b_{ij})$ of the same dimensions $m \times n$ are equal if $a_{ij} = b_{ij}$. The sum of matrices $\boldsymbol{A}$ and $\mathbf{B}$ of the same dimensions is

$$\boldsymbol{C} = \boldsymbol{A} + \boldsymbol{B}, \tag{A.28}$$

whose elements are given by $c_{ij} = a_{ij} + b_{ij}$.

The product of a scalar α by a matrix $\boldsymbol{A}$ is the matrix $\alpha\boldsymbol{A} = (\alpha a_{uj})$.

The product of $\boldsymbol{A} \in \mathbb{R}^{m\times l}$ by $\boldsymbol{B} \in \mathbb{R}^{l\times n}$ is the matrix $\boldsymbol{C} = \boldsymbol{AB} \in \mathbb{R}^{m\times n}$, whose elements are given by

$$c_{ij} = \sum_{k=1}^{l} a_{ik} b_{kj}. \tag{A.29}$$

It should be noted that in general, $\boldsymbol{AB} \neq \boldsymbol{BA}$.

A.3.3. *Determinant*

For a square matrix $\boldsymbol{A} \in \mathbb{R}^{n\times n}$, the **determinant** of $\boldsymbol{A}$ is the scalar given by

$$\det(\boldsymbol{A}) = \sum_{j=1}^{n} a_{ij}(-1)^{i+j} \det(\boldsymbol{A}_{(ij)}), \tag{A.30}$$

where $\boldsymbol{A}_{(ij)}$ is a matrix of dimensions $(n-1) \times (n-1)$ that is obtained by eliminating the i-th row and j-th column of $\boldsymbol{A}$. The determinant has the following properties:

(i) $$\det(\boldsymbol{A}) = \det(\boldsymbol{A}^T). \tag{A.31}$$

(ii) $$\det(\boldsymbol{AB}) = \det(\boldsymbol{A}) \det(\boldsymbol{B}). \tag{A.32}$$

A.3.4. *Inverse*

For a square matrix $\boldsymbol{A} \in \mathbb{R}^{n\times n}$ with $rank(\boldsymbol{A}) = n$, the **inverse** of the matrix is denoted by $\boldsymbol{A}^{-1} \in \mathbb{R}^{n\times n}$ and satisfies

$$\boldsymbol{AA}^{-1} = \boldsymbol{A}^{-1}\boldsymbol{A} = \boldsymbol{E}_{n\times n}. \tag{A.33}$$

The following expressions hold:

(i) $$(\boldsymbol{A}^{-1})^{-1} = \boldsymbol{E}_{n\times n}.$$

(ii) $$(\boldsymbol{A}^T)^{-1} = (\boldsymbol{A}^{-1})^T.$$

(iii) $$(\boldsymbol{AB})^{-1} = \boldsymbol{B}^{-1}\boldsymbol{A}^{-1}.$$

A.3.5. *Differentiation of Matrices*

The derivative of a matrix $\boldsymbol{A}(t) = (a_{ij}(t))$, whose elements $a_{ij}(t)$ are differentiable functions, is the matrix $\dot{\boldsymbol{A}}(t) = (\dot{a}_{ij}(t))$. If a square matrix $\boldsymbol{A}(t) \in \mathbb{R}^{n\times n}$ is such that $rank(\boldsymbol{A}) = n$ for any t, and its elements $a_{ij}(t)$are differentiable functions, then the derivative of the inverse of $\boldsymbol{A}(t)$ is given by

$$\frac{d}{dt}\boldsymbol{A}^{-1}(t) = -\boldsymbol{A}^{-1}(t)\dot{\boldsymbol{A}}(t)\boldsymbol{A}^{-1}(t). \tag{A.34}$$

A.3.6. *Gradient*

Given a scalar function $f(\boldsymbol{x})$, endowed with the partial derivatives with respect to the element x_i of the vector $\boldsymbol{x} \in \mathbb{R}^n$, the **gradient** of the function $f(\boldsymbol{x})$ with respect to vector $\boldsymbol{x}$ is the row vector expressed as

$$\frac{\partial f(\boldsymbol{x})}{\partial \boldsymbol{x}} = \left[\frac{\partial f(\boldsymbol{x})}{\partial x_1} \cdots \frac{\partial f(\boldsymbol{x})}{\partial x_n}\right] \in \mathbb{R}^{1\times n}. \tag{A.35}$$

Further, if $\boldsymbol{x}(t)$ is a differentiable function with respect to t, then

$$\dot{f}(\boldsymbol{x}(t)) = \frac{df(\boldsymbol{x}(t))}{dt} = \frac{\partial f}{\partial \boldsymbol{x}}\dot{\boldsymbol{x}}. \tag{A.36}$$

A.3.7. *Jacobian*

Given a vector function $\boldsymbol{f}(\boldsymbol{x}) \in \mathbb{R}^m$, whose elements f_i are differentiable with respect to the vector $\boldsymbol{x} \in \mathbb{R}^n$, the **Jacobian** matrix of the function $\boldsymbol{f}(\boldsymbol{x})$ is defined as

$$\boldsymbol{J_f}(\boldsymbol{x}) = \frac{\partial \boldsymbol{f}}{\partial \boldsymbol{x}} = \begin{bmatrix} \frac{\partial f_1}{\partial x_1} & \cdots & \frac{\partial f_1}{\partial x_n} \\ \cdots & \cdots & \cdots \\ \frac{\partial f_m}{\partial x_1} & \cdots & \frac{\partial f_m}{\partial x_n} \end{bmatrix} \in \mathbb{R}^{m \times n}. \quad (A.37)$$

If $\boldsymbol{x}(t)$ is a differentiable function with respect to t, then

$$\dot{\boldsymbol{f}}(\boldsymbol{x}) = \frac{\partial \boldsymbol{f}}{\partial \boldsymbol{x}}\dot{\boldsymbol{x}} = \boldsymbol{J_f}(\boldsymbol{x})\dot{\boldsymbol{x}} \in \mathbb{R}^m. \quad (A.38)$$

A.3.8. *Linear Transformation*

Consider a vector space X of dimension n and a vector space Y of dimension m with $m \leq n$. The **linear transformation** (or **linear map**) from a vector $\boldsymbol{x} \in X$ to a vector $\boldsymbol{y} \in Y$ can be defined as

$$\boldsymbol{y} = \boldsymbol{A}\boldsymbol{x}, \quad (A.39)$$

where $\boldsymbol{A}$ is an $m \times n$ matrix. The vector $\boldsymbol{y}$ is called the **image** (or **range**) of $\boldsymbol{x}$ under the transformation $\boldsymbol{A}$. The image space of the transformation is the subspace

$$\mathrm{Im}(\boldsymbol{A}) = \{\boldsymbol{y} : \boldsymbol{y} = \boldsymbol{A}\boldsymbol{x}, \boldsymbol{x} \in X\} \subseteq Y, \quad (A.40)$$

which is the subspace generated by the linearly independent columns of $\boldsymbol{A}$ taken as a basis of Y.

On the other hand, the **kernel** (or **null) space** of the transformation is the subspace

$$Ker(\boldsymbol{A}) = \{\boldsymbol{x} : \boldsymbol{A}\boldsymbol{x} = 0, \boldsymbol{x} \in X\} \subseteq X. \quad (A.41)$$

For a given matrix $\boldsymbol{A}$, the following property holds:

$$\dim(\mathrm{Im}(\boldsymbol{A})) + \dim(Ker(\boldsymbol{A})) = n. \quad (A.42)$$

If $\boldsymbol{x} \in Ker(\boldsymbol{A})$ and $\boldsymbol{y} \in \mathrm{Im}(\boldsymbol{A}^T)$, then $\boldsymbol{y}^T\boldsymbol{x} = 0$; i.e., the vectors in the kernel space of $\boldsymbol{A}$ are orthogonal to each vector in the image space of the transpose of $\boldsymbol{A}$.

A.3.9. *Eigenvalue and Eigenvectors*

Consider a linear transformation of $\boldsymbol{x} \in \mathbb{R}^n$ by a square matrix $\boldsymbol{A} \in \mathbb{R}^{n\times n}$. If the imaged vector is in the same direction of $\boldsymbol{x}$, then

$$\boldsymbol{A}\boldsymbol{x} = \lambda \boldsymbol{x}, \tag{A.43}$$

where λ is a scalar. To have a solution different from the trivial one $\boldsymbol{x} = 0$, it must be

$$\det(\lambda \boldsymbol{E}_{n\times n} - \boldsymbol{A}) = 0, \tag{A.44}$$

which is a polynomial in λ called the **characteristic polynomial** of $\boldsymbol{A}$, and can be written in the following general form:

$$a_n\lambda^n + a_{n-1}\lambda^{n-1} + \cdots a_1\lambda + a_0 = 0, \tag{A.45}$$

where a_i are the coefficients of the polynomial. Its solutions $\lambda_1, \lambda_2, \ldots, \lambda_n$ are **eigenvalues** of matrix $\boldsymbol{A}$. On the assumption of distinct eigenvalues, the n vectors $\boldsymbol{u}_i$ satisfying $\boldsymbol{A}\boldsymbol{u}_i = \lambda_i \boldsymbol{u}_i$ are said to be **eigenvectors** associated with the eigenvalues λ_i. It should be noted that when $\boldsymbol{u}_i$ is an eigenvector and α is an arbitrary scalar, then $\alpha\boldsymbol{u}_i$ is another expression of the eigenvector.

The matrix $\boldsymbol{U} \in \mathbb{R}^{n\times n}$ formed by

$$\boldsymbol{U} = [\boldsymbol{u}_1 \boldsymbol{u}_2 \cdots \boldsymbol{u}_m] \tag{A.46}$$

is invertible and constitutes a basis in the space of dimension n. Furthermore, the following holds:

$$diag(\lambda_1, \lambda_2, \ldots, \lambda_n) = \boldsymbol{U}^{-1}\boldsymbol{A}\boldsymbol{U}. \tag{A.47}$$

It follows that $\det(\boldsymbol{A}) = \prod_{i=1}^{n} \lambda_i$.

If the matrix $\boldsymbol{A} \in \mathbb{R}^{n\times n}$ is symmetric, its eigenvalues are real, then Eq. (A.47) can be written as

$$diag(\lambda_1, \lambda_2, \ldots, \lambda_n) = \boldsymbol{U}^T\boldsymbol{A}\boldsymbol{U}. \tag{A.48}$$

Hence, the matrix $\boldsymbol{U}$ is orthogonal.

A.3.10. *Singular Value Decomposition*

For a non-square matrix, it is not possible to define eigenvalues. An extension of the eigenvalue concept can be obtained by singular values, however. Given a matrix $\boldsymbol{A} \in \mathbb{R}^{n \times n}$, the matrix $\boldsymbol{A}^T\boldsymbol{A}$ has non-negative eigenvalues $\lambda_1 \geq \lambda_2 \geq \cdots \geq \lambda_n \geq 0$, which can be expressed in the form

$$\lambda_i = \sigma_i^2, \sigma_i \geq 0. \tag{A.49}$$

The scalars $\sigma_1 \geq \sigma_2 \geq \cdots \geq \sigma_n \geq 0$ are said to be the **singular values** of matrix $\boldsymbol{A}$. The **singular value decomposition** of matrix $\boldsymbol{A}$ is given by

$$\boldsymbol{A} = \boldsymbol{U}\boldsymbol{\Sigma}\boldsymbol{V}^T, \tag{A.50}$$

where $\boldsymbol{U} \in \mathbb{R}^{m \times m}$, $\boldsymbol{V} \in \mathbb{R}^{n \times n}$ are both orthogonal matrices, and $\boldsymbol{\Sigma} \in \mathbb{R}^{m \times n}$ is

$$\boldsymbol{\Sigma} = \begin{bmatrix} \boldsymbol{S} & 0 \\ 0 & 0 \end{bmatrix}^T, \boldsymbol{S} = diag(\sigma_1, \sigma_2, \ldots, \sigma_r), \tag{A.51}$$

where $\sigma_1 \geq \sigma_2 \geq \cdots \geq \sigma_r > 0$. The number of non-zero singular values is equal to the rank of matrix $\boldsymbol{A}$.

From a relation of $\boldsymbol{A}^T\boldsymbol{A} = \mathbf{V}\boldsymbol{\Sigma}^T\boldsymbol{\Sigma}\mathbf{V}^T$, the column vectors of $\boldsymbol{V}$ are eigenvectors of matrix $\boldsymbol{A}^T\boldsymbol{A}$, whereas from $\boldsymbol{A}\boldsymbol{A}^T = \boldsymbol{U}\boldsymbol{\Sigma}\boldsymbol{\Sigma}^T\boldsymbol{U}^T$ the column vectors of $\boldsymbol{U}$ are eigenvectors of matrix $\boldsymbol{A}\boldsymbol{A}^T$.

The **condition number** of matrix $\boldsymbol{A}$ is defined by

$$cond\boldsymbol{A} = \frac{\sigma_1}{\sigma_n}. \tag{A.52}$$

This provides ill conditioning for a numerical solution of the system established by Eq. (A.39).

A.3.11. *Quadratic Forms*

A **quadratic form** of a square matrix $\boldsymbol{A} \in \mathbb{R}^{n \times n}$ is the scalar given by

$$\boldsymbol{x}^T\boldsymbol{A}\boldsymbol{x} = \sum_{i=1}^{n}\sum_{j=1}^{n} a_{ij}x_ix_j. \tag{A.53}$$

The quadratic form is said to be **positive definite** if

$$\boldsymbol{x}^T\boldsymbol{A}\boldsymbol{x} > 0 \quad \text{for } \forall\, \boldsymbol{x} \neq 0, \quad \boldsymbol{x}^T\boldsymbol{A}\boldsymbol{x} = 0 \quad \text{for } \boldsymbol{x} = 0. \tag{A.54}$$

For any matrix $\boldsymbol{A} \in \mathbb{R}^{n\times n}$, it can be decomposed as

$$\boldsymbol{A} = \boldsymbol{A}_{\text{sym}} + \boldsymbol{A}_{\text{sk}}, \tag{A.55}$$

where $\boldsymbol{A}_{\text{sym}} \in \mathbb{R}^{n\times n}$ is a symmetric matrix, $\boldsymbol{A}_{\text{sk}} \in \mathbb{R}^{n\times n}$ is a skew-symmetric matrix, and these are expressed by

$$\boldsymbol{A}_{\text{sym}} = \frac{\boldsymbol{A} + \boldsymbol{A}^T}{2}, \boldsymbol{A}_{\text{sk}} = \frac{\boldsymbol{A} - \boldsymbol{A}^T}{2}. \tag{A.56}$$

Since $\boldsymbol{x}^T \boldsymbol{A}_{\text{sk}} \boldsymbol{x} = 0$ for arbitrary $\boldsymbol{x}$, this leads to

$$\boldsymbol{x}^T \boldsymbol{A} \boldsymbol{x} = \boldsymbol{x}^T \boldsymbol{A}_{\text{sym}} \boldsymbol{x}. \tag{A.57}$$

The symmetric square matrix $\boldsymbol{A}_{\text{sym}}$ is transformed into a diagonal matrix as shown in Eq. (A.48). Hence, a necessary and sufficient condition for $\boldsymbol{A}_{\text{sym}}$ to be positive definite is that all eigenvalues of $\boldsymbol{A}_{\text{sym}}$ be positive.

The quadratic form is said to be **positive semi-definite** if

$$\boldsymbol{x}^T \boldsymbol{A} \boldsymbol{x} \geq 0 \quad \text{for } \forall\, \boldsymbol{x}. \tag{A.58}$$

The matrix $\boldsymbol{A}$ is also said to be positive definite or positive semi-definite when its quadratic form is positive definite or positive semi-definite.

A.3.12. *Pseudo-Inverse*

From Eq. (A.50) the **pseudo-inverse** of a $m \times n$ matrix $\boldsymbol{A}$ is the $n \times m$ matrix defined as

$$\boldsymbol{A}^+ = \boldsymbol{V} \begin{bmatrix} \boldsymbol{S}^{-1} & 0 \\ 0 & 0 \end{bmatrix}^T \boldsymbol{U}^T. \tag{A.59}$$

Note that $\boldsymbol{A}^+$ has the same dimensions as the transpose of $\boldsymbol{A}$. This matrix has many useful properties:

(i) $\boldsymbol{A}\boldsymbol{A}^+\boldsymbol{A} = \boldsymbol{A}$, $\boldsymbol{A}^+\boldsymbol{A}\boldsymbol{A}^+ = \boldsymbol{A}^+$, $(\mathbf{AA}^+)^T = \boldsymbol{A}\boldsymbol{A}^+$, $(\boldsymbol{A}^+\boldsymbol{A})^T = \boldsymbol{A}^+\boldsymbol{A}$.

(ii) If $\boldsymbol{A}$ is full column rank, meaning $rank(\boldsymbol{A}) = n < m$ (i.e., $\boldsymbol{A}^T\boldsymbol{A}$ is not singular), then $\boldsymbol{A}^+$ is a left inverse of $\boldsymbol{A}$ in the sense that $\boldsymbol{A}^+\boldsymbol{A} = \boldsymbol{E}_{n\times n}$. The closed-form is expressed by

$$\boldsymbol{A}^+ = (\boldsymbol{A}^T\boldsymbol{A})^T \boldsymbol{A}^T. \tag{A.60}$$

(iii) If $\boldsymbol{A}$ is full row rank, meaning $rank(\boldsymbol{A}) = m < n$ (i.e., $\boldsymbol{A}\boldsymbol{A}^T$ is not singular), then $\boldsymbol{A}^+$ is a right inverse of A in the sense that $\boldsymbol{A}\boldsymbol{A}^+ = \boldsymbol{E}_{n\times n}$. The closed-form is expressed by

$$\boldsymbol{A}^+ = \boldsymbol{A}^T(\boldsymbol{A}\boldsymbol{A}^T)^T. \tag{A.61}$$

(iv) If $\boldsymbol{A}$ is square, invertible, then its inverse is $\boldsymbol{A}^+ = \boldsymbol{A}^{-1}$.

(v) Consider a linear equation given by Eq. (A.39). If $\boldsymbol{A}$ is full row rank of $rank(\boldsymbol{A}) = m < n$, then the solution of Eq. (A.39) is not unique; it can be expressed by

$$\boldsymbol{x} = \boldsymbol{A}^+\boldsymbol{y} + (\boldsymbol{E}_{n\times n} - \boldsymbol{A}^+\boldsymbol{A})\boldsymbol{k}, \tag{A.62}$$

where $\boldsymbol{k}$ is an arbitrary n vector and $\boldsymbol{A}^+$ is given by Eq. (A.61). If $\boldsymbol{A}$ is full column rank of $rank(\boldsymbol{A}) = n < m$, the equation has no solution. A solution minimizing $\|\boldsymbol{A}\boldsymbol{x} - \boldsymbol{y}\|$ is given by

$$\boldsymbol{x} = \boldsymbol{A}^+\boldsymbol{y}, \tag{A.63}$$

where $\boldsymbol{A}^+$ is given by Eq. (A.60).

Solutions

Chapter 1

1.1. The German psychologist Weinstein (1968) conducted experiments on the tactile senses. One of his studies examined if subjects could recognize when they were being touched by one or two points of an object as the legs of a drafting compass closed. The least distance was called the two-point threshold, which is shown in Fig. S.1. The tip of tongue had the smallest threshold (1 mm). On the surface of the hand, two separated points are noticeable at as little as 2 mm on the all fingertips except for little finger, 4 mm on the fingertip of little and index finger, and 8–11 mm on the center of palm.

1.2. Pioneering neuroscientist Wilder Penfield (1950) painstakingly mapped out how the body is represented both in the primary motor cortex and the somatosensory cortex by opening up the skulls of 126 patients already undergoing surgery for another procedure (such as the treatment of intractable epilepsy) and electrically stimulating different parts of these regions with a probe. He recorded that simulating the primary motor cortex elicited involuntary movements — hence the name motor cortex: this part of the brain is crucial for controlling movement. About a quarter of the motor cortex in the human brain (the part of the brain that controls all movements in the body) is devoted to the muscles of the hands. This is usually illustrated with a drawing of a human figure draped over the side

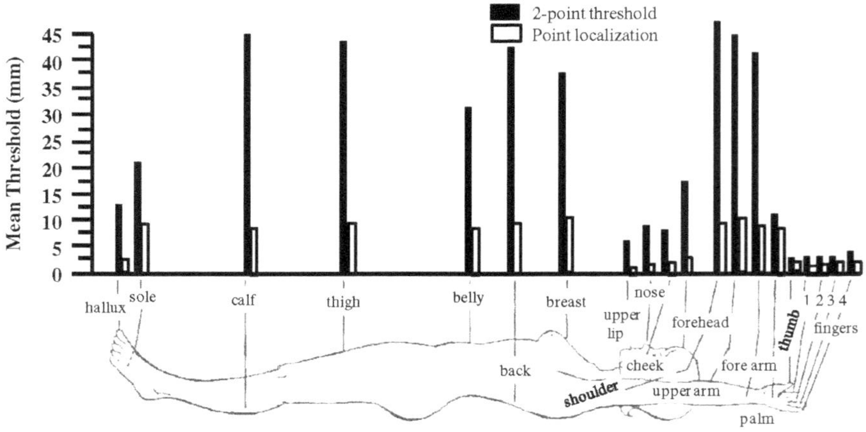

Fig. S.1. Two-point touch and point localization thresholds (Adapted from Lederman and Klatsky, 2009).

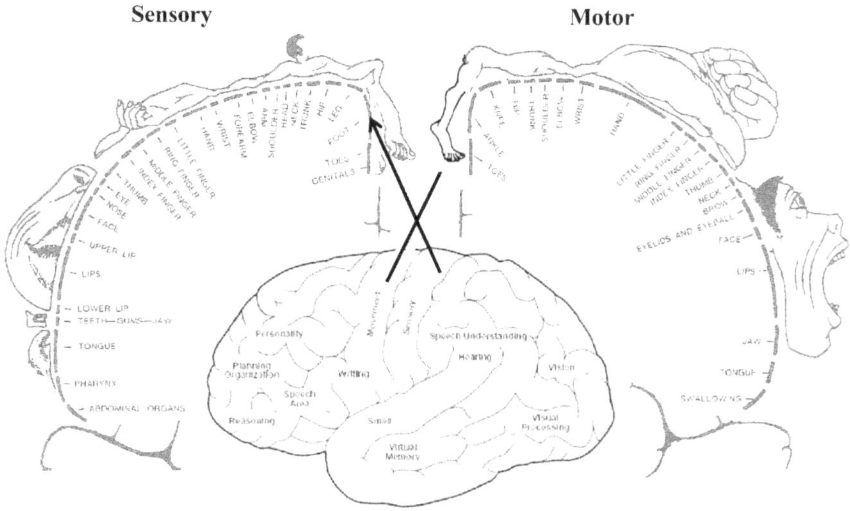

Fig. S.2. The cortical homunculus (Modified from Penfield and Rasmussen, 1950).

of the brain with body parts sized proportionally to the amount of brain devoted to their movement, referred to as a homunculus. See Fig. S.2, which was modified from the original in Dr. Penfield's monograph 'The Cerebral Cortex of Man'.

Chapter 2

2.1. ${}^{c}\boldsymbol{F}_c = {}^{C}\boldsymbol{H}_C \boldsymbol{f}_{FC}$ and $\boldsymbol{f}_{FC} \in FC$ are represented as follows:

(1) Frictionless line contact model

$$
{}^{C}\boldsymbol{H}_C = \begin{bmatrix} 0 & 0 \\ 0 & 0 \\ 1 & 0 \\ 0 & 0 \\ 0 & 1 \\ 0 & 0 \end{bmatrix}, \quad FC = \left\{ \boldsymbol{f} = \begin{bmatrix} f_1 \\ f_2 \end{bmatrix} \in \mathbb{R}^2 : f_1 \geq 0 \right\}.
$$

(2) Line contact model with friction

$$
{}^{C}\boldsymbol{H}_C = \begin{bmatrix} 1 & 0 & 0 & 0 & 0 \\ 0 & 1 & 0 & 0 & 0 \\ 0 & 0 & 1 & 0 & 0 \\ 0 & 0 & 0 & 0 & 0 \\ 0 & 0 & 0 & 1 & 0 \\ 0 & 0 & 0 & 0 & 1 \end{bmatrix},
$$

$$
FC = \left\{ \boldsymbol{f} = \begin{bmatrix} f_1 \\ f_2 \\ f_3 \\ f_4 \\ f_5 \end{bmatrix} \in \mathbb{R}^5, \sqrt{f_1^2 + f_2^2} \leq \mu f_3, f_3 \geq 0, |f_5| \leq \gamma f_3 \right\}.
$$

(3) Frictionless plane contact model

$$
{}^{C}\boldsymbol{H}_C = \begin{bmatrix} 0 & 0 & 0 \\ 0 & 0 & 0 \\ 1 & 0 & 0 \\ 0 & 1 & 0 \\ 0 & 0 & 1 \\ 0 & 0 & 0 \end{bmatrix}, \quad FC = \left\{ \boldsymbol{f} = \begin{bmatrix} f_1 \\ f_2 \\ f_3 \end{bmatrix} \in \mathbb{R}^3 : f_1 \geq 0 \right\}.
$$

(4) Plane contact model with friction

$$
{}^{C}\boldsymbol{H}_C = \begin{bmatrix} 1 & 0 & 0 & 0 & 0 & 0 \\ 0 & 1 & 0 & 0 & 0 & 0 \\ 0 & 0 & 1 & 0 & 0 & 0 \\ 0 & 0 & 0 & 1 & 0 & 0 \\ 0 & 0 & 0 & 0 & 1 & 0 \\ 0 & 0 & 0 & 0 & 0 & 1 \end{bmatrix},
$$

$$
FC = \left\{ \boldsymbol{f} = \begin{bmatrix} f_1 \\ f_2 \\ f_3 \\ f_4 \\ f_5 \\ f_6 \end{bmatrix} \in \mathbb{R}^6 : \sqrt{f_1^2 + f_2^2} \le \mu f_3, f_3 \ge 0, |f_6| \le \gamma f_3 \right\}.
$$

2.2. The rotation matrix $\boldsymbol{R}_{C_1}$ and the position from the origin of the object coordinate system to the contact point C_i, $\boldsymbol{p}_{OC_1}$, are represented in Exercise 2.2. Since the contact point C_1 is the soft finger contact, and the contact point C_2 is the frictionless point contact, the wrench base at each contact point ${}^{C_i}\boldsymbol{H}_{C_i} (i = 1, 2)$ is obtained using the relation of Eq. (2.6) and Eq. (2.5) and Eq. (2.1), respectively, by

$$
{}^{C_1}\boldsymbol{H}_{C_1} = \begin{bmatrix} 1 & 0 & 0 & 0 \\ 0 & 1 & 0 & 0 \\ 0 & 0 & 1 & 0 \\ 0 & 0 & 0 & 0 \\ 0 & 0 & 0 & 0 \\ 0 & 0 & 0 & 1 \end{bmatrix}, \quad {}^{C_2}\boldsymbol{H}_{C_2} = \begin{bmatrix} 0 \\ 0 \\ 1 \\ 0 \\ 0 \\ 0 \end{bmatrix}. \tag{1}
$$

The friction cone is represented by

$$
\begin{aligned}
FC_{C_1} &= \{\boldsymbol{f}_{FC_1} \in \mathbb{R}^4 : \sqrt{f_1^2 + f_2^2} \le \mu f_3, f_3 \ge 0, |f_4| \le \gamma f_3\}, \\
FC_2 &= \{\boldsymbol{f}_{FC_2} \in \mathbb{R}^1 : f_1' \ge 0\}.
\end{aligned} \tag{2}
$$

From Eq. (2.15), the matrix $\boldsymbol{W}_{OC_i}$ $(i = 1, 2)$ are given by

$$\boldsymbol{W}_{OC_1} = \begin{bmatrix} 1 & 0 & 0 & 0 & 0 & 0 \\ 0 & 1 & 0 & 0 & 0 & 0 \\ 0 & 0 & 1 & 0 & 0 & 0 \\ 0 & 0 & -a & 1 & 0 & 0 \\ 0 & 0 & 0 & 0 & 1 & 0 \\ a & 0 & 0 & 0 & 0 & 1 \end{bmatrix},$$

$$\boldsymbol{W}_{OC_2} = \begin{bmatrix} 1 & 0 & 0 & 0 & 0 & 0 \\ 0 & 1 & 0 & 0 & 0 & 0 \\ 0 & 0 & 1 & 0 & 0 & 0 \\ 0 & 0 & a & 1 & 0 & 0 \\ 0 & 0 & 0 & 0 & 1 & 0 \\ -a & 0 & 0 & 0 & 0 & 1 \end{bmatrix}. \tag{3}$$

From the relation of $\boldsymbol{F}_{C_1} = \Re(\boldsymbol{R}_{C_1})^{C_1}\mathbf{H}_{C_1}\boldsymbol{f}_{FC_i}$, the wrench referred to the reference coordinate system is represented as

$$\boldsymbol{F}_{C_1} = \begin{bmatrix} 1 & 0 & 0 & 0 \\ 0 & 0 & 1 & 0 \\ 0 & -1 & 0 & 0 \\ 0 & 0 & 0 & 0 \\ 0 & 0 & 0 & 1 \\ 0 & 0 & 0 & 0 \end{bmatrix} \boldsymbol{f}_{FC_1}, \quad \boldsymbol{F}_{C_2} = \begin{bmatrix} 0 \\ -1 \\ 0 \\ 0 \\ 0 \\ 0 \end{bmatrix} \boldsymbol{f}_{FC_2}. \tag{4}$$

From the relation of $\boldsymbol{G}_C = \boldsymbol{W}_{OC}\boldsymbol{H}_C = [\boldsymbol{W}_{C_1}\boldsymbol{H}_{C_1} \quad \boldsymbol{W}_{C_2}\boldsymbol{H}_{C_2}]$, the grasp map is represented as

$$\boldsymbol{G}_C = \begin{bmatrix} 1 & 0 & 0 & 0 & 0 \\ 0 & 0 & 1 & 0 & -1 \\ 0 & -1 & 0 & 0 & 0 \\ 0 & a & 0 & 0 & 0 \\ 0 & 0 & 0 & 1 & 0 \\ a & 0 & 0 & 0 & 0 \end{bmatrix}. \tag{5}$$

Here we set $\boldsymbol{f}_{FC} = [\boldsymbol{f}_{FC_1}^T, \boldsymbol{f}_{FC_1}^T]^T = [f_1, f_2, f_3, f_4, f_1']^T$; the total of the wrenches acting on the origin of the object coordinate system from fingertips is then represented as

$$\boldsymbol{F}_O = \mathbf{G}_C \mathbf{f}_{FC} = [f_1, f_3 - f_1', -f_2, af_2, f_4, af_1]^T. \tag{6}$$

2.3. Rotation matrix from the object coordinate system to the contact coordinate system, $\boldsymbol{R}_{C_i}(i = 1, 2)$, and the position vector of contact points $\boldsymbol{p}_{OC_i}(i = 1, 2)$ are represented as

$$\boldsymbol{R}_{C_1} = \begin{bmatrix} 1 & 0 & 0 \\ 0 & 0 & 1 \\ 0 & -1 & 0 \end{bmatrix}, \quad \boldsymbol{R}_{C_2} = \begin{bmatrix} 0 & -1 & 0 \\ 0 & 0 & -1 \\ 1 & 0 & 0 \end{bmatrix},$$

$$\boldsymbol{p}_{OC_1} = \begin{bmatrix} 0 \\ -a \\ 0 \end{bmatrix}, \quad \boldsymbol{p}_{OC_2} = \begin{bmatrix} 0 \\ a \\ 0 \end{bmatrix}. \tag{1}$$

From $\boldsymbol{H}_C = \Re(\boldsymbol{R}_C)^C \boldsymbol{H}_C$, the wrench bases are represented as

$$\boldsymbol{H}_{C_1} = \begin{bmatrix} 1 & 0 & 0 & 0 \\ 0 & 0 & 1 & 0 \\ 0 & -1 & 0 & 0 \\ 0 & 0 & 0 & 0 \\ 0 & 0 & 0 & 1 \\ 0 & 0 & 0 & 0 \end{bmatrix}, \quad \boldsymbol{H}_{C_2} = \begin{bmatrix} 0 & -1 & 0 & 0 \\ 0 & 0 & -1 & 0 \\ 1 & 0 & 0 & 0 \\ 0 & 0 & 0 & 0 \\ 0 & 0 & 0 & -1 \\ 0 & 0 & 0 & 0 \end{bmatrix}. \tag{2}$$

From these equations, we get

$$\boldsymbol{H}_C = \begin{bmatrix} \boldsymbol{H}_{C_1} & 0 \\ 0 & \boldsymbol{H}_{C_2} \end{bmatrix} = \begin{bmatrix} 1 & 0 & 0 & 0 & 0 & 0 & 0 & 0 \\ 0 & 0 & 1 & 0 & 0 & 0 & 0 & 0 \\ 0 & -1 & 0 & 0 & 0 & 0 & 0 & 0 \\ 0 & 0 & 0 & 0 & 0 & 0 & 0 & 0 \\ 0 & 0 & 0 & 1 & 0 & 0 & 0 & 0 \\ 0 & 0 & 0 & 0 & 0 & 0 & 0 & 0 \\ 0 & 0 & 0 & 0 & 0 & -1 & 0 & 0 \\ 0 & 0 & 0 & 0 & 0 & 0 & -1 & 0 \\ 0 & 0 & 0 & 0 & 1 & 0 & 0 & 0 \\ 0 & 0 & 0 & 0 & 0 & 0 & 0 & 0 \\ 0 & 0 & 0 & 0 & 0 & 0 & 0 & -1 \\ 0 & 0 & 0 & 0 & 0 & 0 & 0 & 0 \end{bmatrix}. \tag{3}$$

Next, using $\boldsymbol{W}_{OC_1} = \begin{bmatrix} \boldsymbol{I}_{3\times 3} & 0 \\ [\boldsymbol{P}_{OC_1}\times] & \boldsymbol{I}_{3\times 3} \end{bmatrix}$, we get

$$\boldsymbol{W}_{OC_1} = \begin{bmatrix} 1 & 0 & 0 & 0 & 0 & 0 \\ 0 & 1 & 0 & 0 & 0 & 0 \\ 0 & 0 & 1 & 0 & 0 & 0 \\ 0 & 0 & -a & 1 & 0 & 0 \\ 0 & 0 & 0 & 0 & 1 & 0 \\ a & 0 & 0 & 0 & 0 & 1 \end{bmatrix},$$

$$\boldsymbol{W}_{OC_2} = \begin{bmatrix} 1 & 0 & 0 & 0 & 0 & 0 \\ 0 & 1 & 0 & 0 & 0 & 0 \\ 0 & 0 & 1 & 0 & 0 & 0 \\ 0 & 0 & a & 1 & 0 & 0 \\ 0 & 0 & 0 & 0 & 1 & 0 \\ -a & 0 & 0 & 0 & 0 & 1 \end{bmatrix}. \quad (4)$$

These relations derive

$$\begin{aligned} \boldsymbol{W}_{OC} &= \begin{bmatrix} \boldsymbol{W}_{OC_1} & \boldsymbol{W}_{OC_2} \end{bmatrix} \\ &= \begin{bmatrix} 1 & 0 & 0 & 0 & 0 & 0 & 1 & 0 & 0 & 0 & 0 & 0 \\ 0 & 1 & 0 & 0 & 0 & 0 & 0 & 1 & 0 & 0 & 0 & 0 \\ 0 & 0 & 1 & 0 & 0 & 0 & 0 & 0 & 1 & 0 & 0 & 0 \\ 0 & 0 & -a & 1 & 0 & 0 & 0 & 0 & a & 1 & 0 & 0 \\ 0 & 0 & 0 & 0 & 1 & 0 & 0 & 0 & 0 & 0 & 1 & 0 \\ a & 0 & 0 & 0 & 0 & 1 & -a & 0 & 0 & 0 & 0 & 1 \end{bmatrix}. \end{aligned} \quad (5)$$

Grasp map $\boldsymbol{G}_C = \boldsymbol{W}_{OC}\boldsymbol{H}_C$ is calculated as

$$\boldsymbol{G}_C = \begin{bmatrix} 1 & 0 & 0 & 0 & 0 & -1 & 0 & 0 \\ 0 & 0 & 1 & 0 & 0 & 0 & -1 & 0 \\ 0 & -1 & 0 & 0 & 1 & 0 & 0 & 0 \\ 0 & a & 0 & 0 & a & 0 & 0 & 0 \\ 0 & 0 & 0 & 1 & 0 & 0 & 0 & -1 \\ a & 0 & 0 & 0 & 0 & a & 0 & 0 \end{bmatrix}. \quad (6)$$

It is evident that $rank(\boldsymbol{G_C}) = 6$. The grasp is a force closure because a solution of Eq. (7) satisfying $f_{FC} \in FC$ for arbitral external force $F_e \in \mathbb{R}^6$ exists.

$$\boldsymbol{G_C f_{FC}} = -\boldsymbol{F_e} \tag{7}$$

Next, we derive the Jacobian. The position vectors $\boldsymbol{p}_{F_i}$ $(i = 1, 2)$ are represented as

$$\begin{aligned}\boldsymbol{p}_{F_1} &= \begin{bmatrix} -l_1 S\theta_{11} + l_2 S(\theta_{11} + \theta_{12}) \\ l_1 C\theta_{11} + (l_1 + l_2) C\theta_{12} \\ -d_{13} \end{bmatrix}, \\ \boldsymbol{p}_{F_2} &= \begin{bmatrix} l_1 S\theta_{21} + l_2 S(\theta_{21} + \theta_{22}) \\ -l_1 C\theta_{21} - l_2 S(\theta_{21} + \theta_{22}) \\ -d_{23} \end{bmatrix},\end{aligned} \tag{8}$$

where

$$\dot{\boldsymbol{q}}_1 = \begin{bmatrix} \dot{\theta}_{11} \\ \dot{\theta}_{12} \\ \dot{d}_{13} \end{bmatrix}, \quad \dot{\boldsymbol{q}}_2 = \begin{bmatrix} \dot{\theta}_{21} \\ \dot{\theta}_{22} \\ \dot{d}_{23} \end{bmatrix}. \tag{9}$$

Hence, $\dot{\boldsymbol{r}}_{F_i} = [\dot{\boldsymbol{p}}_{F_i}^T, \boldsymbol{\omega}_{F_i}^T]^T (i = 1, 2)$ are given as

$$\dot{\boldsymbol{r}}_{F_1} = \begin{bmatrix} -\{l_1 C\theta_{11} + l_2 C(\theta_{11} + \theta_{12})\}\dot{\theta}_{11} - l_2 C(\theta_{11} + \theta_{12})\dot{\theta}_{12} \\ -\{l_1 S\theta_{11} + l_2 S(\theta_{11} + \theta_{12})\}\dot{\theta}_{11} - l_2 S(\theta_{11} + \theta_{12})\dot{\theta}_{12} \\ -\dot{d}_{13} \\ 0 \\ 0 \\ \dot{\theta}_{11} + \dot{\theta}_{12} \end{bmatrix},$$

$$\dot{\boldsymbol{r}}_{F_2} = \begin{bmatrix} \{l_1 C\theta_{21} + l_2 C(\theta_{21} + \theta_{22})\}\dot{\theta}_{21} + l_2 C(\theta_{21} + \theta_{22})\dot{\theta}_{22} \\ \{l_1 S\theta_{21} + l_2 S(\theta_{21} + \theta_{22})\}\dot{\theta}_{21} + l_2 S(\theta_{21} + \theta_{22})\dot{\theta}_{22} \\ -\dot{d}_{23} \\ 0 \\ 0 \\ \dot{\theta}_{21} + \dot{\theta}_{22} \end{bmatrix} \tag{10}$$

From these equations and $\dot{\boldsymbol{r}}_F = \boldsymbol{J}(\boldsymbol{q})\dot{\boldsymbol{q}}$, Jacobian $\boldsymbol{J}_{\boldsymbol{F}_i}(i = 1, 2)$ are represented as

$$\boldsymbol{J}_{\boldsymbol{F}_1} = \frac{\partial \boldsymbol{r}_{F_1}}{\partial \boldsymbol{q}_1} = \begin{bmatrix} -l_1C\theta_{11} - l_2C(\theta_{11}+\theta_{12}) & -l_2C(\theta_{11}+\theta_{12}) & 0 \\ -l_1S\theta_{11} - l_2S(\theta_{11}+\theta_{12}) & -l_2S(\theta_{11}+\theta_{12}) & 0 \\ 0 & 0 & -1 \\ 0 & 0 & 0 \\ 0 & 0 & 0 \\ 1 & 1 & 0 \end{bmatrix},$$

$$\boldsymbol{J}_{\boldsymbol{F}_2} = \frac{\partial \boldsymbol{r}_{F_2}}{\partial \boldsymbol{q}_2} = \begin{bmatrix} l_1C\theta_{21} + l_2C(\theta_{21}+\theta_{22}) & l_2C(\theta_{21}+\theta_{22}) & 0 \\ l_1S\theta_{21} + l_2S(\theta_{21}+\theta_{22}) & l_2S(\theta_{21}+\theta_{22}) & 0 \\ 0 & 0 & -1 \\ 0 & 0 & 0 \\ 0 & 0 & 0 \\ 1 & 1 & 0 \end{bmatrix} \tag{11}$$

The position vector from the fingertip coordinate origin to the object coordinate origin, $\boldsymbol{p}_{\boldsymbol{FC}_i} = \boldsymbol{R}_{Fi}{}^{Fi}\boldsymbol{p}_{\boldsymbol{FC}_i}$ $(i = 1, 2)$, are represented as

$$\boldsymbol{p}_{\boldsymbol{FC}_1} = \begin{bmatrix} C(\theta_{11}+\theta_{12}) & 0 & -S(\theta_{11}+\theta_{12}) \\ S(\theta_{11}+\theta_{12}) & 0 & C(\theta_{11}+\theta_{12}) \\ 0 & -1 & 0 \end{bmatrix} \begin{bmatrix} 0 \\ 0 \\ l_3 \end{bmatrix}$$

$$= \begin{bmatrix} -l_3S(\theta_{11}+\theta_{12}) \\ l_3C(\theta_{11}+\theta_{12}) \\ 0 \end{bmatrix}, \quad \boldsymbol{p}_{\boldsymbol{FC}_2} = \begin{bmatrix} l_3S(\theta_{21}+\theta_{22}) \\ -l_3C(\theta_{21}+\theta_{22}) \\ 0 \end{bmatrix}. \tag{12}$$

From $\boldsymbol{W}_{\boldsymbol{FC}_i} = \begin{bmatrix} \boldsymbol{I}_{3\times 3} & 0 \\ [\boldsymbol{p}_{\boldsymbol{FC}_i}\times] & \boldsymbol{I}_{3\times 3} \end{bmatrix}$, we get

$$\boldsymbol{W}_{\boldsymbol{FC}_1}^T = \begin{bmatrix} 1 & 0 & 0 & 0 & 0 & -l_3C(\theta_{11}+\theta_{12}) \\ 0 & 1 & 0 & 0 & 0 & -l_3S(\theta_{11}+\theta_{12}) \\ 0 & 0 & 1 & l_3C(\theta_{11}+\theta_{12}) & l_3S(\theta_{11}+\theta_{12}) & 0 \\ 0 & 0 & 0 & 1 & 0 & 0 \\ 0 & 0 & 0 & 0 & 1 & 0 \\ 0 & 0 & 0 & 0 & 0 & 1 \end{bmatrix},$$

$$\boldsymbol{W}_{\boldsymbol{FC_2}}^T = \begin{bmatrix} 1 & 0 & 0 & 0 & 0 & l_3C(\theta_{21}+\theta_{22}) \\ 0 & 1 & 0 & 0 & 0 & l_3S(\theta_{21}+\theta_{22}) \\ 0 & 0 & 1 & -l_3C(\theta_{21}+\theta_{22}) & -l_3S(\theta_{21}+\theta_{22}) & 0 \\ 0 & 0 & 0 & 1 & 0 & 0 \\ 0 & 0 & 0 & 0 & 1 & 0 \\ 0 & 0 & 0 & 0 & 0 & 1 \end{bmatrix}. \tag{13}$$

From the relation of $\boldsymbol{J}_{\boldsymbol{FC_i}} = \boldsymbol{W}_{\boldsymbol{FC_i}}^T \boldsymbol{J}_{\boldsymbol{F_i}}$, we get

$$\boldsymbol{J}_{\boldsymbol{FC_1}} = \begin{bmatrix} -l_1C\theta_{11} - l_2C(\theta_{11}+\theta_{12}) & -l_2C(\theta_{11}+\theta_{12}) & 0 \\ \quad -l_3C(\theta_{11}+\theta_{12}) & \quad -l_3C(\theta_{11}+\theta_{12}) & \\ -l_1S\theta_{11} - l_2S(\theta_{11}+\theta_{12}) & -l_2S(\theta_{11}+\theta_{12}) & 0 \\ \quad -l_3S(\theta_{11}+\theta_{12}) & \quad -l_3S(\theta_{11}+\theta_{12}) & \\ 0 & 0 & -1 \\ 0 & 0 & 0 \\ 0 & 0 & 0 \\ 1 & 1 & 0 \end{bmatrix} \tag{14}$$

$$\boldsymbol{J}_{\boldsymbol{FC_2}} = \begin{bmatrix} l_1C\theta_{21} + l_2C(\theta_{21}+\theta_{22}) & l_2C(\theta_{21}+\theta_{22}) & 0 \\ \quad +l_3C(\theta_{21}+\theta_{22}) & \quad +l_3C(\theta_{21}+\theta_{22}) & \\ l_1S\theta_{21} + l_2S(\theta_{21}+\theta_{22}) & l_2S(\theta_{21}+\theta_{22}) & 0 \\ \quad +l_3S(\theta_{21}+\theta_{22}) & \quad +l_3S(\theta_{21}+\theta_{22}) & \\ 0 & 0 & -1 \\ 0 & 0 & 0 \\ 0 & 0 & 0 \\ 1 & 1 & 0 \end{bmatrix} \tag{15}$$

From the relation of $\boldsymbol{J}_{\boldsymbol{C}} = \boldsymbol{H}_{\boldsymbol{C}}^T \boldsymbol{J}_{\boldsymbol{FC}} = \begin{bmatrix} \boldsymbol{H}_{\boldsymbol{C_1}}^T \boldsymbol{J}_{\boldsymbol{FC_1}} & 0 \\ 0 & \boldsymbol{H}_{\boldsymbol{C_2}}^T \boldsymbol{J}_{\boldsymbol{FC_2}} \end{bmatrix}$, we get

$$\boldsymbol{H}_{\boldsymbol{C_1}}^T \boldsymbol{J}_{\boldsymbol{FC_1}} = \begin{bmatrix} -l_1C\theta_{11} - l_2C(\theta_{11}+\theta_{12}) & -l_2C(\theta_{21}+\theta_{22}) & 0 \\ \quad -l_3C(\theta_{11}+\theta_{12}) & \quad -l_3C(\theta_{21}+\theta_{22}) & \\ 0 & 0 & 1 \\ -l_1S\theta_{11} - l_2S(\theta_{11}+\theta_{12}) & -l_2S(\theta_{11}+\theta_{12}) & 0 \\ \quad -l_3S(\theta_{11}+\theta_{12}) & \quad -l_3S(\theta_{11}+\theta_{12}) & \\ 0 & 0 & 0 \end{bmatrix} \tag{16}$$

$$
H_{C_2}^T J_{FC_2} = \begin{bmatrix} 0 & 0 & -1 \\ -l_1C\theta_{21} - l_2C(\theta_{21}+\theta_{22}) & -l_2C(\theta_{21}+\theta_{22}) & 0 \\ \quad - l_3C(\theta_{21}+\theta_{22}) & \quad - l_3C(\theta_{21}+\theta_{22}) & \\ -l_1S\theta_{21} - l_2S(\theta_{21}+\theta_{22}) & -l_2S(\theta_{21}+\theta_{22}) & 0 \\ \quad - l_3S(\theta_{21}+\theta_{22}) & \quad - l_3S(\theta_{21}+\theta_{22}) & \\ 0 & 0 & 0 \end{bmatrix} \tag{17}
$$

When θ_1 and θ_2 are zero, then we get

$$
J_C = \begin{bmatrix} -l_1-l_2-l_3 & -l_2-l_3 & 0 & 0 & 0 & 0 \\ 0 & 0 & 1 & 0 & 0 & 0 \\ 0 & 0 & 0 & 0 & 0 & 0 \\ 0 & 0 & 0 & 0 & 0 & 0 \\ 0 & 0 & 0 & 0 & 0 & -1 \\ 0 & 0 & 0 & -l_1-l_2-l_3 & -l_2-l_3 & 0 \\ 0 & 0 & 0 & 0 & 0 & 0 \\ 0 & 0 & 0 & 0 & 0 & 0 \end{bmatrix} \tag{18}
$$

In this case, $rank(J_C) = 4$. This means that the grasp is not a manipulable grasp because the condition of $\mathrm{Im}(G_C) \subset \mathrm{Im}(\mathbf{J}_C)$ is not satisfied. This is also shown from the fact that the object cannot be rotated around the line that connects two contact points.

2.4. By setting the elliptic coordinate system, the position on the surface of the ellipsoid is represented as

$$
c(\alpha) = \begin{bmatrix} a\cos u \cos v \\ a\cos u \sin v \\ c\sin u \end{bmatrix}, \tag{1}
$$

where $-\pi/2 < u < \pi/2$ and $-\pi < v < \pi$. From this, we get

$$
c_u = \begin{bmatrix} -a\sin u\cos v \\ -a\sin u \sin v \\ c\cos u \end{bmatrix}, \quad c_v = \begin{bmatrix} -a\cos u\sin v \\ a\cos u\cos v \\ 0 \end{bmatrix},
$$

$$
c_v \times c_u = \begin{bmatrix} -ac\cos^2 u\cos v \\ ac\cos^2 u\sin v \\ a^2\sin u\cos u \end{bmatrix} \tag{2}
$$

The norm of the above vectors are represented as

$$\|\boldsymbol{c}_u\| = a\sqrt{\sin(u)^2 + \alpha^2\cos(u)^2}, \quad \|\boldsymbol{c}_v\| = a\cos(u),$$
$$\|\boldsymbol{c}_u \times \boldsymbol{c}_v\| = a^2\cos(u)\sqrt{\sin^2(u) + \alpha^2\cos^2(u)}. \tag{3}$$

By defining $\alpha = c/a$, we can derive

$$\sum_C = [\boldsymbol{x}, \boldsymbol{y}, \boldsymbol{z}] = \left[\frac{\boldsymbol{c}_u}{\|\boldsymbol{c}_v\|}, \frac{\boldsymbol{c}_v}{\|\boldsymbol{c}_v\|}, \frac{\boldsymbol{c}_u \times \boldsymbol{c}_v}{\|\boldsymbol{c}_u \times \boldsymbol{c}_v\|}\right]$$
$$= \begin{bmatrix} -\dfrac{\sin(u)\cos(v)}{\sqrt{\sin^2(u) + \alpha^2\cos^2(u)}} & -\sin(v) & \dfrac{\alpha\cos(u)\cos(v)}{\sqrt{\sin^2(u) + \alpha^2\cos^2(u)}} \\ -\dfrac{\sin(u)\sin(v)}{\sqrt{\sin^2(u) + \alpha^2\cos^2(u)}} & \cos(v) & \dfrac{\alpha\cos(u)\sin(v)}{\sqrt{\sin^2(u) + \alpha^2\cos^2(u)}} \\ \dfrac{\alpha\cos(u)}{\sqrt{\sin^2(u) + \alpha^2\cos^2(u)}} & 0 & \dfrac{\sin(u)}{\sqrt{\sin^2(u) + \alpha^2\cos^2(u)}} \end{bmatrix}. \tag{4}$$

The geometric parameters of the ellipsoid are derived as

$$\boldsymbol{M}_f = \begin{bmatrix} \|\boldsymbol{c}_u\| & 0 \\ 0 & \|\boldsymbol{c}_v\| \end{bmatrix} = \begin{bmatrix} a\sqrt{\sin(u)^2 + \alpha^2\cos(u)^2} & 0 \\ 0 & a\cos(u) \end{bmatrix}, \tag{5}$$

$$\boldsymbol{K}_f = \begin{bmatrix} \boldsymbol{x}^T \\ \boldsymbol{y}^T \end{bmatrix} \begin{bmatrix} \dfrac{\boldsymbol{z}_u}{\|\boldsymbol{c}_u\|} & \dfrac{\boldsymbol{z}_v}{\|\boldsymbol{c}_v\|} \end{bmatrix}$$
$$= \begin{bmatrix} \frac{\alpha}{a(\sin^2(u)+\alpha^2\cos^2(u))\sqrt{\sin^2(u)+\alpha^2\cos^2(u)}} & 0 \\ 0 & \frac{\alpha}{a\sqrt{\sin^2(u)+\alpha^2\cos^2(u)}} \end{bmatrix}, \tag{6}$$

$$T_f = \boldsymbol{y}^T \begin{bmatrix} \dfrac{\boldsymbol{x}_u}{\|\boldsymbol{c}_u\|} & \dfrac{\boldsymbol{x}_v}{\|\boldsymbol{c}_v\|} \end{bmatrix} = \begin{bmatrix} 0 & -\dfrac{\tan(u)}{a\sqrt{\sin^2(u) + \alpha^2\cos^2(u)}} \end{bmatrix}. \tag{7}$$

2.5. The relative translation velocity between the contact point of the object surface and that of the fingertips is defined by Eq. (2.37). By omitting suffix i for the sake of simplicity, and multiplying ${}^{FC}\boldsymbol{R}$

by both sides of this equation from the left, we get

$$ {}^{FC}\dot{\boldsymbol{p}}_S = {}^{FC}\boldsymbol{R}_F {}^F\dot{\boldsymbol{p}}_{FC} - {}^{FC}\boldsymbol{R}_{OC} {}^{OC}\boldsymbol{R}_O {}^O\dot{\boldsymbol{p}}_{OC}. \tag{1} $$

When the object contact coordinate system is rotated π around the x-axis, and then rotated φ around the z-axis, it coincides with the fingertip contact coordinate system. Hence ${}^{FC}\boldsymbol{R}_{OC}$ can be represented as

$$ {}^{FC}\boldsymbol{R}_{OC} = \begin{bmatrix} \boldsymbol{R}_\varphi & 0 \\ 0 & -1 \end{bmatrix}. \tag{2} $$

By using Eq. (2.47) and defining $\boldsymbol{c}_{u_o} = \partial \boldsymbol{c}_o / \partial u_o$ and $\boldsymbol{c}_{v_o} = \partial \boldsymbol{c}_o / \partial v_o$, we get

$$ {}^O\boldsymbol{R}_{OC} = [{}^O\boldsymbol{x}_{OC}, {}^O\boldsymbol{y}_{OC}, {}^O\boldsymbol{z}_{OC}] = \left[\frac{\boldsymbol{c}_{u_o}}{||\boldsymbol{c}_{u_o}||}, \frac{\boldsymbol{c}_{v_o}}{||\boldsymbol{c}_{v_o}||}, \frac{\boldsymbol{c}_{u_o} \times \boldsymbol{c}_{v_o}}{||\boldsymbol{c}_{u_o} \times \boldsymbol{c}_{v_o}||} \right]. $$

The vector ${}^O\dot{\boldsymbol{p}}_{OC}$ is represented from Eq. (2.51) by

$$ {}^O\dot{\boldsymbol{p}}_{OC} = \frac{\partial \boldsymbol{c}_o}{\partial \boldsymbol{\alpha}_o} \dot{\boldsymbol{\alpha}}_o = [{}^O\boldsymbol{x}_{OC}, {}^O\boldsymbol{y}_{OC}] \boldsymbol{M}_o \dot{\boldsymbol{\alpha}}_o. $$

These relations derive

$$ \begin{aligned} {}^{OC}\boldsymbol{R}_O {}^O\dot{\boldsymbol{p}}_{OC} &= [{}^O\boldsymbol{x}_{OC}, {}^O\boldsymbol{y}_{OC}, {}^O\boldsymbol{z}_{OC}]^T \, [{}^O\boldsymbol{x}_{OC}, {}^O\boldsymbol{y}_{OC}] \boldsymbol{M}_o \dot{\boldsymbol{\alpha}}_o \\ &= \begin{bmatrix} \boldsymbol{M}_o \\ 0 \end{bmatrix} \dot{\boldsymbol{\alpha}}_o. \end{aligned} $$

Similarly, we get

$$ {}^{FC}\boldsymbol{R}_F {}^F\dot{\boldsymbol{p}}_{FC} = \begin{bmatrix} \boldsymbol{M}_f \\ 0 \end{bmatrix} \dot{\boldsymbol{\alpha}}_f. $$

By substituting these relations into Eq. (1), we get

$$ \begin{bmatrix} v_x \\ v_y \end{bmatrix} = \boldsymbol{M}_f \dot{\boldsymbol{\alpha}}_f - \boldsymbol{R}_\varphi \boldsymbol{M}_o \dot{\boldsymbol{\alpha}}_o. \tag{3} $$

Next, $\boldsymbol{R}_{FC}$ is represented as

$$ \boldsymbol{R}_O {}^O\boldsymbol{R}_{OC} {}^{OC}\boldsymbol{R}_{FC} = \boldsymbol{R}_F {}^F\boldsymbol{R}_{FC}, $$

and its time derivative is represented as

$$ \begin{aligned} &[\boldsymbol{\omega}_O \times] \boldsymbol{R}_{FC} + \boldsymbol{R}_O {}^O\dot{\boldsymbol{R}}_{OC} {}^{OC}\boldsymbol{R}_{FC} + \boldsymbol{R}_O {}^O\boldsymbol{R}_{OC} {}^{OC}\dot{\boldsymbol{R}}_{FC} \\ &\quad = [\boldsymbol{\omega}_F \times] \boldsymbol{R}_{FC} + \boldsymbol{R}_F {}^F\dot{\boldsymbol{R}}_{FC}, \end{aligned} $$

where $\boldsymbol{\omega}_O$ and $\boldsymbol{\omega}_F$ are angular velocities of the object and the finger, respectively. By multiplying ${}^{FC}\boldsymbol{R}$ by this equation from the left and using the relation of ${}^{FC}\boldsymbol{R}[\boldsymbol{\omega}_O\times]\boldsymbol{R}_{FC} = [{}^{FC}\boldsymbol{\omega}_O\times]$, we get

$$[{}^{FC}\boldsymbol{\omega}_s\times] - {}^{FC}\boldsymbol{R}_O{}^{O}\dot{\boldsymbol{R}}_{OC}{}^{OC}\boldsymbol{R}_{FC} - {}^{FC}\boldsymbol{R}_{OC}{}^{OC}\dot{\boldsymbol{R}}_{FC} + {}^{FC}\boldsymbol{R}_F{}^{F}\dot{\boldsymbol{R}}_{FC} = 0, \tag{4}$$

where $\boldsymbol{\omega}_s = \boldsymbol{\omega}_F - \boldsymbol{\omega}_O$ is the relative angular velocity from the object contact point to the fingertip contact point. ${}^{FC}\boldsymbol{R}_F{}^{F}\dot{\boldsymbol{R}}_{FC}$ is obtained as follows by using the relations such that $({}^{F}\boldsymbol{x}_{FC})^T(\partial^F\boldsymbol{x}_{FC}/\partial\boldsymbol{\alpha}_f)$ $\dot{\boldsymbol{\alpha}}_f = 0$:

$$\begin{aligned}
{}^{FC}\boldsymbol{R}_F{}^{F}\dot{\boldsymbol{R}}_{FC} &= [{}^{F}\boldsymbol{x}_{FC}, {}^{F}\boldsymbol{y}_{FC}, {}^{F}\boldsymbol{z}_{FC}]^T \\
&\quad \times \left[\frac{\partial^F\boldsymbol{x}_{FC}}{\partial\boldsymbol{\alpha}_f}\dot{\boldsymbol{\alpha}}_f, \frac{\partial^F\boldsymbol{y}_{FC}}{\partial\boldsymbol{\alpha}_f}\dot{\boldsymbol{\alpha}}_f, \frac{\partial^F\boldsymbol{z}_{FC}}{\partial\boldsymbol{\alpha}_f}\dot{\boldsymbol{\alpha}}_f\right] \\
&= \left[\begin{bmatrix} -{}^{F}\boldsymbol{y}_{FC}^T\frac{\partial^F\boldsymbol{z}_{FC}}{\partial\boldsymbol{\alpha}_f}\dot{\boldsymbol{\alpha}}_f \\ {}^{F}\boldsymbol{x}_{FC}^T\frac{\partial^F\boldsymbol{z}_{FC}}{\partial\boldsymbol{\alpha}_f}\dot{\boldsymbol{\alpha}}_f \\ {}^{F}\boldsymbol{y}_{FC}^T\frac{\partial^F\boldsymbol{x}_{FC}}{\partial\boldsymbol{\alpha}_f}\dot{\boldsymbol{\alpha}}_f \end{bmatrix}\times\right].
\end{aligned}$$

Similarly, we get

$$ {}^{OC}\boldsymbol{R}_O{}^{O}\dot{\boldsymbol{R}}_{OC} = \left[\begin{bmatrix} -{}^{O}\boldsymbol{y}_{OC}^T\dfrac{\partial^O\boldsymbol{z}_{OC}}{\partial\boldsymbol{\alpha}_o}\dot{\boldsymbol{\alpha}}_o \\ {}^{O}\boldsymbol{x}_{OC}^T\dfrac{\partial^O\boldsymbol{z}_{OC}}{\partial\boldsymbol{\alpha}_o}\dot{\boldsymbol{\alpha}}_o \\ {}^{O}\boldsymbol{y}_{OC}^T\dfrac{\partial^O\boldsymbol{x}_{OC}}{\partial\boldsymbol{\alpha}_o}\dot{\boldsymbol{\alpha}}_o \end{bmatrix}\times\right].$$

Furthermore, Eq. (2) leads to,

$$ {}^{FC}\boldsymbol{R}_{OC}{}^{OC}\dot{\boldsymbol{R}}_{FC} = \begin{bmatrix} 0 & -\dot{\varphi} & 0 \\ \dot{\varphi} & 0 & 0 \\ 0 & 0 & 0 \end{bmatrix} = \left[\begin{bmatrix} 0 \\ 0 \\ \dot{\varphi} \end{bmatrix}\times\right].$$

By substituting these equations into Eq. (4) and focusing attention on ensuring that all terms are described using the vector product, we get

$$
{}^{FC}\boldsymbol{\omega}_s - {}^{FC}\boldsymbol{R}_{OC}\begin{bmatrix} -{}^{O}\boldsymbol{y}_{OC}^T \dfrac{\partial^O \boldsymbol{z}_{OC}}{\partial \boldsymbol{\alpha}_o}\dot{\boldsymbol{\alpha}}_o \\ {}^{O}\boldsymbol{x}_{OC}^T \dfrac{\partial^O \boldsymbol{z}_{OC}}{\partial \boldsymbol{\alpha}_o}\dot{\boldsymbol{\alpha}}_o \\ {}^{O}\boldsymbol{y}_{OC}^T \dfrac{\partial^O \boldsymbol{x}_{OC}}{\partial \boldsymbol{\alpha}_o}\dot{\boldsymbol{\alpha}}_o \end{bmatrix}
$$

$$
+ \begin{bmatrix} -{}^{F}\boldsymbol{y}_{FC}^T \dfrac{\partial^F \boldsymbol{z}_{FC}}{\partial \boldsymbol{\alpha}_f}\dot{\boldsymbol{\alpha}}_f \\ {}^{F}\boldsymbol{x}_{FC}^T \dfrac{\partial^F \boldsymbol{z}_{FC}}{\partial \boldsymbol{\alpha}_f}\dot{\boldsymbol{\alpha}}_f \\ {}^{F}\boldsymbol{y}_{FC}^T \dfrac{\partial^F \boldsymbol{x}_{FC}}{\partial \boldsymbol{\alpha}_f}\dot{\boldsymbol{\alpha}}_f \end{bmatrix} - \begin{bmatrix} 0 \\ 0 \\ \dot{\varphi} \end{bmatrix} = 0. \tag{5}
$$

Now, an orthogonal matrix is defined as,

$$
\boldsymbol{\Omega} = \begin{bmatrix} 0 & -1 & 0 \\ 1 & 0 & 0 \\ 0 & 0 & 1 \end{bmatrix}.
$$

By multiplying $\boldsymbol{\Omega}$ with Eq. (5) from the left and using the relation of $\boldsymbol{\Omega}^{FC}\boldsymbol{R}_{OC}\boldsymbol{\Omega}^T = \begin{bmatrix} -\boldsymbol{R}_\varphi & 0 \\ 0 & -1 \end{bmatrix}$, we get

$$
\begin{bmatrix} -\omega_y \\ \omega_x \end{bmatrix} - \boldsymbol{R}_\varphi \boldsymbol{K}_o \boldsymbol{M}_o \dot{\boldsymbol{\alpha}}_o - \boldsymbol{K}_f \boldsymbol{M}_f \dot{\boldsymbol{\alpha}}_f = 0 \tag{6}
$$

$$
\omega_z + \boldsymbol{T}_o \boldsymbol{M}_o \dot{\boldsymbol{\alpha}}_o + \mathbf{T}_f \boldsymbol{M}_f \dot{\boldsymbol{\alpha}}_f - \dot{\varphi} = 0 \tag{7}
$$

By getting $\dot{\boldsymbol{\alpha}}_o$ from Eq. (6) and substituting it into Eq. (3), we get

$$
\dot{\boldsymbol{\alpha}}_f = \boldsymbol{M}_f^{-1}(\boldsymbol{K}_f + \tilde{\mathbf{K}}_o)^{-1}\left(\begin{bmatrix} -\omega_y \\ \omega_x \end{bmatrix} - \tilde{\boldsymbol{K}}_o \begin{bmatrix} v_x \\ v_y \end{bmatrix}\right). \tag{8}
$$

Similarly, by getting $\dot{\boldsymbol{\alpha}}_f$ from Eq. (6) and substituting it into Eq. (3), we get

$$
\dot{\boldsymbol{\alpha}}_o = \boldsymbol{M}_o^{-1}\boldsymbol{R}_\varphi(\boldsymbol{K}_f + \tilde{\boldsymbol{K}}_o)^{-1}\left(\begin{bmatrix} -\omega_y \\ \omega_x \end{bmatrix} + \boldsymbol{K}_f \begin{bmatrix} v_x \\ v_y \end{bmatrix}\right). \tag{9}
$$

Equation (2.53) is derived by summing up Eq. (7)–Eq. (9).

Chapter 3

3.1

(1)
$$L_p\left(\varphi_1(\boldsymbol{q})+\varphi_2(\boldsymbol{q})\right)=\frac{\partial\left(\varphi_1(\boldsymbol{q})+\varphi_2(\boldsymbol{q})\right)}{\partial \boldsymbol{q}}p(\boldsymbol{q})$$
$$=\frac{\partial\left(\varphi_1(\boldsymbol{q})\right)}{\partial \boldsymbol{q}}p(\boldsymbol{q})+\frac{\partial\left(\varphi_2(\boldsymbol{q})\right)}{\partial \boldsymbol{q}}p(\boldsymbol{q})$$
$$=L_p\varphi_1(\boldsymbol{q})+L_p\varphi_2(\boldsymbol{q})$$

(2)
$$L_{\alpha p1+\alpha p2}\left(\varphi_1(\boldsymbol{q})\right)=\frac{\partial\left(\varphi_1(\boldsymbol{q})\right)}{\partial \boldsymbol{q}}\left(\alpha_1\boldsymbol{p}_1+\alpha_2\boldsymbol{p}_2\right)(\boldsymbol{q})$$
$$=\alpha_1\frac{\partial\left(\varphi_1(\boldsymbol{q})\right)}{\partial \boldsymbol{q}}\boldsymbol{p}_1(\boldsymbol{q})+\alpha_2\frac{\partial\left(\varphi_2(\boldsymbol{q})\right)}{\partial \boldsymbol{q}}\boldsymbol{p}_2(\boldsymbol{q})$$
$$=\alpha_1L_{p1}\varphi_1(\boldsymbol{q})+\alpha_2L_{p2}\varphi_1(\boldsymbol{q})$$

(3)
$$L_p\left(\varphi_1(\boldsymbol{q})\,\varphi_2(\boldsymbol{q})\right)=\frac{\partial\left(\varphi_1(\boldsymbol{q})\,\varphi_2(\boldsymbol{q})\right)}{\partial \boldsymbol{q}}p(\boldsymbol{q})$$
$$=\varphi_2(\boldsymbol{q})\frac{\partial\left(\varphi_1(\boldsymbol{q})\right)}{\partial \boldsymbol{q}}p(\boldsymbol{q})$$
$$+\varphi_1(\boldsymbol{q})\frac{\partial\left(\varphi_2(\boldsymbol{q})\right)}{\partial \boldsymbol{q}}p(\boldsymbol{q})$$
$$=\varphi_2(\boldsymbol{q})\,L_p\varphi_1(\boldsymbol{q})+\varphi_1(\boldsymbol{q})\,L_p\varphi_2(\boldsymbol{q}).$$

3.2.

(1)
$$[\alpha_1\boldsymbol{p}_1+\alpha_2\boldsymbol{p}_2,\boldsymbol{r}_1]=\frac{\partial\boldsymbol{r}_1}{\partial\boldsymbol{q}}(\alpha_1\boldsymbol{p}_1+\alpha_2\boldsymbol{p}_2)-\frac{\partial(\alpha_1\boldsymbol{p}_1+\alpha_2\boldsymbol{p}_2)}{\partial\boldsymbol{q}}\boldsymbol{r}_1$$
$$=\alpha_1\frac{\partial\boldsymbol{r}_1}{\partial\boldsymbol{q}}\boldsymbol{p}_1+\alpha_2\frac{\partial\boldsymbol{r}_1}{\partial\boldsymbol{q}}\boldsymbol{p}_2-\alpha_1\frac{\partial\boldsymbol{p}_1}{\partial\boldsymbol{q}}\boldsymbol{r}_1-\alpha_2\frac{\partial\boldsymbol{p}_2}{\partial\boldsymbol{q}}\boldsymbol{r}_1$$
$$=\alpha_1\left(\frac{\partial\boldsymbol{r}_1}{\partial\boldsymbol{q}}\boldsymbol{p}_1-\frac{\partial\boldsymbol{p}_1}{\partial\boldsymbol{q}}\boldsymbol{r}_1\right)$$
$$+\ \alpha_2\left(\frac{\partial\boldsymbol{r}_1}{\partial\boldsymbol{q}}\boldsymbol{p}_2-\frac{\partial\boldsymbol{p}_2}{\partial\boldsymbol{q}}\boldsymbol{r}_1\right)$$
$$=\alpha_1[\boldsymbol{p}_1,\boldsymbol{r}_1]+\alpha_2[\boldsymbol{p}_2,\boldsymbol{r}_1]$$

(2)
$$
\begin{aligned}
[\boldsymbol{p}_1, \alpha_1 \boldsymbol{r}_1 + \alpha_2 \boldsymbol{r}_2] &= \frac{\partial(\alpha_1 \boldsymbol{r}_1 + \alpha_2 \boldsymbol{r}_2)}{\partial \boldsymbol{q}} \boldsymbol{p}_1 - \frac{\partial \boldsymbol{p}_1}{\partial \boldsymbol{q}} (\alpha_1 \boldsymbol{r}_1 + \alpha_2 \boldsymbol{r}_2) \\
&= \alpha_1 \frac{\partial \boldsymbol{r}_1}{\partial \boldsymbol{q}} \boldsymbol{p}_1 + \alpha_2 \frac{\partial \boldsymbol{r}_2}{\partial \boldsymbol{q}} \boldsymbol{p}_1 - \alpha_1 \frac{\partial \boldsymbol{p}_1}{\partial \boldsymbol{q}} \boldsymbol{r}_1 - \alpha_2 \frac{\partial \boldsymbol{p}_1}{\partial \boldsymbol{q}} \boldsymbol{r}_2 \\
&= \alpha_1 \left(\frac{\partial \boldsymbol{r}_1}{\partial \boldsymbol{q}} \boldsymbol{p}_1 - \frac{\partial \boldsymbol{p}_1}{\partial \boldsymbol{q}} \boldsymbol{r}_1 \right) \\
&\quad + \alpha_2 \left(\frac{\partial \boldsymbol{r}_2}{\partial \boldsymbol{q}} \boldsymbol{p}_1 - \frac{\partial \boldsymbol{p}_1}{\partial \boldsymbol{q}} \boldsymbol{r}_2 \right) \\
&= \alpha_1 [\boldsymbol{p}_1, \boldsymbol{r}_1] + \alpha_2 [\boldsymbol{p}_1, \boldsymbol{r}_2].
\end{aligned}
$$

(3)
$$
\begin{aligned}
[\boldsymbol{p}_1, \boldsymbol{p}_2] &= \frac{\partial \boldsymbol{p}_2}{\partial \boldsymbol{q}} \boldsymbol{p}_1 - \frac{\partial \boldsymbol{p}_1}{\partial \boldsymbol{q}} \boldsymbol{p}_2 \\
&= - \left(\frac{\partial \boldsymbol{p}_1}{\partial \boldsymbol{q}} \boldsymbol{p}_2 - \frac{\partial \boldsymbol{p}_2}{\partial \boldsymbol{q}} \boldsymbol{p}_1 \right) = - [\boldsymbol{p}_2, \boldsymbol{p}_1]
\end{aligned}
$$

(4)
$$
\begin{aligned}
&[\boldsymbol{p}_1, [\boldsymbol{p}_2, \boldsymbol{p}_3]] + [\boldsymbol{p}_2, [\boldsymbol{p}_3, \boldsymbol{p}_1]] + [\boldsymbol{p}_3, [\boldsymbol{p}_1, \boldsymbol{p}_2]] \\
&= \left[\boldsymbol{p}_1, \left(\frac{\partial \boldsymbol{p}_3}{\partial \boldsymbol{q}} \boldsymbol{p}_2 - \frac{\partial \boldsymbol{p}_2}{\partial \boldsymbol{q}} \boldsymbol{p}_3 \right) \right] + \left[\boldsymbol{p}_2, \left(\frac{\partial \boldsymbol{p}_1}{\partial \boldsymbol{q}} \boldsymbol{p}_3 - \frac{\partial \boldsymbol{p}_3}{\partial \boldsymbol{q}} \boldsymbol{p}_1 \right) \right] \\
&\quad + \left[\boldsymbol{p}_3, \left(\frac{\partial \boldsymbol{p}_2}{\partial \boldsymbol{q}} \boldsymbol{p}_1 - \frac{\partial \boldsymbol{p}_1}{\partial \boldsymbol{q}} \boldsymbol{p}_2 \right) \right] \\
&= \frac{\partial \left(\frac{\partial \boldsymbol{p}_3}{\partial \boldsymbol{q}} \boldsymbol{p}_2 - \frac{\partial \boldsymbol{p}_2}{\partial \boldsymbol{q}} \boldsymbol{p}_3 \right)}{\partial \boldsymbol{q}} \boldsymbol{p}_1 - \frac{\partial \boldsymbol{p}_1}{\partial \boldsymbol{q}} \left(\frac{\partial \boldsymbol{p}_3}{\partial \boldsymbol{q}} \boldsymbol{p}_2 - \frac{\partial \boldsymbol{p}_2}{\partial \boldsymbol{q}} \boldsymbol{p}_3 \right) \\
&\quad + \frac{\partial \left(\frac{\partial \boldsymbol{p}_1}{\partial \boldsymbol{q}} \boldsymbol{p}_3 - \frac{\partial \boldsymbol{p}_3}{\partial \boldsymbol{q}} \boldsymbol{p}_1 \right)}{\partial \boldsymbol{q}} \boldsymbol{p}_2 - \frac{\partial \boldsymbol{p}_2}{\partial \boldsymbol{q}} \left(\frac{\partial \boldsymbol{p}_1}{\partial \boldsymbol{q}} \boldsymbol{p}_3 - \frac{\partial \boldsymbol{p}_3}{\partial \boldsymbol{q}} \boldsymbol{p}_1 \right) \\
&\quad + \frac{\partial \left(\frac{\partial \boldsymbol{p}_2}{\partial \boldsymbol{q}} \boldsymbol{p}_1 - \frac{\partial \boldsymbol{p}_1}{\partial \boldsymbol{q}} \boldsymbol{p}_2 \right)}{\partial \boldsymbol{q}} \boldsymbol{p}_3 - \frac{\partial \boldsymbol{p}_3}{\partial \boldsymbol{q}} \left(\frac{\partial \boldsymbol{p}_2}{\partial \boldsymbol{q}} \boldsymbol{p}_1 - \frac{\partial \boldsymbol{p}_1}{\partial \boldsymbol{q}} \boldsymbol{p}_2 \right) \\
&= \left\{ \left(\frac{\partial^2 \boldsymbol{p}_3}{\partial \boldsymbol{q}^2} \boldsymbol{p}_2 - \frac{\partial^2 \boldsymbol{p}_2}{\partial \boldsymbol{q}^2} \boldsymbol{p}_3 \right) + \left(\frac{\partial \boldsymbol{p}_3}{\partial \boldsymbol{q}} \frac{\partial \boldsymbol{p}_2}{\partial \boldsymbol{q}} - \frac{\partial \boldsymbol{p}_2}{\partial \boldsymbol{q}} \frac{\partial \boldsymbol{p}_3}{\partial \boldsymbol{q}} \right) \right\} \boldsymbol{p}_1 \\
&\quad - \frac{\partial \boldsymbol{p}_1}{\partial \boldsymbol{q}} \frac{\partial \boldsymbol{p}_3}{\partial \boldsymbol{q}} \boldsymbol{p}_2 + \frac{\partial \boldsymbol{p}_1}{\partial \boldsymbol{q}} \frac{\partial \boldsymbol{p}_2}{\partial \boldsymbol{q}} \boldsymbol{p}_3
\end{aligned}
$$

$$+\left\{\left(\frac{\partial^2 \boldsymbol{p}_1}{\partial \boldsymbol{q}^2}\boldsymbol{p}_3 - \frac{\partial^2 \boldsymbol{p}_3}{\partial \boldsymbol{q}^2}\boldsymbol{p}_1\right) + \left(\frac{\partial \boldsymbol{p}_1}{\partial \boldsymbol{q}}\frac{\partial \boldsymbol{p}_3}{\partial \boldsymbol{q}} - \frac{\partial \boldsymbol{p}_3}{\partial \boldsymbol{q}}\frac{\partial \boldsymbol{p}_1}{\partial \boldsymbol{q}}\right)\right\}\boldsymbol{p}_2$$

$$-\frac{\partial \boldsymbol{p}_2}{\partial \boldsymbol{q}}\frac{\partial \boldsymbol{p}_1}{\partial \boldsymbol{q}}\boldsymbol{p}_3 + \frac{\partial \boldsymbol{p}_2}{\partial \boldsymbol{q}}\frac{\partial \boldsymbol{p}_3}{\partial \boldsymbol{q}}\boldsymbol{p}_1$$

$$+\left\{\left(\frac{\partial^2 \boldsymbol{p}_2}{\partial \boldsymbol{q}^2}\boldsymbol{p}_1 - \frac{\partial^2 \boldsymbol{p}_1}{\partial \boldsymbol{q}^2}\boldsymbol{p}_2\right) + \left(\frac{\partial \boldsymbol{p}_2}{\partial \boldsymbol{q}}\frac{\partial \boldsymbol{p}_1}{\partial \boldsymbol{q}} - \frac{\partial \boldsymbol{p}_1}{\partial \boldsymbol{q}}\frac{\partial \boldsymbol{p}_2}{\partial \boldsymbol{q}}\right)\right\}\boldsymbol{p}_3$$

$$-\frac{\partial \boldsymbol{p}_3}{\partial \boldsymbol{q}}\frac{\partial \boldsymbol{p}_2}{\partial \boldsymbol{q}}\boldsymbol{p}_1 + \frac{\partial \boldsymbol{p}_3}{\partial \boldsymbol{q}}\frac{\partial \boldsymbol{p}_1}{\partial \boldsymbol{q}}\boldsymbol{p}_2 = 0.$$

3.3 The generalized coordinates can be given by $\boldsymbol{q} = [x, y, \theta]^T$. The traveling direction of the vehicle cannot be changed because there are no handles. The traveling velocity is $\dot{\boldsymbol{q}} = [\cos\theta, \sin\theta, 0]^T$. A vector field for the constraint is represented by

$$\boldsymbol{p} = \begin{bmatrix} \cos\theta \\ \sin\theta \\ 0 \end{bmatrix}, \tag{1}$$

and a Pfaffian constraint is represented by

$$\boldsymbol{A}(\boldsymbol{q})\boldsymbol{p} = 0 \tag{2}$$

where

$$\boldsymbol{A}(\boldsymbol{q}) = \begin{bmatrix} \sin\theta & -\cos\theta & 0 \end{bmatrix} \tag{3}$$

Since the distribution $\Delta(\boldsymbol{q})$ spanned by a vector field $\boldsymbol{p}$ is involutive, the constraint is holonomic. Therefore, this system is not controllable.

Chapter 4

4.1. Eq. (4.4) can be expanded as

$$\boldsymbol{f} = m_O \boldsymbol{I}_{3\times3}\ddot{\boldsymbol{p}} - m_O\tilde{\boldsymbol{g}}, \tag{1}$$

$$\boldsymbol{n} = \boldsymbol{R}_C{}^C\hat{\mathbf{I}}_O\boldsymbol{R}_O^T\dot{\boldsymbol{\omega}}_O + \boldsymbol{\omega}_O \times \boldsymbol{R}_C{}^C\hat{\mathbf{I}}_O\boldsymbol{R}_C^T\boldsymbol{\omega}_O. \tag{2}$$

First, by substituting Eq. (1) and Eq. (4.8) into Eq. (4.7), it is expressed by

$$\begin{aligned}\boldsymbol{f}_O &= m_O\ddot{\boldsymbol{p}}_O - m_O[\boldsymbol{R}_O{}^O\boldsymbol{s}_{OC}\times]\dot{\boldsymbol{\omega}}_O \\ &\quad -[(\boldsymbol{\omega}_O \times m_O\boldsymbol{R}_O{}^O\boldsymbol{s}_{OC})\times]\boldsymbol{\omega}_O - m_O\tilde{\boldsymbol{g}}.\end{aligned} \tag{3}$$

Next, by substituting Eq. (4.9) and Eq. (3) into Eq. (2), it is expressed by

$$\begin{aligned}\boldsymbol{n}_O = \boldsymbol{R}_O{}^O\boldsymbol{s}_{OC} \times (m_O\ddot{\boldsymbol{p}}_O \underline{- m_O[\boldsymbol{R}_O{}^C\boldsymbol{s}_{OC}\times]\dot{\omega}_O} \\ - \underline{\underline{[(\boldsymbol{\omega}_O \times m_O\boldsymbol{R}_O{}^O\boldsymbol{s}_{OC})\times]\boldsymbol{\omega}_O}} - m_O\tilde{\boldsymbol{g}}) \\ + \boldsymbol{R}_C\left\{{}^o\boldsymbol{I}_o - \underline{m_O({}^O\boldsymbol{s}_{OC}^T{}^O\boldsymbol{s}_{OC}\boldsymbol{I}_{3\times3} - {}^O\boldsymbol{s}_{OC}{}^O\boldsymbol{s}_{OC}^T)}\right\}\boldsymbol{R}_C^T\dot{\boldsymbol{\omega}}_C \\ + \boldsymbol{\omega}_O \times \boldsymbol{R}_C\{{}^o\boldsymbol{I}_o - \underline{\underline{m_O({}^O\boldsymbol{s}_{OC}^T{}^O\boldsymbol{s}_{OC}I_{3\times3} - {}^O\boldsymbol{s}_{OC}{}^O\boldsymbol{s}_{OC}^T)}}\}\boldsymbol{R}_C^T\boldsymbol{\omega}_O.\end{aligned} \tag{4}$$

Now, let us prove that the underlined parts of above equation are equal. By using a property of vector product $\boldsymbol{a}\times(\boldsymbol{b}\times\boldsymbol{c}) = (\boldsymbol{c}^T\boldsymbol{a}I_{3\times3} - \boldsymbol{c}\boldsymbol{a}^T)\boldsymbol{b}$ and the relation of $\boldsymbol{R}_C = \boldsymbol{R}_O$, because Σ_O is parallel to Σ_C, the single-underline part of Eq. (4) is represented by

$$\begin{aligned}& m_O(\boldsymbol{R}_O{}^O\boldsymbol{s}_{OC}) \times (\dot{\boldsymbol{\omega}}_O \times \boldsymbol{R}_O{}^O\boldsymbol{s}_{OC}) \\ &= m_O({}^O\boldsymbol{s}_{OC}^T\boldsymbol{R}_O^T\boldsymbol{R}_O{}^O\boldsymbol{s}_{OC}\boldsymbol{I}_{3\times3} - \boldsymbol{R}_O{}^O\boldsymbol{s}_{OC}{}^O\boldsymbol{s}_{OC}^T\boldsymbol{R}_O^T)\dot{\boldsymbol{\omega}}_O. \\ &= m_O\boldsymbol{R}_O({}^O\boldsymbol{s}_{OC}^T{}^O\boldsymbol{s}_{OC}\boldsymbol{I}_{3\times3} - {}^O\boldsymbol{s}_{OC}{}^O\boldsymbol{s}_{OC}^T)\boldsymbol{R}_O^T\dot{\boldsymbol{\omega}}_O\end{aligned} \tag{5}$$

Furthermore, by using a property of vector product $\boldsymbol{a}^T(\boldsymbol{b} \times \boldsymbol{c}) = \boldsymbol{b}^T(\boldsymbol{c} \times \boldsymbol{a}) = \boldsymbol{c}^T(\boldsymbol{a} \times \boldsymbol{b})$, the double-underline part of Eq. (4) is represented by

$$\begin{aligned}& m_O(\boldsymbol{R}_O{}^O\boldsymbol{s}_{OC}) \times \{\boldsymbol{\omega}_O \times (\boldsymbol{\omega}_O \times \boldsymbol{R}_O{}^O\boldsymbol{s}_{OC})\} \\ &= m_O\Big[\{\boldsymbol{\omega}_O \times (\boldsymbol{R}_O{}^O\boldsymbol{s}_{OC})\}^T(\boldsymbol{R}_O{}^O\boldsymbol{s}_{OC})\boldsymbol{I}_{3\times3} \\ &\quad - \{\boldsymbol{\omega}_O \times (\boldsymbol{R}_O{}^O\boldsymbol{s}_{OC})\}(\boldsymbol{R}_O{}^O\boldsymbol{s}_{OC})^T\Big]\boldsymbol{\omega}_O. \\ &= m_O\boldsymbol{\omega}_O \times \boldsymbol{R}_O(-{}^O\boldsymbol{s}_{OC}{}^O\boldsymbol{s}_{OC}^T)\boldsymbol{R}_O^T\boldsymbol{\omega}_O\end{aligned} \tag{6}$$

By substituting these relations into Eq. (4), it is then represented by

$$\begin{aligned}\boldsymbol{n}_O = m_O[\boldsymbol{R}_O{}^O\boldsymbol{s}_{OC}\times]\ddot{\boldsymbol{p}}_O - \boldsymbol{R}_O{}^O\boldsymbol{I}_O\boldsymbol{R}_O^T\dot{\boldsymbol{\omega}}_O \\ -[\boldsymbol{R}_O{}^O\boldsymbol{I}_O\boldsymbol{R}_O^T\boldsymbol{\omega}_O\times]\boldsymbol{\omega}_O - m_O[\boldsymbol{R}_O{}^O\boldsymbol{s}_{OC}\times]\tilde{\boldsymbol{g}}.\end{aligned} \tag{7}$$

The matrix representation of Eq. (3) and Eq. (7) is

$$\boldsymbol{M}_O\dot{\boldsymbol{v}}_O + \boldsymbol{C}_O\boldsymbol{v}_C + \boldsymbol{g}_O = \boldsymbol{F}_O, \tag{8}$$

where

$$\boldsymbol{v}_O = \begin{bmatrix} \dot{\boldsymbol{p}}_O \\ \boldsymbol{\omega}_O \end{bmatrix}, \quad \boldsymbol{M}_O = \begin{bmatrix} m_O \boldsymbol{I}_{3\times 3} & -m_O[\boldsymbol{R}_O{}^O\boldsymbol{s}_{OC}\times] \\ m_O[\boldsymbol{R}_O{}^O\boldsymbol{s}_{OC}\times] & \boldsymbol{R}_O{}^O\boldsymbol{I}_O\boldsymbol{R}_O^T \end{bmatrix},$$

$$\boldsymbol{C}_o = \begin{bmatrix} 0 & [(\boldsymbol{\omega}_O \times m_O\boldsymbol{R}_O{}^O\boldsymbol{s}_{OC})\times] \\ 0 & [(\boldsymbol{R}_O{}^O\boldsymbol{I}_O\boldsymbol{R}_O^T\boldsymbol{\omega}_O)\times] \end{bmatrix},$$

$$\boldsymbol{g}_O = \begin{bmatrix} -m_O\tilde{\boldsymbol{g}} \\ -m_O[\boldsymbol{R}_O{}^O\boldsymbol{s}_{OC}\times]\tilde{\boldsymbol{g}} \end{bmatrix}, \quad \text{and} \quad \boldsymbol{F}_O = \begin{bmatrix} \boldsymbol{f}_O \\ \boldsymbol{n}_O \end{bmatrix}.$$

Equation (4.11) is thus proven.

4.2. By substituting Eq. (4.53) into Eq. (4.52), it is expressed by

$$C_{ij} = \sum_{k=1}^{n} \frac{1}{2}\left(\frac{\partial M_{ij}}{\partial q_k} + \frac{\partial M_{ik}}{\partial q_j} - \frac{\partial M_{jk}}{\partial q_i}\right)\dot{q}_k. \qquad (1)$$

Multiplying $\dot{q}_j$ to Eq. (1) from the right side, the summation of it for j from 1 to n is expressed by

$$\sum_{j=1}^{n} C_{ij}\dot{q}_j = \sum_{j=1}^{n}\sum_{k=1}^{n} \frac{1}{2}\left(\frac{\partial M_{ij}}{\partial q_k} + \frac{\partial M_{ik}}{\partial q_j} - \frac{\partial M_{jk}}{\partial q_i}\right)\dot{q}_k\dot{q}_j. \qquad (2)$$

From Eq. (4.40), the following is obtained:

$$\frac{dM_{ij}}{dt} = \sum_{k=1}^{n} \frac{\partial M_{ij}}{\partial q_k}\dot{q}_k.$$

Similarly, by changing to, it is expressed by

$$\frac{dM_{ik}}{dt} = \sum_{j=1}^{n} \frac{\partial M_{ik}}{\partial q_j}\dot{q}_j.$$

By substituting these equations into Eq. (2), it is then expressed by

$$\begin{aligned} \sum_{j=1}^{n} C_{ij}\dot{q}_j &= \frac{1}{2}\left(\sum_{j=1}^{n}\frac{dM_{ij}}{dt}\dot{q}_j + \sum_{k=1}^{n}\frac{dM_{ik}}{dt}\dot{q}_k - \sum_{j=1}^{n}\sum_{k=1}^{n}\frac{\partial M_{jk}}{\partial q_i}\dot{q}_j\dot{q}_k\right) \\ &= \sum_{j=1}^{n}\sum_{k=1}^{n}\frac{\partial M_{ij}}{\partial q_k}\dot{q}_j\dot{q}_k - \frac{1}{2}\sum_{j=1}^{n}\sum_{k=1}^{n}\frac{\partial M_{jk}}{\partial q_i}\dot{q}_j\dot{q}_k \end{aligned}$$

$$= \sum_{j=1}^{n}\sum_{k=1}^{n}\left(\frac{\partial M_{ij}}{\partial q_k} - \frac{1}{2}\frac{\partial M_{jk}}{\partial q_i}\right)\dot{q}_j\dot{q}_k$$

$$= \sum_{j=1}^{n}\sum_{k=1}^{n} h_{ijk}\dot{q}_j\dot{q}_k$$

where h_{ijk} is given by Eq. (4.45).

4.3. The mass center of Link 1 and Link 2 are expressed by

$$\boldsymbol{s}_1 = \begin{bmatrix} l_{g1}S_1 \\ l_{g1}C_1 \end{bmatrix}, \quad \boldsymbol{s}_2 = \begin{bmatrix} l_1S_1 + l_{g2}S_{12} \\ l_1C_1 + l_{g2}C_{12} \end{bmatrix}, \tag{1}$$

where $S_1 = \sin\theta_1$, $C_1 = \cos\theta_1$, $S_{12} = \sin(\theta_1+\theta_2)$, and $C_{12} = \cos(\theta_1+\theta_2)$. The velocities of the mass centers are represented by

$$\hat{\boldsymbol{v}}_1 = \boldsymbol{J}_p^{(1)}\dot{\boldsymbol{q}}, \quad \hat{\boldsymbol{v}}_2 = \boldsymbol{J}_p^{(2)}\dot{\boldsymbol{q}}, \tag{2}$$

where $\dot{\boldsymbol{q}} = [\dot{\theta}_1 \quad \dot{\theta}_2]$ and $\boldsymbol{J}_p^{(i)}$ $(i = 1, 2)$ are given by

$$\boldsymbol{J}_p^{(1)} = \begin{bmatrix} l_{g1}C_1 & 0 \\ -l_{g1}S_1 & 0 \end{bmatrix}, \quad \boldsymbol{J}_p^{(2)} = \begin{bmatrix} -l_1S_1 + l_{g2}C_{12} & l_{g2}C_{12} \\ l_1C_1 - l_{g2}S_{12} & -l_{g2}S_{12} \end{bmatrix}. \tag{3}$$

The angular viscosities of Link 1 and Link 2 are represented by

$$\boldsymbol{\omega}_1 = \boldsymbol{J}_o^{(1)}\dot{\boldsymbol{q}}, \quad \boldsymbol{\omega}_2 = \boldsymbol{J}_o^{(2)}\dot{\boldsymbol{q}}, \tag{4}$$

where

$$\boldsymbol{J}_o^{(1)} = [1\ 0], \quad \boldsymbol{J}_o^{(2)} = [1\ 1]. \tag{5}$$

By substituting $\boldsymbol{J}_p^{(i)}$ and $\boldsymbol{J}_o^{(i)}(i = 1, 2)$ into Eq. (4.37), the robotic inertia matrix $\boldsymbol{M} = (M_{ij})$ is represented by

$$\boldsymbol{M} = \begin{bmatrix} M_{11} & M_{12} \\ M_{21} & M_{22} \end{bmatrix}, \tag{6}$$

where

$$M_{11} = I_1 + I_2 + 2m_2l_1l_{g2}C_2 + m_2l_1^2,$$
$$M_{12} = M_{21} = I_2 + m_2l_1l_{g2}C_2, \quad M_{22} = I_2,$$

and

$$I_i = \hat{I}_i + m_il_{gi}^2 \quad (i = 1, 2).$$

The I_i is the inertia moment about the joint axis of Link i. From $q_1 = \theta_1, q_2 = \theta_2$, Eq. (4.45) and Eq. (4.47), the quadratic velocity term $\mathbf{C\dot{q}} = [C_1 \quad C_2]^T$ is represented by

$$\boldsymbol{C}\dot{\boldsymbol{q}} = \begin{bmatrix} -m_2 l_1 l_{g2} S_2 \left(2\dot{\theta}_1\dot{\theta}_2 + \dot{\theta}_2^2\right) \\ m_2 l_1 l_{g2} S_2 \dot{\theta}_1^2 \end{bmatrix}. \tag{7}$$

From Eq. (4.48), the gravitational term is represented by

$$\boldsymbol{g} = \begin{bmatrix} m_1 l_{g1} C_1 + m_2 \left(l_{g2} C_{12} + l_1 C_1\right) \\ m_2 l_{g2} C_{12} \end{bmatrix} \tilde{g}, \tag{8}$$

where $\tilde{g}$ is the gravitational acceleration. Since there is no external force, the equation of motion is obtained by substituting Eqs. (6)–(7) into Eq. (4.46).

4.4. The constraint of the fingertip is represented by

$$\phi(\boldsymbol{q}) = l_1 S_1 + l_2 S_{12} - l_0 = 0. \tag{1}$$

Its time derivative is expressed by

$$\boldsymbol{A}\dot{\boldsymbol{q}} = 0, \tag{2}$$

where

$$\boldsymbol{A} = \begin{bmatrix} l_1 C_1 + l_2 C_{12} & l_2 C_{12} \end{bmatrix}. \tag{3}$$

The equation of motion is represented by

$$\boldsymbol{M}\ddot{\boldsymbol{q}} + \boldsymbol{C}\dot{\boldsymbol{q}} + \boldsymbol{g} = \boldsymbol{\tau} + \boldsymbol{A}^T\boldsymbol{\lambda}, \tag{4}$$

where $\boldsymbol{M}$, $\boldsymbol{C}\dot{\boldsymbol{q}}$, and $\boldsymbol{g}$ are given by Eqs. (6)–(8) of the solution of Exercise 4.3. The partial derivative of Eq. (1) of the solution of Exercise 4.4 by joint variables is represented by

$$\frac{\partial\phi(\theta)}{\partial\theta_1} = l_1 C_1 + l_2 C_{12}, \quad \frac{\partial\phi(\theta)}{\partial\theta_2} = l_2 C_{12}.$$

From these relations we get

$$-\left(\frac{\partial\phi(\theta)}{\partial\theta_2}\right)^{-1} \frac{\partial\phi(\theta)}{\partial\theta_1} = -\frac{l_1 C_1 + l_2 C_{12}}{l_2 C_{12}}.$$

Hence, Eq. (4.92) is represented by

$$\boldsymbol{L}(\boldsymbol{q}) = \begin{bmatrix} 1 \\ -\dfrac{l_1 C\theta_1 + l_2 C_{12}}{l_2 C_{12}} \end{bmatrix}.$$

By substituting this into Eq. (4.95), the low-dimensional model of the equation of motion is represented by

$$M_1(\theta_1)\ddot{\theta}_1 + C_1(\theta_1, \dot{\theta}_1)\dot{\theta}_1 + g_1(\theta_1) = \tau_1,$$

where $M_1 = \boldsymbol{L}^T \boldsymbol{M}\boldsymbol{L}, C_1 = \boldsymbol{L}^T(\boldsymbol{M}\dot{\boldsymbol{L}} + \boldsymbol{C}\boldsymbol{L}), g_1 = \boldsymbol{L}^T\boldsymbol{g}$, and $\tau_1 = \boldsymbol{L}^T\tau$.

4.5. Abbreviation.

Chapter 5

5.1. By variable transformation of $z = x - 1$, the system is written by

$$\dot{z} = -(1 + z)z. \tag{1}$$

Consider a Lyapunov function candidate:

$$V(z) = \frac{1}{2}z^2. \tag{2}$$

This satisfies $V(0) = 0$ and $V(z) > 0$ (for $z \neq 0$). A time derivative of $V(z)$ along the trajectory is represented by:

$$\dot{V}(z) = z\dot{z} = -z^2(z + 1). \tag{3}$$

Since $\dot{V}(z) \leq 0$ in the region $z + 1 \geq 0$, $V(z)$ is a Lyapunov function in the region. Therefore, in the region $x \geq 0$, the equilibrium of $x = 1$ is stable. Furthermore, in the region of $z + 1 > 0$, the equilibrium is asymptotically stable because $\dot{V}(z) < 0$. Actually, the solution is represented by

$$x(t) = \frac{a}{a + (1 - a)e^{-t}},$$

where $x(0) = a$. It is evident that $x(t) \to 1$ as $t \to \infty$ for $x(0) \neq 0$ and $x(t) = 0$ for $x(0) = 0$.

5.2. Consider a Lyapunov function candidate:

$$V(x,t) = x^2 + \dot{x}^2. \tag{1}$$

Since $V(0,t) = 0$ and $V(x,t) > 0$ for $x \neq 0$, $V(x,t)$ is positive definite. A time derivative of $V(x,t)$ along the trajectory is represented as

$$\dot{V}(x,t) = 2x\dot{x} + 2\dot{x}\ddot{x} = -2\dot{x}^2(1+t) \leq 0, \tag{2}$$

and it is positive semi-definite. Furthermore, since there exists a time invariant positive definite function $V_1(x) = x^2 + \dot{x}^2$ such that $V(x,t) \leq V_1(x)$, $V(x,t)$ is decrescent. From Theorem 5.9, the origin is therefore uniformly continuous.

5.3. Consider the lower bounded function

$$V = e^2 + \sigma^2.$$

The derivative of the function is:

$$\dot{V} = 2e(-2e + \sigma \sin t) + 2\sigma(-e \sin t) = -4e^2 \leq 0.$$

This implies $V(t) \leq V(0)$. Therefore, e and σ are bounded. Furthermore, the derivative of $\dot{V}$ is represented by

$$\ddot{V} = -8e(-2e + \sigma \sin t),$$

and $\ddot{V}$ is bounded. This leads us to conclude that $\dot{V}$ is uniformly continuous and, from Barbalet's lemma, $e \to 0$ as $t \to \infty$.

5.4. The derivative of quadratic form $\boldsymbol{x}^T \boldsymbol{M} \boldsymbol{x}$ is:

$$\frac{d}{dt}(\boldsymbol{x}^T \mathbf{M} \mathbf{x}) = \dot{\boldsymbol{x}}^T \mathbf{M} \mathbf{x} + \boldsymbol{x}^T \dot{\boldsymbol{M}} x + \boldsymbol{x}^T \boldsymbol{M} \dot{\boldsymbol{x}}.$$

Since the quadratic form is scalar, $\dot{\boldsymbol{x}}^T \boldsymbol{M} \boldsymbol{x} = \boldsymbol{x}^T \boldsymbol{M} \dot{\boldsymbol{x}}$. Therefore, the above equation is

$$\frac{d}{dt}\left(\boldsymbol{x}^T \boldsymbol{M} \boldsymbol{x}\right) = 2\dot{\boldsymbol{x}}^T \boldsymbol{M} \boldsymbol{x} + \boldsymbol{x}^T \dot{\boldsymbol{M}} \boldsymbol{x}.$$

5.5. Consider a Lyapunov function candidate:

$$V = \boldsymbol{x}^T \boldsymbol{P}\boldsymbol{x}.$$

The derivative of the function is

$$\dot{V} = \dot{\boldsymbol{x}}^T \boldsymbol{P}\boldsymbol{x} + \boldsymbol{x}^T \boldsymbol{P}\dot{\boldsymbol{x}} = \boldsymbol{x}^T(\boldsymbol{A}^T\boldsymbol{P} + \boldsymbol{P}\boldsymbol{A})\boldsymbol{x} = -\boldsymbol{y}^T\boldsymbol{y} \le 0.$$

The function V is a Lyapunov function and the system is stable. From the invariant set theorem, $\boldsymbol{y} \to 0$ as $t \to \infty$ and from the observability condition of the system, $\boldsymbol{x} \to 0$ as $t \to \infty$.

5.6. Consider a Lyapunov function candidate:

$$V(t) = \frac{1}{2}\left(\dot{\boldsymbol{e}}_r^T \boldsymbol{M}_r \dot{\mathrm{e}}_r + \boldsymbol{e}_r^T \boldsymbol{K}_p \boldsymbol{e}_r\right). \quad (1)$$

The time derivative of the function is:

$$\dot{V}(t) = \frac{1}{2}\left(2\dot{\boldsymbol{e}}_r^T \boldsymbol{M}_r \ddot{\mathrm{e}}_r + \dot{\boldsymbol{e}}_r^T \dot{\boldsymbol{M}}_r \dot{\boldsymbol{e}}_r + 2\dot{\mathrm{e}}_r^T \boldsymbol{K}_p \boldsymbol{e}_r\right). \quad (2)$$

Since $\dot{\boldsymbol{M}}_r - 2\boldsymbol{C}_r$ is a skew-symmetric matrix, $\dot{\boldsymbol{e}}_r^T(\dot{\boldsymbol{M}}_r - 2\boldsymbol{C}_r)\dot{\boldsymbol{e}}_r = 0$. Substituting this equation into Eq. (1), it can be written by

$$\dot{V}(t) = \dot{\boldsymbol{e}}_r^T\left(\boldsymbol{M}_r \ddot{\boldsymbol{e}}_r + \boldsymbol{C}_r \dot{\boldsymbol{e}}_r + \boldsymbol{K}_p \boldsymbol{e}_r\right). \quad (3)$$

Now, using relations of $\boldsymbol{e}_r = \boldsymbol{r}_d - \boldsymbol{r}$, $\dot{\boldsymbol{e}}_r = -\dot{\boldsymbol{r}}$, and $\ddot{\boldsymbol{e}}_r = -\ddot{\boldsymbol{r}}$, we get:

$$\dot{V}(t) = -\dot{\mathrm{e}}_r^T \boldsymbol{K}_v \dot{\mathrm{e}}_r \le 0.$$

Therefore, $V(t)$ is a Lyapunov function. A set that satisfies $\dot{V}(t) = 0$ is $R = \{\dot{\boldsymbol{e}}_r : \dot{\boldsymbol{e}}_r = 0\}$, then $\boldsymbol{M}_r\ddot{\mathrm{e}} + \boldsymbol{K}_p\boldsymbol{e}_r = 0$. This leads to the conclusion that the invariant set is $M = \{(\boldsymbol{e}_r, \dot{\boldsymbol{e}}_r) : \boldsymbol{e}_r = 0, \dot{\boldsymbol{e}}_r = 0\}$. Therefore, the origin in Cartesian space is locally asymptotically stable.

Chapter 6

6.1. Consider a candidate Lyapunov function given by

$$V = \frac{1}{2}\left(\boldsymbol{s}_O^T \boldsymbol{M}_O \boldsymbol{s}_O + \boldsymbol{s}^T \boldsymbol{M}_F \boldsymbol{s} + \frac{1}{\alpha}\boldsymbol{v}_C^T \boldsymbol{\Omega} \boldsymbol{v}_C + \boldsymbol{\eta}_C^T \boldsymbol{\Psi} \boldsymbol{\eta}_C\right), \quad (1)$$

where $\boldsymbol{v}_C = [\boldsymbol{v}_{C_1}^T, \ldots, \boldsymbol{v}_{C_k}^T]^T$, $\boldsymbol{\eta}_C = [\boldsymbol{\eta}_{C_1}^T, \ldots, \boldsymbol{\eta}_{C_k}^T]^T$, $\boldsymbol{\Omega} = block\ diag[\boldsymbol{\Omega}_1, \ldots, \boldsymbol{\Omega}_k] > 0$, and $\boldsymbol{\Psi} = block\ diag[\boldsymbol{\Psi}_1, \ldots, \boldsymbol{\Psi}_k] > 0$. Since $\dot{\boldsymbol{M}}_O - 2\boldsymbol{C}_O$ and $\dot{\boldsymbol{M}}_F - 2\boldsymbol{C}_F$ are skew-symmetric, a time derivative of V along the trajectory using these properties and Eq. (6.24) is given by

$$\begin{aligned}\dot{V} &= \frac{1}{2}\bigg(\boldsymbol{s}_O^T\dot{\boldsymbol{M}}_O\boldsymbol{s}_O + 2\boldsymbol{s}_O^T\boldsymbol{M}_O\dot{\boldsymbol{s}}_O + \boldsymbol{s}^T\dot{\boldsymbol{M}}_F\boldsymbol{s} + 2\boldsymbol{s}^T\boldsymbol{M}_F\dot{\boldsymbol{s}} \\ &\quad + \frac{2}{\alpha}\boldsymbol{v}_C^T\boldsymbol{\Omega}\dot{\boldsymbol{v}}_C + 2\boldsymbol{\eta}_C^T\boldsymbol{\Psi}\Delta\boldsymbol{F}_C\bigg) \\ &= \boldsymbol{s}_O^T(\boldsymbol{M}_O\dot{\boldsymbol{s}}_O + \boldsymbol{C}_O\boldsymbol{s}_O) + \boldsymbol{s}^T(\boldsymbol{M}_F\dot{\boldsymbol{s}} + \boldsymbol{C}_F\boldsymbol{s}) \\ &\quad + \boldsymbol{v}_C^T\boldsymbol{\Omega}(\Delta\boldsymbol{F}_C - \boldsymbol{v}_C) + \boldsymbol{\eta}_C^T\boldsymbol{\Psi}\Delta\boldsymbol{F}_C, \end{aligned} \tag{2}$$

where, $\Delta\boldsymbol{F}_C = \boldsymbol{F}_C^d - \boldsymbol{F}_C$. By substituting Eq. (6.14), Eq. (6.25), Eqs. (4.109) and (4.111) of Chapter 4, into $\boldsymbol{M}_O\dot{\boldsymbol{s}}_O + \boldsymbol{C}_O\boldsymbol{s}_O$ in order, and defining $\Delta\boldsymbol{F}_O = \boldsymbol{F}_O^d - \boldsymbol{F}_O$, it is represented by

$$\boldsymbol{M}_O\dot{\boldsymbol{s}}_O + \boldsymbol{C}_O\boldsymbol{s}_O = -\Delta\boldsymbol{F}_O - \boldsymbol{K}_O\boldsymbol{s}_O. \tag{3}$$

In the same way, by substituting Eq. (6.21), Eq. (4.110) of Chapter 4, Eq. (6.19), and Eq. (6.20) into $\boldsymbol{M}_F\dot{\boldsymbol{s}} + \boldsymbol{C}_F\boldsymbol{s}$ in order, it is then represented as

$$\boldsymbol{M}_F\dot{\boldsymbol{s}} + \boldsymbol{C}_F\boldsymbol{s} = \boldsymbol{J}_C^T\Delta\boldsymbol{F}_C - \boldsymbol{K}\boldsymbol{s}. \tag{4}$$

Now, by substituting Eq. (3) and Eq. (4) into Eq. (2), and using the relationship of $\boldsymbol{G}_C\boldsymbol{\Delta}\boldsymbol{F}_C = \Delta\boldsymbol{F}_O$, Eq. (2) can be represented by

$$\dot{V} = -\boldsymbol{s}_O^T\boldsymbol{K}_O\boldsymbol{s}_O - \boldsymbol{s}^T\boldsymbol{K}\boldsymbol{s} - \boldsymbol{v}_C^T\boldsymbol{\Omega}\boldsymbol{v}_C \leq 0. \tag{5}$$

Therefore, V is a Lyapunov function and $\boldsymbol{s}_O, \boldsymbol{s}, \boldsymbol{v}_C$, and $\boldsymbol{\eta}_C$ are bounded. From Eq. (6.14), $\boldsymbol{v}_O, \boldsymbol{v}_{Or}, \dot{\boldsymbol{e}}_O$, and $\boldsymbol{e}_O$ are bounded, and so $\dot{\boldsymbol{r}}_O, \boldsymbol{r}_O, \boldsymbol{G}_C$, and $\boldsymbol{J}_O$ are bounded. $\dot{\boldsymbol{v}}_{Or}$ is bounded from the time differentiation of Eq. (6.13), and $\boldsymbol{F}_O^d$ and $\boldsymbol{Y}_O$ are bounded from Eq. (6.25). In addition, $\boldsymbol{F}_C^d$ is bounded from Eq. (6.16), $\dot{\boldsymbol{q}}_{Fr}$ is bounded from Eq. (6.18), and $\dot{\boldsymbol{q}}_F$ is bounded from Eq. (6.17).

Next, the following is derived from Eq. (3):

$$\dot{\boldsymbol{s}}_O = -\boldsymbol{M}_O^{-1}(\boldsymbol{G}_C\Delta\boldsymbol{F}_C + (\boldsymbol{K}_O + \boldsymbol{C}_O)\boldsymbol{s}_O). \tag{6}$$

Then the time derivative of Eq. (6.28) is given as

$$\begin{aligned}\dot{\boldsymbol{s}} &= \dot{\boldsymbol{J}}_C^{-1}(\boldsymbol{G}_C^T\boldsymbol{s}_O - \boldsymbol{\Omega}\boldsymbol{v}_C - \boldsymbol{\Psi}\eta_C)\\ &\quad + \boldsymbol{J}_C^{-1}(\dot{\boldsymbol{G}}_C^T\boldsymbol{s}_O + \boldsymbol{G}_C^T\dot{\boldsymbol{s}}_O - \alpha\boldsymbol{\Omega}(\Delta\boldsymbol{F}_C - \boldsymbol{v}_C) - \boldsymbol{\Psi}\Delta\boldsymbol{F}_C)\\ &= -\boldsymbol{J}_C^{-1}(\boldsymbol{G}_C^T\boldsymbol{M}_O^{-1}\boldsymbol{G}_C + \alpha\boldsymbol{\Omega} - \boldsymbol{\Psi})\Delta\boldsymbol{F}_C + \boldsymbol{b},\end{aligned} \tag{7}$$

where $\boldsymbol{b}$ is a bounded vector given by

$$\begin{aligned}\boldsymbol{b} &= \dot{\boldsymbol{J}}_C^{-1}(\boldsymbol{G}_C^T\boldsymbol{s}_O - \boldsymbol{\Omega}_C - \boldsymbol{\Psi}\boldsymbol{\eta}_C)\\ &\quad + \boldsymbol{J}_C^{-1}(\dot{\boldsymbol{G}}_C^T\boldsymbol{s}_O - \boldsymbol{G}_C^T\boldsymbol{M}_O^{-1}(\boldsymbol{K}_O + \boldsymbol{C}_O)\boldsymbol{s}_O + \alpha\boldsymbol{\Omega}\boldsymbol{v}_C).\end{aligned} \tag{8}$$

By substituting Eq. (7) into Eq. (4), it shows

$$\begin{aligned}&(\boldsymbol{J}_C^T + \boldsymbol{M}_F\boldsymbol{J}_C^{-1}(\boldsymbol{G}_C^T\boldsymbol{M}_O^{-1}\boldsymbol{G}_C + \alpha\boldsymbol{\Omega} + \boldsymbol{\Psi}))\Delta\boldsymbol{F}_C\\ &\quad = \boldsymbol{M}_F\boldsymbol{b} + (\boldsymbol{C}_F + \boldsymbol{K})\boldsymbol{s}.\end{aligned} \tag{9}$$

Multiply $\boldsymbol{J}_C\boldsymbol{M}_F^{-1}$ from left in this equation, and it is represented as

$$\boldsymbol{U}\Delta\boldsymbol{F}_C = \boldsymbol{J}_C(\boldsymbol{b} + \boldsymbol{M}_F^{-1}(\boldsymbol{C}_F + \boldsymbol{K})\boldsymbol{s}), \tag{10}$$

where

$$\boldsymbol{U} = \boldsymbol{J}_C\boldsymbol{M}_F^{-1}\boldsymbol{J}_C^T + \boldsymbol{G}_C^T\boldsymbol{M}_O^{-1}\boldsymbol{G}_C + \alpha\boldsymbol{\Omega} + \boldsymbol{\Psi}. \tag{11}$$

Obviously, $\boldsymbol{U}$ is a positive and symmetric matrix. Therefore, $\Delta\boldsymbol{F}_C$ is bounded. As a result, $\Delta\boldsymbol{F}_O$ is bounded, $\dot{\mathbf{s}}_O$ is bounded from Eq. (6), $\dot{\boldsymbol{v}}_C$ is bounded from Eq. (6.24), and $\dot{\boldsymbol{s}}$ is bounded from Eq. (4). From these relations, the time derivative of Eq. (5) is given as

$$\ddot{V} = -2(\boldsymbol{s}_O^T\boldsymbol{K}_O\dot{\boldsymbol{s}}_O + \boldsymbol{s}^T\mathbf{K}\dot{\mathbf{s}} + \boldsymbol{v}_C^T\boldsymbol{\Omega}\dot{v}_C), \tag{12}$$

and it is bounded. Hence, $\dot{V}$ is uniformly continuous. According to the Lyapunov-like lemma, $\dot{V} \to 0$ as $t \to \infty$. Therefore, $\boldsymbol{s}_O \to 0$ and $\boldsymbol{s} \to 0$ as $t \to \infty$. As a result, from Eq. (6.14), $\dot{\boldsymbol{e}}_O, \boldsymbol{e}_O$ converges to zero and the object trajectory converges to the desired trajectory as $t \to \infty$. Furthermore, since it can be shown that $\boldsymbol{U}$ and the right side of Eq. (10) are uniformly continuous, then $\Delta\boldsymbol{F}_C$, $\dot{\boldsymbol{s}}$, $\dot{\boldsymbol{s}}_O$, and $\dot{\boldsymbol{v}}_O$ are uniformly continuous. According to the Barbalat's lemma, it is derived that $\dot{\boldsymbol{s}} \to 0$, $\dot{\boldsymbol{s}}_O \to 0$, and $\dot{\boldsymbol{v}}_O \to 0$ as $t \to \infty$. Also, $\Delta \boldsymbol{F}_O \to 0$ and $\Delta \boldsymbol{F}_C \to 0$ as $t \to \infty$. From Eq. (3), $\boldsymbol{F}_C \to \boldsymbol{F}_C^d$ and $\boldsymbol{F}_O \to \boldsymbol{F}_O^d$ as $t \to \infty$; i.e., the contact forces converge to the desired values.

6.2. Abbreviation.

6.3. Abbreviation.

6.4. Consider a candidate Lyapunov function given by

$$V = \frac{1}{2}\Bigg(\boldsymbol{s}_O^T \boldsymbol{M}_O \boldsymbol{s}_O + \Delta\boldsymbol{\sigma}_O^T \boldsymbol{\Gamma}_O^{-1} \Delta\boldsymbol{\sigma}_O + \frac{1}{\alpha}\boldsymbol{v}_C^T \boldsymbol{\Omega} \boldsymbol{v}_C + \boldsymbol{\eta}_C^T \boldsymbol{\Psi} \boldsymbol{\eta}_C \\ + \boldsymbol{s}^T \boldsymbol{M}_F \boldsymbol{s} + \Delta\boldsymbol{\sigma}_F^T \boldsymbol{\Gamma}^{-1} \Delta\boldsymbol{\sigma}_F \Bigg). \tag{1}$$

Since $\dot{\boldsymbol{M}}_O - 2\boldsymbol{C}_O$ and $\dot{\boldsymbol{M}}_F - 2\boldsymbol{C}_F$ are skew-symmetric, a time derivative of V along the trajectory using these properties, Eq. (6.44), Eq. (6.45), and Eq. (6.50) are given by

$$\begin{aligned} \dot{V} &= \frac{1}{2}\Bigg(\boldsymbol{s}_O^T \dot{\boldsymbol{M}}_O \boldsymbol{s}_O + 2\boldsymbol{s}_O^T \boldsymbol{M}_O \dot{\boldsymbol{s}}_O + 2\Delta\boldsymbol{\sigma}_O^T \boldsymbol{\Gamma}_O^{-1} \Delta\boldsymbol{\sigma}_O + \frac{2}{\alpha}\boldsymbol{v}_C^T \boldsymbol{\Omega} \dot{\boldsymbol{v}}_C \\ &\quad + 2\boldsymbol{\eta}_C^T \boldsymbol{\Psi} \Delta\boldsymbol{F}_C + \boldsymbol{s}^T \dot{\boldsymbol{M}}_F \boldsymbol{s} + 2\boldsymbol{s}^T \boldsymbol{M}_F \dot{\boldsymbol{s}} + 2\Delta\boldsymbol{\sigma}_F^T \boldsymbol{\Gamma}^{-1} \Delta\boldsymbol{\sigma}_F \Bigg) \\ &= \boldsymbol{s}_O^T (\boldsymbol{M}_O \dot{\boldsymbol{s}}_O + \boldsymbol{C}_O \boldsymbol{s}_O - \boldsymbol{Y}_O \Delta\boldsymbol{\sigma}_O) \\ &\quad + \boldsymbol{s}^T (\boldsymbol{M}_F \dot{\boldsymbol{s}} + \boldsymbol{C}_F \boldsymbol{s} - \boldsymbol{Y}_F \Delta\boldsymbol{\sigma}_F) \\ &\quad + (\boldsymbol{v}_C^T \boldsymbol{\Omega} (\Delta\boldsymbol{F}_C - \boldsymbol{v}_C) + \boldsymbol{\eta}_C^T \boldsymbol{\Psi} \Delta\boldsymbol{F}_C). \end{aligned} \tag{2}$$

By substituting Eq. (6.14), Eq. (6.43) and Eq. (4.100) of Chapter 4, into $\boldsymbol{M}_O \dot{\boldsymbol{s}}_O + \boldsymbol{C}_O \boldsymbol{s}_O$, in order, and defining $\Delta\boldsymbol{F}_O = \boldsymbol{F}_O^d - \boldsymbol{F}_O$, it is represented as

$$\boldsymbol{M}_O \dot{\boldsymbol{s}}_O + \boldsymbol{C}_O \boldsymbol{s}_O = -\Delta\boldsymbol{F}_O + \boldsymbol{Y}_O \Delta\boldsymbol{\sigma}_O - \boldsymbol{K}_O \boldsymbol{s}_O. \tag{3}$$

In the same way, by substituting Eq. (6.21), Eq. (6.19), Eq. (4.110) of Chapter 4, and Eq. (6.50) into $\boldsymbol{M}_F \dot{\boldsymbol{s}} + \boldsymbol{C}_F \boldsymbol{s}$, in order, and defining

$\Delta \boldsymbol{F}_C = \boldsymbol{F}_C^d - \boldsymbol{F}_C$, then it is represented as

$$\boldsymbol{M}_F \dot{\boldsymbol{s}} + \boldsymbol{C}_F \boldsymbol{s} = \boldsymbol{J}_C^T \Delta \boldsymbol{F}_C - \boldsymbol{K}\boldsymbol{s} + \boldsymbol{Y}_F \Delta \boldsymbol{\sigma}_F. \tag{4}$$

Now, by substituting Eq. (3) and Eq. (4) into Eq. (2) and using the relationship of $\boldsymbol{G}_C \boldsymbol{\Delta}\mathbf{F}_C = \Delta \boldsymbol{F}_O$, Eq. (2) is represented by

$$\dot{V} = -\boldsymbol{s}_O^T \boldsymbol{K}_O \boldsymbol{s}_O - \boldsymbol{s}^T \boldsymbol{K} \boldsymbol{s} - \boldsymbol{v}_C^T \boldsymbol{\Omega} \boldsymbol{v}_C \leq 0. \tag{5}$$

Therefore, V is a Lyapunov function and $\boldsymbol{s}_O, \boldsymbol{s}, \boldsymbol{v}_C, \boldsymbol{\eta}_C$, $\Delta \boldsymbol{\sigma}_O$, and $\Delta \sigma_F$ are bounded. From Eq. (6.14), $\boldsymbol{v}_O, \boldsymbol{v}_{Or}$, $\dot{\boldsymbol{e}}_O$, $\boldsymbol{e}_O$, $\hat{\boldsymbol{\sigma}}_O$, and $\hat{\sigma}_F$ are bounded, and so $\dot{\boldsymbol{r}}_O, \boldsymbol{r}_O$, $\boldsymbol{G}_C$, and $\boldsymbol{J}_O$ are bounded. $\dot{\boldsymbol{v}}_{Or}$ is bounded from the time differentiation of Eq. (6.13), and $\boldsymbol{F}_O^d$ and $\boldsymbol{Y}_O$ are bounded from Eq. (6.42). In addition, $\boldsymbol{F}_C^d$ is bounded from Eq. (6.16), $\dot{\boldsymbol{q}}_{Fr}$ is bounded from Eq. (6.45), $\dot{\boldsymbol{q}}_F$ is bounded from Eq. (6.21), and $\dot{\boldsymbol{G}}_C$ is bounded.

Next, the following is derived from Eq. (3):

$$\dot{\boldsymbol{s}}_O = -\boldsymbol{M}_O^{-1}(\boldsymbol{G}_C \Delta \boldsymbol{F}_C + (\boldsymbol{K}_O + \boldsymbol{C}_O)\boldsymbol{s}_O - \boldsymbol{Y}_O \Delta \boldsymbol{\sigma}_O). \tag{6}$$

Then the time derivative of Eq. (6.46) is derived from Eq. (3) as

$$\begin{aligned}\dot{\boldsymbol{s}} &= \dot{\boldsymbol{J}}_C^{-1}(\boldsymbol{G}_C^T \boldsymbol{s}_O - \boldsymbol{\Omega}\boldsymbol{v}_C - \boldsymbol{\Psi}\boldsymbol{\eta}_C) \\ &\quad + \boldsymbol{J}_C^{-1}(\dot{\boldsymbol{G}}_C^T \boldsymbol{s}_O + \boldsymbol{G}_C^T \dot{\boldsymbol{s}}_O - \alpha\boldsymbol{\Omega}(\Delta \boldsymbol{F}_C - \boldsymbol{v}_C) - \boldsymbol{\Psi}\Delta \boldsymbol{F}_C) \\ &= -\boldsymbol{J}_C^{-1}(\boldsymbol{G}_C^T \boldsymbol{M}_O^{-1} \boldsymbol{G}_C + \alpha\boldsymbol{\Omega} + \boldsymbol{\Psi})\Delta \boldsymbol{F}_C + \boldsymbol{b},\end{aligned} \tag{7}$$

where $\boldsymbol{b}$ is a bounded vector given by

$$\begin{aligned}\boldsymbol{b} &= \dot{\boldsymbol{J}}_C^{-1}(\boldsymbol{G}_C^T \boldsymbol{s}_O - \boldsymbol{\Omega}\boldsymbol{v}_C - \boldsymbol{\Psi}\boldsymbol{\eta}_C) \\ &\quad + \boldsymbol{J}_C^{-1}(\dot{\boldsymbol{G}}_C^T \boldsymbol{s}_O - \boldsymbol{G}_C^T \boldsymbol{M}_O^{-1}((\boldsymbol{K}_O + \boldsymbol{C}_O)\boldsymbol{s}_O - \boldsymbol{Y}_O \Delta \boldsymbol{\sigma}_O) + \alpha\boldsymbol{\Omega}_C).\end{aligned} \tag{8}$$

By substituting Eq. (7) into Eq. (4), it shows:

$$(\boldsymbol{J}_C^T + \boldsymbol{M}_F \boldsymbol{J}_C^{-1}(\boldsymbol{G}_C^T \boldsymbol{M}_O^{-1} \boldsymbol{G}_C + \alpha\boldsymbol{\Omega} + \boldsymbol{\Psi}))\Delta \boldsymbol{F}_C = \boldsymbol{M}_F \boldsymbol{b} + (\boldsymbol{C}_F + \boldsymbol{K})\boldsymbol{s}. \tag{9}$$

By multiplying $\boldsymbol{J}_C \boldsymbol{M}_F^{-1}$ from the left of this equation, it is represented as

$$\boldsymbol{U}\Delta \boldsymbol{F}_C = \boldsymbol{J}_C(\boldsymbol{b} + \boldsymbol{M}_F^{-1}(\boldsymbol{C}_F + \boldsymbol{K})\boldsymbol{s}), \tag{10}$$

where

$$\boldsymbol{U} = \boldsymbol{J}_C \boldsymbol{M}_F^{-1} \boldsymbol{J}_C^T + \boldsymbol{G}_C^T \boldsymbol{M}_O^{-1} \boldsymbol{G}_C + \alpha\boldsymbol{\Omega} + \boldsymbol{\Psi}. \tag{11}$$

Obviously, $\boldsymbol{U}$ is a positive and symmetric matrix. Therefore, $\Delta \boldsymbol{F}_C$ is bounded. As a result, $\Delta \boldsymbol{F}_O$ is bounded, $\dot{\boldsymbol{s}}_O$ is bounded from Eq.

(6), $\dot{\boldsymbol{v}}_C$ is bounded from Eq. (6.24), and $\dot{\boldsymbol{s}}$ is bounded from Eq. (4). From these relations, the time derivative of Eq. (5) is given as

$$\ddot{V} = -2(\boldsymbol{s}_O^T \boldsymbol{K}_O \dot{\boldsymbol{s}}_O + \boldsymbol{s}^T \boldsymbol{K} \dot{\boldsymbol{s}} + \boldsymbol{v}_C^T \boldsymbol{\Omega} \dot{\boldsymbol{v}}_C), \tag{12}$$

and it is bounded. Hence, $\dot{V}$ is uniformly continuous. According to the Lyapunov-like lemma, $\dot{V} \to 0$ as $t \to \infty$. Therefore, $\boldsymbol{s}_O \to 0$, $\boldsymbol{s} \to 0$, and $\boldsymbol{v}_C \to 0$ as $t \to \infty$. As a result, from Eq. (6.14), $\dot{\boldsymbol{e}}_O$ and $\boldsymbol{e}_O$ converge to zero and the object trajectory converges to the desired trajectory as $t \to \infty$. Also, $\boldsymbol{b} \to 0$ as $t \to \infty$. From Eq. (10), $\boldsymbol{F}_C \to \boldsymbol{F}_C^d$ and $\boldsymbol{F}_O \to \boldsymbol{F}_O^d$ as $t \to \infty$.

Furthermore, since it can be shown that $\boldsymbol{U}$ and the right side of Eq. (10) are uniformly continuous, then $\Delta\boldsymbol{F}_C$, $\dot{\boldsymbol{s}}$, $\dot{\boldsymbol{s}}_O$, and $\dot{\boldsymbol{v}}_O$ are uniformly continuous. According the Barbalat's lemma, it is derived that $\dot{\boldsymbol{s}} \to 0$, $\dot{\boldsymbol{s}}_O \to 0$, and $\dot{v}_O \to 0$ as $t \to \infty$. Also, $\Delta\boldsymbol{F}_O \to 0$ and $\Delta\boldsymbol{F}_C \to 0$ as $t \to \infty$. From Eq. (3), $\boldsymbol{F}_C \to \boldsymbol{F}_C^d$ and $\boldsymbol{F}_O \to \boldsymbol{F}_O^d$ as $t \to \infty$; i.e., the contact forces converge to the desired values.

Chapter 7

7.1. Abbreviation.

Chapter 8

8.1. Abbreviation.

References

Chapter 1

1. Adachi, Y., Kumano T., Ikemoto A., Hattori A. and Suzuki M. (2002). Development of a haptic device for multi fingers by macro-micro structure. *Journal of the Robotics Society of Japan*, 20(7), 725–733.
2. Cutkosky, M.R. (1989). On grasp choice, grasp models, and the design of hands for manufacturing tasks. *IEEE Transactions on Robotics and Automation*, 5(3), 269–279.
3. CyberGrasp (2014). CyberGlove system., http://www.cyberglovesystems.com/products/cybergrasp/overview (Octorber 1. 2014 visited).
4. Endo, T., Kawachi, Y., Kawasaki, H. and Mouri, T. (2008). FPGA-based control for the wire-saving of five-fingered haptic interface. *Haptics: Perception, Devices and Scenarios*, Proceedings of the Sixth International Conference: EuroHaptics 2008, pp. 536–542.
5. Endo, T., Kanno, T., Kobayashi, M. and Kawasaki, H. (2010). Human perception test of discontinuous force and a trial of skill transfer using a five-fingered haptic interface. *Journal of Robotics*, Vol. 2010, Article ID 542360, 14 pages. doi:10.1155/2010/542360.
6. Endo, T., Kawasaki, H., Mouri, T., Ishigure, Y., Shimomura, H., Matsumura, M. and Koketsu. K. (2011). Five-fingered haptic interface robot: HIRO III. *IEEE Transactions on Haptics*, 4(1), 14–27.
7. Endo, T. and Kawasaki, H. (2013). Force perception of human finger using a multi-fingered haptic interface. In R. Balasubramanian and V.J. Santos (Eds.), *The Human Hand as an Inspiration for Robot Hand Development* (pp. 345–361). New York: Springer Verlag.
8. Fearing, R.S. (1990). Tactile sensing mechanisms. *International Journal of Robotics Research*, 9(3), 3–23.
9. Grunwald, M. (Ed.) (2008). *Human Haptic Perception:Basics and Applications*. Basel, Switzerland: Birkhauser Verlag.
10. Howe, RD. (1994). Tactile sensing and control of robotic manipulation. *Advanced Robotics*, 8(3), 245–261.
11. Isobe, Y. (1985). Hand Position. In Murota (Ed.), *Orthopedics Mook* (39, pp. 24–34) (in Japanese). Japan: K. Kanehara Co. Ltd.

12. Jacobsen, S.C., Wood, D.F., Knutti, D.F. and Biggers, K.B. (1984). The Utah/MIT dexterous hand: Work in progress. *International Journal of Robotics Research*, 3(4), 21–50.
13. Johansson, R.S. and Vallbo, Å.B. (1983). Tactile sensory coding in the glabrous skin of the human hand. *Trends in NeuroSciences*, 6(1), 27–32.
14. Johnston, D., Zhang, P., Hollerbach, J. and Jacobsen, S. (1996). A full tactile sensing suite for dexterous robot hands and use in contact force control. Proceedings of IEEE International Conference on Robotics and Automation, Minneapolis, USA, pp. 3222–3227.
15. Kamakura, Y., Nakata, M. and Yamazaki, S. (1998). Fundamental of movement patterns in human hands. *Journal of Advances in Neurological Sciences*, 42(1), 7–17 (in Japanese).
16. Kawasaki, H. and Komatsu, T. (1999). Mechanism design of anthropomorphic robot hand: Gifu hand I. *Journal of Robotics and Mechatronics*, 11(4), 269–273.
17. Kawasaki, H., Komatsu, T. and Uchiyama, K. (2002). Dexterous anthropomorphic robot hand with distributed tactile sensor: Gifu hand II, *IEEE/ASME Transactions on Mechatronics*, 7(3), 296–303.
18. Kawasaki, H., Takai, J., Tanaka, Y., Mrad, C. and Mouri, T. (2003). Control of multi-fingered haptic interface opposite to human hand. Proceedings of IEEE/RSJ International Conference on Intelligent Robots and Systems (IROS '03), Las Vegas, Nevada, USA, pp. 2707–2712.
19. Kawasaki, H. and Mouri, T. (2007). Design and control of five-fingered haptic interface opposite to human hand. *IEEE Transactions on Robotics*, 23(5), 909–918.
20. Kawasaki, H., Otsuka, Y., Koide, S. and Mouri, T. (2011). Perception and haptic rendering of friction moments. *IEEE Transactions on Haptics*, 4(1), 28–38.
21. Kyriakopoulos, K.J., Zink, A. and Stephanou, H.E. (1997). Kinematic analysis and position/force control of the anthrobot dexterous hand. *Transactions on System, Man, and Cybernetics–Part B: Cybernetics*, 27(1), 95–104.
22. Lederman, S.L. and Klatzky, R.L. (2009). Haptic perception: a tutorial. *Attention, Perception, & Psychophysics*, 71(7), 1439–1459.
23. Lin, L.R. and Huang, H.P. (1996). Integrating fuzzy control of the dexterous National Taiwan University (NTU) hand. *IEEE/ASME Transactions on Mechatronics*, 1(3), 216–229.
24. Liu, H., Butterfass, J., Knoch, S., Meusel, P. and Hirzinger, G. (1999). Multisensory articulated hand. *IEEE Control Systems*, 19(2), 47–54.
25. Montoy, M., Oyarzabal, M., Ferre, M., Campos, A. and Barrio, J. (2008). MasterFinger: multi-finger haptic interface for collaborative environment. Proceedings of EuroHaptics, Madrid, Spain, pp. 411–419.
26. Mouri, T., Kawasaki H., Yoshikawa, K., Takai, J. and Ito, S. (2002). Anthropomorphic robot hand: Gifu hand III. Proceedings of International Conference on Control, Automation and Systems (ICCAS '02), Jeonbuk, Korea, pp. 1288–1293.

27. Mouri, T. and Kawasaki, H. (2007). A novel anthropomorphic robot hand and its master slave system. In H. Hackel (Ed.), *Humanoid Robots, Human-like Machines* (pp. 29–42). Vienna, Austria: I-Tech Education and Publishing.
28. Napier, J.R. (1956). The prehensile movements of the human hand. *The Journal of Bone and Joint Surgery,* 38B(4), 902–913.
29. Penfield, W. and Rasmussen, T. (1950). *The Cerebral Cortex of Man.* New York: The Macmillan Company.
30. Rosheim, M. (1994). *Robot Evolution* (pp. 216–224). Hoboken, New Jersey: John Wiley & Sons, Inc.
31. Salisbury, J.K. and Craig, J.J. (1982). Articulated hands: force control and kinematic issues. *International Journal of Robotics Research,* 1(1), 4–17.
32. Walairacht, S., Ishii, M., Koike, Y. and Sato, M. (2001). Two-handed multi-fingers string-based haptic interface device. *IEICE Transactions on Information and Systems*, E84D(3), 365–373.
33. Weinstein, S. (1968). Intensive and extensive aspects of tactile sensitivity as a function of body part, sex, and laterally. In D.R. Kenshalo, R. Springfield and C. Thomas (Eds.), Proceedings of the First International Symposium on The Skin Senses, Thomas, Springfield, Illinois, USA (pp. 195–218).

Chapter 2

34. Cole, B.A., Hauser, J.E. and Sastry, S.S. (1989). Kinematics and control of multifingered hands with rolling contact. *IEEE Transactions on Automatic Control,* 34(4), 398–404.
35. Montana, D.J. (1988). The kinematics of contact and grasp. *The International Journal of Robotics Research,* 7(3), 17–32.
36. Murray, R.M., Li, Z. and Sastry, S.S. (1994). *A Mathematical Introduction to Robotic Manipulation.* Boca Raton Florida: CRC Press.
37. Siciliano, B., Sciavicco, L., Villani, L. and Oriolo, G. (2009). *Robotics, Modeling, Planning and Control.* Berlin: Springer.
38. Spong, M.W., Mutchinson, S. and Vidyasagar, M. (2006). *Robot Modeling and Control.* Hoboken, New Jersey: John Willey & Sons, Inc.
39. Yoshikawa, T. (1995). Foundations of grasping and manipulation, 1. passive closure and active closure. *Journal of Robotics Society of Japan,* 13(7), 950–957 (in Japanese).
40. Yoshikawa, T. (1996a). Foundations of grasping and manipulation, 2. Fingertip force. *Journal of Robotics Society of Japan,* 14(1), 48–54 (in Japanese).
41. Yoshikawa, T. (1996b). Foundations of grasping and manipulation, 3. Control. *Journal of Robotics Society of Japan,* 14(4), 505–511 (in Japanese).
42. Zhang, X.Y., Nakamura, Y. and Yoshimoto, K. (1997). Statics of grasping and manipulation that is applicable to power grasp. *Journal of Robotics Society of Japan,* 15(2), 45–53 (in Japanese).

Chapter 3

43. Bloch, A.M. (2010). *Nonholonomic Mechanics and Control.* New York: Springer Verlag.
44. Kwatny, H.G. and Blankenship, G.L. (2000). *Nonlinear Control and Analytical Mechanics:A Computational Approach.* Boston: Birkhäuser.
45. Mason, M.T. (1998). *Mechanics of Robotic Manipulation.* Cambridge, MA, USA: The MIT Press.
46. Murray, R.M., Li, Z. and Sastry S.S. (1993). *A Mathematical Introduction to Robotic Manipulation.* Boca Raton, Florida: CRC Press.
47. Nakamura, Y. (1993a). Nonholonomic robot systems, part 1: what is a nonholonomic robot? *Journal of Robotics Society of Japan,* 11(4), 521–528 (in Japanese).
48. Nakamura, Y. (1993b). Nonholonomic robot systems, part 2: motion planning under kinematic nonholonomic constraints. *Journal of Robotics Society of Japan,* 11(5), 655–662 (in Japanese).
49. Nakamura, Y. (1993c). Nonholonomic robot systems, part 3: motion control under kinematic nonholonomic constraints. *Journal of Robotics Society of Japan,* 11(6), 837–844 (in Japanese).
50. Nakamura, Y. (1993d). Nonholonomic robot systems, part 4: motion under dynamical nonholonomic constraints. *Journal of Robotics Society of Japan,* 11(7), 999–1005 (in Japanese).
51. Nakamura, Y. (1994). Non-holonomic robot systems: part 5 motion control under dynamical non-holonomic constraints. *Journal of Robotics Society of Japan,* 12(2), 231–239 (in Japanese).
52. Neimark, J.I. and Fufaev, A.N. (1972). Dynamics of nonholonomic systems. In *Translations of Mathematical Monographs.* Providence, Rhode Island: American Mathematical Society.
53. Soltakhanov, S., Yushkov, M. and Zegzhda, S. (2009). *Mechanics of Non-Holonomic Systems.* New York: Springer Verlag.

Chapter 4

54. Büchberger, B. (1988). Application of Grobner bases in non-linear computational geometry. *Lecture Notes in Computer Science,* 296, 52–80.
55. Ghorbel, F., Srinivasan, B. and Spong, M.W. (1998). On the uniform boundedness of the inertia matrix of serial robot manipulators. *Journal of Robotics Systems,* 15(1), 17–28.
56. Kawasaki, H. and Nishimura, K. (1986). Parameter identification of mechanical manipulators. *Transactions of Society of Instrument and Control Engineering,* 22(1), 76–83 (in Japanese).
57. Kawasaki, H. and Nishimura, K. (1988). Terminal-Link parameter estimation of robotic manipulators. *IEEE Journal of Robotics and Automation,* 4(5), 485–490.
58. Kawasaki, H., Shimizu, T. and Kanzaki, K. (1996). Symbolic analysis of the base parameters for closed-chain robots based on the completion

procedure. Proceedings of 1996 IEEE International Conference on Robotics and Automation, Minneapolis, Minnesota, USA, pp. 1781–1786.

59. Kawasaki, H. and Shimizu, T. (2000). *Robot Symbolic Analysis.* Shokodo Co., LTD. (in Japanese).
60. Mayeda, H., Yoshida, K. and Osuka, K. (1990). Base parameters of manipulator dynamic models. *IEEE Transactions on Robotics and Automation*, 6(3), 312–321.
61. Nakamura, Y. and Ghodoussi, M. (1989). Dynamic computation of closed-link robot mechanisms with nonredundant and redundant actuators, *IEEE Transactions on Robotics and Automation*, 5(3), 294–302.

Chapter 5

62. Arimoto, S. and Miyazaki, F. (1984). Stability and robustness of PID feedback control for robot manipulators of sensory capability. In M. Brady and R.P. Paul (Eds.), *Robotics Research, First International Symposium* (pp. 783–799). Cambridge, MA, USA: The MIT Press.
63. Bacciotti, A. and Rosier, L. (2005). *Liapunov Functions and Stability in Control Theory*, 2nd Ed. New York: Springer Verlag.
64. Boyd S. and Sastry S.S. (1986). Necessary and sufficient conditions for parameter convergence in adaptive control. *Automatica,* 22(6), 629–639.
65. Imura, J. (2000). *Stability Theory in Systems and Control.* Tokyo: Corona Publishing Co. Ltd. (in Japanese).
66. La Salle J. and Leffschtz, S. (1961). *Stability by Liapunov's Direct Method with Applications.* New York: Academic Press Inc.
67. Morgan, A. P. and K. S. Narendra (1977). On the uniform asymptotic stability of certain linear non-autonomous differential equations. *SIAM Journal on Control and Optimization*, 15, 5–24.
68. Queiros, M.S., Dawson, D.M., Nagarkatti, S.P. and Zhang, F. (2000). *Lyapunov-Based Control of Mechanical Systems.* Basel, Switzerland: Birkhauser Verlag.
69. Slotine J.J.E. and Li, W. (1991). *Applied Nonlinear Control.* Englewood Cliffs, New Jersey: Prentice-Hall.

Chapter 6

70. Anderson, B.O.D. (1977). Exponential stability of linear systems arising from adaptive identification. *IEEE Transactions on Automatic Control*, 22(2), 83–88.
71. Birglen, L., Lalibert, T. and Gosselin, C.M. (2010). *Underactuated Robotic Hands.* New York: Springer Verlag.
72. Cutkosky, M.R. and Kao, I. (1989). Computing and controlling the compliance of a robot hand. *IEEE Transactions on Robotics and Automation*, 5(2), 151–165.

73. Hayati, S. (1986). Hybrid position/force control of multi-arm cooperating robots. Proceedings of IEEE International Conference on Robotics and Automation, San Francisco, CA, USA, pp. 82–89.
74. Hwang, Y.Y. and Toda, I. (1994). Coordinated control of two direct-drive robots using neural networks, *JSME International Journal Series C, Mechanical Systems*, 37, 335–341.
75. Kawasaki, H., Mouri, T., Takai, J. and Ito, S. (2002). Grasping of un 79.74. Hwang known object imitating human grasping reflex. Proceedings of 15th Triennial World Congress of the IFAC, Barcelona, Spain, 15(1), pp. 851–856.
76. Kawasaki, H., Ueki, S. and Ito, S. (2006). Decentralized adaptive coordinated control of multiple robot arms without using force sensor. *Automatica*, 42(3), 481–488.
77. Kosuge, K., Yoshida, H., Fukuda, T., Kanitani, K., Sakai, M. and Hariki, K. (1995). Coordinated motion control of manipulators based on impedance control. *Journal of the Robotics Society of Japan*, 13(3), 404–410 (in Japanese).
78. Mason, M.T. and Salisbury, J.K. (1985). *Robot Hands and Mechanics of Manipulation*. Cambridge, MA, USA: The MIT Press.
79. Mouri, T., Kawasaki, H. and Ito, S. (2007). Unknown object grasping strategy imitating human grasping reflex for anthropomorphic robot hand. *Journal of Advanced Mechanical Design, Systems, and Manufacturing*, 1(1), 1–11.
80. Nakamura, Y. (1991). Controlling multifingered hands. *Journal of the Society of Instrument and Control Engineers*, 30(5), 395–399 (in Japanese).
81. Naniwa, T. and Arimoto, S. (1995). Learning control for robot tasks under geometric endpoint constraints. *IEEE Transactions on Robotics and Automation*, 2(3), 432–441.
82. Ozawa, R., Arimoto, S., Nakamura, S. and Bae, J. (2005). Control of an object with parallel surfaces by a pair of finger robots without object sensing. *IEEE Transactions. on Robotics and Automation*, 21(5), 965–976.
83. Su, Y. and Stepanenco, Y. (1995). Adaptive sliding mode coordinated control of multiple robot arms attached to a constrained object. *IEEE Transactions on Systems, Man, and Cybernetics*, 25, 871–877.
84. Ueki, S., Kawasaki, H. and Mouri, T. (2006). Adaptive coordinated control of multi-fingered hands with sliding contact. Proceedings of SICE-ICASE International Joint Conference, pp. 5893–5898. Busan, South Korea.
85. Ueki, S., Kawasaki, H. and Mouri, T. (2009). Adaptive coordinated control of multi-fingered robot hand. *Journal of Robotics and Mechatronics*, 21(1), 36–43.
86. Yoshikawa, T. and Zheng, X. (1990). Coordinated dynamic hybrid position/force control for multiple robot manipulators handling one constrained object. Proceedings of IEEE International Conference on Robotics and Automation, Cincinnati, OH, USA, pp. 1178–1183.
87. Zheng, X., Nakashima, R. and Yoshikawa, T. (2000). On dynamic control of finger sliding and object motion in manipulation with mutifingered hands. *IEEE Transactions on Robotics and Automation*, 16(5), 469–481.

Chapter 7

88. Adams, R.J. and Hannaford, B. (2002). Control law design for haptic interfaces to virtual reality. *IEEE Transactions on Control System Technology*, 10(1), 3–13.
89. Allart, J., Courtecuisse, H. and Faure, F. (2011). Implicit FEM solver on GPU for interactive deformation simulation. In Jade (Ed.), *GPU Computing Gems* (pp. 281–294). Amsterdam: Elsevier B.V.
90. Alharabi, M.O., Daniulaitis, V., Kawasaki, H. and Hori, T. (2005). Medical training simulation for palpation of subsurface tumor using HIRO. Proceedings of WorldHaptics2005, Pisa, Italy pp.623–624.
91. Burdea, G.C. (1996). *Force and Touch Feedback for Virtual Reality*. New York: Wiley-Interscience Publication.
92. Colgate, J.E., Grafing, P.E., Stanley, M.C. and Schenkel, G. (1993). Implementation of stiff virtual walls in force-reflecting interfaces. Proceedings of IEEE Virtual Reality Annual International Symposium, Seattle, WA, USA, pp. 202–208.
93. Daniulaitis, V., Alhalabi, M.O., Kawasaki, H., Tanaka, Y. and Hori, T. (2004). Medical palpation of deformable tissue using physics-based model for haptic interface robot (HIRO). Proceedings of IROS2004, Sendai, Japan, pp. 3907–3911.
94. Endo, T., Kanno, T., Kobayashi, M. and Kawasaki, H. (2010). Human perception test of discontinuous force and a trial of skill transfer using a five-fingered haptic interface. *Journal of Robotics*, Vol. 2010, Article ID 542360, 14 pages. doi:10.1155/2010/542360.
95. Endo, T., Kawasaki, H., Mouri, T. Ishigure, Y., Shimomura, H., Matsumura, M. and Koketsu. K. (2011). Five-fingered haptic interface robot: HIRO III. *IEEE Transactions on Haptics*, 4(1), 14–27.
96. Endo, T., Tanimura, S. and Kawasaki, H. (2013a). Development of tool-type devices for a multi-fingered haptic interface robot, *IEEE Transactions on Robotics*, 29(1), 68–81.
97. Endo, T., Kobayashi, M. and Kawasaki, H. (2013b). A finger skill transfer system using a multi-fingered haptic interface robot and a hand motion image. *Robotica*, 31(8), 1251–1261.
98. Hioki, M., Kawasaki, H., Sakaeda, H., Nishimoto, Y. and Mouri, T. (2011). Finger rehabilitation support system using a multifingered haptic interface controlled by a surface electromyogram. *Journal of Robotics*, Article ID 167516, 10 pages.
99. Honma, S. and Wakamatsu, H. (2004). Cutting moment analysis of materials by the saw for force display system. *Transactions of the Virtual Reality Society of Japan*, 9(3), 319–326 (in Japanese).
100. Kawai, M. and Yoshikawa, T. (2004). Haptic display with an interface device capable of continuous-time impedance display within a sampling period, *IEEE/ASME Transactions on Mechatronics*, 9(1), 58–64.
101. Kawai, T., Mouri, T., Endo, T. and Kawasaki, H. (2013). Breast cancer palpation training system using five-fingered haptic interface robot and

GPGPU. Proceedings of the 18th International Symposium on Artificial Life and Robotics (AROB '13), Daejeon, Korea, pp. 547–550.
102. Kawasaki, H., Mouri, T. and Ikenohata, S. (2007a). Multi-fingered haptic interface robot handling plural tool devices. Proceedings of World Haptics 2007, Tsukuba, Japan, pp. 397–402.
103. Kawasaki, H. and Mouri, T. (2007b). Design and control of five-fingered haptic interface opposite to human hand. *IEEE Transactions on Robotics*, 23(5), 909–918.
104. Mouri, T., Kawasaki, H., Kigaku, K. and Ohtsuka, Y. (2006). Novel control methods for multi-fingered haptic interface robot. Proceedings of the IEEE/RSJ International Conference on Intelligent Robots and Systems (IROS '06), Beijing, China, pp. 1576–1581.
105. Mouri, T., Imada, Y., Ishigure, Y. and Kawasaki, H. (2012). Haptic interface robotic measurement system for breast engorgement. Proceedings of 2012 5th International Conference on BioMedical Engineering and Informatics (BMEI '12), Chongqing, China, pp. 670–675.
106. Ueki, S., Kawasaki, H., Ito, S., Nishimoto, Y., Abe, M., Aoki, T., Ishigure, Y., Ojika, T. and Mouri, T. (2012). Development of a hand-assist robot with multi-degrees-of-freedom for rehabilitation therapy. *IEEE/ASME Transactions on Mechatronics*, 17(1), 136–146.

Chapter 8

107. Arvetti, M., Gini, G. and Folgheraiter, M. (2007). Classification of EMG signals through wavelet analysis and neural networks for controlling an active hand prosthesis. Proceedings of the 10th International Conference on Rehabilitation Robotics (ICORR 2007), Noordwijk, Netherlands, pp. 531–536.
108. Chopra, N., Spong, M.W., Ortega, R. and Barabanov, N.E. (2006). On tracking performance in bilateral teleoperation. *IEEE Transactions on Robotics*, 22(4), 861–886.
109. Cortesao, R., Park, J. and Khatib, O. (2006). Real-time adaptive control for haptic telemanipulation with Kalman active observers. *IEEE Transactions on Robotics*, 22(5), 987–999.
110. Ebisawal, S., Mouril, T., Nakagawa, S., Adachi, M., Sakaeda, H. and Kawasaki, H. (2013). Myoelectric prosthetic hand with exteriority and functionality. Proceedings of SICE Annual Conference, Nagoya, Japan, pp. 2688–2693.
111. Hioki, M., Ebisawa, S., Sakaeda, H., Mouri, T., Nakagawa, S., Uchida, Y. and Kawasaki, H. (2011). Design and control of electromyogram prosthetic hand with high grasping force and 2 degrees of freedom at wrist. Proceedings of 3rd IEEE RAS & EMBS International Conference on Biomedical Robotics and Biomechatronics (BIOROBO2011), Tokyo, Japan, pp. 1128–1133.
112. Hioki, M. and Kawasaki, H. (2012). Estimation of finger joint angles from sEMG using a neural network including time delay factor and recurrent structure. *ISRN Rehabilitation,* Article ID 604314, 13 pages.

113. Hokayem, P.F. and Spong, M.W. (2006). Bilateral teleoperation: an historical survey, *Automatica*, 42(12), 2035–2057.
114. Kawai, Y. and Fujita, M. (2008). Bilateral teleoperation of constrained system with time delay. *IEEJ Transactions on Electronics, Information and Systems*, 128(1), 1–6.
115. Kawasaki, H., Koide, S., Endo, T. and Mouri, T. (2012). Development of a hand haptic interface and its basic experimental evaluation. Proceedings of International Symposium on Innovations in Intelligent Systems and Applications (INISTA '12), Trabzon, Turkey.
116. Lawrence, D.A. (1993). Stability and transparency in bilateral teleoperation. *IEEE Transactions on Robotics and Automation*, 9(5), 624–637.
117. Lee, D. and Li, P.Y. (2005). Passive bilateral control and tool dynamics rendering for nonlinear mechanical teleoperators. *IEEE Transactions on Robotics*, 21(5), 936–951.
118. Mishima, M., Kawasaki, H., Mouri, T. and Endo, T. (2009). Haptic teleoperation of humanoid robot hand using three-dimensional force feedback. Reprints of the 9th International Symposium on Robot Control (SYROCO '09), Gifu, Japan, pp. 565–570.
119. Mouri, T. and Kawasaki, H. (2007). A novel anthropomorphic robot hand and its master slave system. In M. Hackel (Ed.), *Humanoid Robots:Human-like Machines* (pp. 29–42). Vienna, Austria: I-Tech Education and Publishing.

Index

G

H

I

J

K

L

M

N

P

Q

R

S

T

U

V

W

www.ingramcontent.com/pod-product-compliance
Ingram Content Group UK Ltd.
Pitfield, Milton Keynes, MK11 3LW, UK
UKHW020400150926
8624IPUK00004B/254

9 789814 635608